*Press, Revolution,
and Social Identities in France
1830-1835*

Press, Revolution, and Social Identities in France 1830–1835

Jeremy D. Popkin

The Pennsylvania State University Press
University Park, Pennsylvania

Library of Congress Cataloging-in-Publication Data

Popkin, Jeremy D., 1948–
    Press, revolution, and social identities in France, 1830–1835 / Jeremy D.
    Popkin.
    p.   cm.
    Includes bibliographical references and index.
    ISBN 0-271-02152-7 (cloth : alk. paper)
    ISBN 0-271-02153-5 (pbk. : alk. paper)
    1. Press—France—History—19th century. 2. Journalism—Social
    aspects—France—History—19th century. 3. France—History—July
    Revolution, 1830.

PN5177.P67 2001
074'.09'034—dc21

                                                                                                 2001021545

Copyright © 2002 The Pennsylvania State University
All rights reserved
Printed in the United States of America
Published by The Pennsylvania State University Press,
University Park, PA 16802-1003

It is the policy of The Pennsylvania State University Press to use acid-free paper for the first printing of all clothbound books. Publications on uncoated stock satisfy the minimum requirements of American National Standard for Information Sciences—Permanence of Paper for Printed Library Materials, ANSI Z39.48–1992.

Contents

List of Illustrations   vii

Acknowledgments   ix

Abbreviations   xi

Introduction   1

1   *Newspapers, Journalists, and Public Space*   23

2   *The Press, Liberal Society, and Bourgeois Identity*   67

3   *Reshaping Journalistic Discourse*   The Alternative Press in Lyon   105

4   *Echoes of the Working Classes*   135

5   *Creating Events*
Press Banquets and Press Trials in the July Monarchy   167

6   *Textualizing Insurrection*
The Press and the Lyon Revolts of 1831 and 1834   193

7   *From Newspapers to Books*
The Recasting of Revolutionary Narrative   229

Conclusion   263

Appendix 1: Sophie Grangé, "Moi" and "A la femme"   271

Appendix 2: The *Écho de la fabrique*'s Anniversary Salute to the Victims of the 1831 Workers' Insurrection   277

Notes   279

Bibliography   307

Index   321

List of Illustrations

1. "La Revue de Lyon," portraying Lyon's "journalistic field" in early 1835 (Courtesy of Bibliothèque municipale de Lyon, Fonds Coste no. 775) (Photo: Cliché Bibliothèque municipale, Didier Nicole)   27
2. "Plan de la Ville de Leon 1835." Map of Lyon, 1835 (Courtesy of Archives municipales de Lyon)   53
3. Public Newspaper-Reading as a Bourgeois Ritual, from *Lyon vu de Fourvières*, 1833 (Courtesy of McCormick Library of Special Collections, Northwestern University Library)   58
4. "Assemblée d'Actionnaires," 1831 (Courtesy of Bibliothèque municipale de Lyon, Fonds Coste no. 768) (Photo: Cliché Bibliothèque municipale, Didier Nicole)   75
5. "Grande Croisade Contre la Liberté," Grandville's drawing from *La Caricature*, 1834 (Courtesy of the Harvard College Library)   96
6. Front page of *L'Écho des travàilleurs*, 25 November 1833, commemorating the anniversary of the November 1831 insurrection (Courtesy of Bibliothèque municipale de Lyon) (Photo: Cliché Bibliothèque municipale, Didier Nicole)   162
7. *L'Indicateur*, 2 May 1835, comments on the silencing of the opposition press (Courtesy of Bibliothèque municipale de Lyon) (Photo: Cliché Bibliothèque municipale, Didier Nicole)   190
8. "Serment de l'Hôtel-de-Ville (de Lyon)," in which Michel-Ange Pèrier, one of the editors of the *Glaneuse*, leads the insurgents who had seized Lyon's Hôtel-de-Ville in a revolutionary oath on 23 November 1831 (Courtesy of Bibliothèque municipale de Lyon, Fonds Coste no. 717) (Photo: Cliché Bibliothèque municipale, Didier Nicole)   200
9. "Événements de Lyon, Combat du Pont Morand," celebrating the workers' November 1831 victory in the Lyon suburb of

Brotteaux (Courtesy of Bibliothèque municipale de Lyon,
　　　Fonds Coste no. 716) (Photo: Cliché Bibliothèque municipale,
　　　Didier Nicole) 223
10. Jean-Baptiste Monfalcon (1792–1874) (Courtesy of
　　　Bibliothèque municipale de Lyon, Fonds Coste) (Photo: Cliché
　　　Bibliothèque municipale, Didier Nicole) 240

Acknowledgments

This relatively short book has been a long time in the making, and I owe thanks to many people and institutions for their assistance and encouragement. I began the research on this project during a sabbatical year in Lyon, France, where I was a guest researcher at the Centre d'études du dix-huitième siècle, housed in the Maison des Sciences de l'Homme–Rhône-Alpes. I would like to thank my good friend Pierre Rétat, director of the Centre d'études du dix-huitième siècle, and Yves Lequin, director of the Centre Pierre Léon, for their hospitality and encouragement. I have completed the final revisions on the manuscript in the wonderful working conditions provided by a fellowship year at the National Humanities Center, for whose support I am also grateful. Other support for this research came from the John Simon Guggenheim Foundation, the Fulbright Foundation, and the University of Kentucky Research Foundation. Over the years, I received valuable assistance from the staffs of the Bibliothèque municipale de Lyon, the Archives municipales de Lyon, the Archives départementales du Rhône, and the Musée Gadagne in Lyon, and from the Archives nationales, the Bibliothèque nationale (especially the now-closed Annexe de Versailles), the Bibliothèque de l'Arsenal, the Bibliothèque Marguerite-Durand, and the Bibliothèque historique de la ville de Paris in Paris.

I have had the opportunity to present portions of this work to audiences at the New York University Institute for French Studies; to the history departments at Indiana University, the University of Illinois, and the University of Kentucky; to the North Carolina French History group; at the Washington, D.C., Old Regime Seminar; at a conference on "Presse et événement" at the Universität von Saarbrücken; and at meetings of the Society for French Historical Studies and the Western Society for French

History; and I have benefited from lively discussions on all these occasions. By inviting me to contribute to the volume they edited, *Making the News,* Dean de la Motte and Jeannene Przyblyski provided a much-needed incentive to develop ideas that are presented here at greater length.

I am grateful to the University of Massachusetts Press for permission to reuse material in Chapters 1, 3, and 4 that appeared in another form in my article "Press and 'Counter-Discourse' in the Early July Monarchy," in Dean de la Motte and Jeannene M. Przyblyski, eds., *Making the News: Modernity and the Mass Press in Nineteenth-Century France* (Amherst: University of Massachusetts Press, 1999).

James Albisetti, James W. Carey, Jack Censer, Phil Harling, and Steven Vincent took time from their own busy schedules to critique the entire manuscript, and Elinor Accampo, Ellen Furlough, Sarah Maza, Jeffrey Peters, Sandy Petrey, and Karen Petrone offered helpful comments on parts of it. The ideas developed here owe much to discussions over the years with leading French and German specialists on the history of the press, especially Gilles Feyel, Claude Labrosse, Hans-Jürgen Lüsebrink, Marc Martin, Rolf Reichardt, and Pierre Rétat. Alex Popkin compiled the index.

I would like to dedicate this book to my wife, Beate, and my sons, Gabriel and Alex, who accompanied me on the year-long visit that made this immersion in the history of Lyon possible, and to the many friends who have made it such a pleasure to return to *l'ancien capitale des Gaules* over the years.

Abbreviations

ADR   Archives départementales du Rhône (Lyon)
AML   Archives municipales de Lyon
AN   Archives nationales (Paris)
BML   Bibliothèque municipale de Lyon

Introduction

The Press and Revolution in 1830

The Paris journalist Saint-Marc Girardin's report on the Lyon silk workers' insurrection of November 1831 is one of the most famous newspaper articles ever printed in France. Girardin, a writer for the Paris *Journal des débats*, a paper devoted to the "bourgeois monarchy" brought to power by the Revolution of 1830, happened to pass through France's second-largest city a few days after the insurrection, the largest workers' uprising in France since 1789. The article he wrote commenting on the significance of what had occurred was reprinted all over Europe and has continued to be quoted regularly down to the present day as a description of the central problem facing nineteenth-century European civilization. Saint-Marc Girardin insisted that "the Lyon uprising has revealed a fundamental secret, that of the internal social conflict between the class that owns things and the one that does not." His warning that "the barbarians who threaten society are not in the Caucasus or the steppes of Tartary; they are in the suburbs of our manufacturing cities" dramatized that conflict for the express purpose, as Saint-Marc Girardin later wrote, of inspiring his middle-class readers with a clear consciousness of the dangers facing them. The newspaper press, he knew, was a uniquely powerful instrument for this purpose. Read as soon as it appeared, all over the country, the *Journal des débats* helped tie its audience together and give them a common sense of identity.[1]

In his highly colored depiction of Lyon, Saint-Marc Girardin failed to comment on one of the most significant aspects of the city's situation: the "barbarians" of whom he spoke had armed themselves with a printing press. Three weeks before the insurrection, a newspaper claiming to speak for them, the *Écho de la fabrique*, had circulated its first issue, and two weeks before the fighting broke out that paper had printed a clear and detailed warning of the looming confrontation, publishing what it labeled a "rumor" that "the masses were ready to rise up, that the Croix-Rousse [the working-class suburb where most of the silk weavers lived] was on the march with a black flag to assault the Hotel-de-Ville," but blaming *agents provocateurs* for creating the fear of an imminent class war.[2] Saint-Marc Girardin's famous article was an attempt to give the Lyon insurrection a particular construction, one that is easily categorized as a classic example of bourgeois ideology, but the *Écho de la fabrique* had already taken the initiative in constructing that event and showing that it was not only bourgeois interests that could use the press for their purposes. This periodical was not a clandestine publication challenging the legal order from outside; it had been duly registered with the authorities,[3] and its founders intended to take advantage of the press freedom that was one of the defining characteristics of the liberal society inaugurated in 1789 and consolidated in 1830. The threat they posed to the bourgeois world was not one of blind violence, as Saint-Marc Girardin's rhetoric suggested; it was the more insidious threat of using the communications media of that society to transform it. The same medium that Saint-Marc Girardin used to summon a self-conscious bourgeoisie into existence could be used to help a working-class identity coalesce.

The two very different treatments of the Lyon insurrection in the *Journal des débats* and the *Écho de la fabrique* amply illustrate the importance of the periodical press during the revolutionary crisis opened by overthrow of the Restoration monarchy in the July Days of 1830. In 1830, as in 1789, the press played a major role in the revolutionary process. The press was one of the major forces transforming politics and society, even as newspapers themselves were transformed by the new circumstances created by the revolution. As the two articles about Lyon just cited demonstrate, the papers published during the early 1830s played an especially important role in the definition of the new social identities that would characterize what both contemporaries and historians have called bourgeois society. In a period when political parties and organized mass movements were still in their infancy, newspapers were the most effective means for the articulation of collective interests. They operated on the boundary between a political system limited to a small, exclusively male elite and an increasingly democratic print culture, and served simultaneously as instruments for the construction of bourgeois cultural hegemony and for its subversion.

That newspapers and journalism were central to the revolutionary crisis that began with the "Three Glorious Days" in July 1830 has always been acknowledged. The July Days were set off by the Restoration government's attack on the opposition press, and the revolutionary cycle begun in 1830 came to an end five years later with the imposition of a new set of press controls, the "September Laws" of 1835, that finally tamed the period's unruly texts. To date, however, there has been no detailed study of the French press in these years. The scholarship that does exist has generally considered the press only from the point of view of its role in national politics. Charles Ledré's standard work, *La presse à l'assaut de la monarchie,* and the survey in the second volume of the *Histoire générale de la presse française* both exemplify this approach.[4] Charles Philipon's famous satirical journals have been a perennial subject of interest, but they too have been analyzed primarily from the point of view of their political content.[5] Women's historians have highlighted the importance of the period's feminist journals,[6] demonstrating that the impact of the periodical press went well beyond the realm of traditional politics, but the full range of the press has not yet been treated as it deserves to be: as one of the period's most important cultural institutions and as a prime force in defining the conflicts that shaped the early years of nineteenth-century bourgeois society. The main reasons for this are the long-standing tendency to consider the Revolution of 1830 as a minor episode in comparison with 1789, 1848, and 1871, the other high points of France's century of upheavals, and the neglect of newspapers that has characterized the cultural history of the last few decades.

Certainly, the French Revolution of 1830 does not rank as one of the "great" revolutions of modern times, but it was a more significant event than the modest place it occupies in most comparative studies would suggest. The three days of fighting in Paris to which the revolution is often reduced were in fact only the start of a revolutionary cycle whose outcome remained in doubt throughout the early 1830s. Although the structure of French government was not greatly changed, David Pinkney, author of the most detailed history of the event, claims that it resulted in the most sweeping changeover of administrative personnel in the country's history.[7] Recent scholarship has also shown that the 1830 revolution was not, as is often claimed, a purely Parisian or even a purely urban phenomenon; like 1789, it also involved a concatenation of popular protests in some rural areas where the incomplete penetration of capitalist practices and the liberal state caused resentment, as well as movements in most of France's other major cities.[8] Nor was its relatively peaceful outcome necessarily inevitable. In June 1832, Louis-Philippe's regime faced a genuine crisis reminiscent of 1793. As a republican demonstration involving perhaps 100,000 people filled the streets of Paris, eleven western *départements* rose in revolt on behalf of the exiled Bourbons. If the republican leaders had shown a little more determination, they might well have toppled the constitutional monarchy and inaugurated a radical revolutionary process reminiscent of the 1790s.[9] Only the defeat of the republican insurrections in Lyon and Paris in April 1834 and the draconian September Laws enacted in 1835 really demonstrated the regime's capacity to defend itself.

There are therefore compelling arguments for defining the Revolution of 1830 not as a minor or insignificant revolution but as a rare example of a major revolution in which the elites who seized power in the immediate aftermath of the collapse of an old regime succeeded in retaining control despite significant challenges from both right and left, rather than going the way of the liberal reformers of 1789 or the idealistic republicans of 1848. Seen in this light, the Revolution of 1830 becomes a challenge to general theories of revolution, such as Crane Brinton's

Anatomy of Revolution, that suggest a certain inevitability in the development of revolutionary crises. If there is much to be learned even from failed revolutions, as Ronald Aminzade has recently argued in his study of local insurrections in nineteenth-century France, there is certainly something important to be gained from studying one that might be said to have stopped halfway.[10]

The Revolution of 1830 is also significant, however, because it signaled a major change in the nature of French society. From the time it occurred, 1830 has been identified as the time when the bourgeoisie definitively took over political power in France, finally completing the process of rupture with the past begun so spectacularly in 1789. As recent scholars have increasingly argued that the modern French bourgeoisie was more the product than the cause of that revolution, and that it was, in David Garrioch's words, not the 1790s but the 1820s that "saw the formation of a politically and ideologically united bourgeoisie,"[11] they have underlined the importance of the Revolution of 1830 as a historical watershed. A whole complex of processes that were to structure French society well into the twentieth century have their roots in the years around that time. Long recognized as the beginning of French industrialization, these years are now also seen as the period when a new kind of consumer economy took root. A distinctive "bourgeois" culture and lifestyle achieved hegemony, marked symbolically by the adoption of an almost uniform code of masculine dress and a sharp distinction in gender roles as bourgeois women began to retreat from their earlier active participation in family businesses.[12]

The crystallization of a modern bourgeois identity in the 1830s was closely linked to the emergence of new identities among workers and women. The years after 1830 saw a sudden upsurge in the use of the terms *prolétaire* and *prolétariat*, and the articulation of the notion that all those who worked for a living shared common interests distinct from those of the possessing classes.[13] France had a long tradition of workers' organizations, such as the *compagnonnages* formed by journeymen artisans in many trades, but only after 1830 did the idea of a "working class" begin to take root.[14] Socialist doctrines had begun to develop in France before 1830, but only in the wake of the July Revolution did they begin to penetrate the working classes and create the potent identification between proletarian identity and collectivism that was to endure down to the late twentieth century. The early 1830s were also critical years for the emergence of French feminism. Proponents of women's rights during the revolutionary decade of the 1790s had emphasized what women shared with men; in the 1830s, a new form of feminism developed that urged women to construct their identity around notions of difference instead.[15] The ideological issues raised by the Revolution of 1830 generated new debates about women's rights that had profound implications for how women understood their own social position and how men regarded them.[16] The Revolution of 1830 may have been less violent and less dramatic than its successor of 1848, but it was in many ways an equally significant turning point.

Nowhere in France were the fundamental changes taking place in the early 1830s more evident than in the country's second-largest city. During this period, Lyon, normally lost in the shadow of Paris, was suddenly thrust into the national and international spotlight. In 1830, Lyon and its working-class suburbs formed an urban concentration of nearly 180,000 people, whose economy was vitally dependent on a single industry, the silk trade. Like Manchester in the same period, Lyon became a "shock city," a place where the future of European society could be glimpsed with special clarity.[17] Proposing to transfer the headquarters of the Saint-Simonian movement from Paris to Lyon in 1832, Michel Chevalier wrote, "We will go seek the air they breathe and the wind that blows in the greatest center of production . . . on which the European continent can pride itself. We will go there where a million arms move fourteen hours a day with a single goal, to produce. . . . Last year, Lyon, this great worker, stamped its foot, and the earth shook."[18] The stamping of the foot to which Chevalier was referring was the first of the two revolutionary insurrections in the city in the early 1830s: the massive silk workers' uprising of November 1831, which forced the local army garrison to evacuate Lyon and leave it under the control of armed workers for more than a week. This event, which claimed some 170 lives, demonstrated the potential power of an urban proletariat organized in opposition to its bourgeois employers. Lyon's second insurrection in April 1834, with its overt republican coloration, cost even more lives—more than 300—and seemed to show that radical political ideas had penetrated into the city's working classes and created a threat to the liberal political order proclaimed in 1830.[19] These spectacular explosions were classic newspaper stories, extensively covered in Lyon's local press as well as in France's national papers. They made the city an obligatory stop for anyone seeking to understand the new configuration of social and political forces of the nineteenth century. Not only French social experts such as Louis-René Villermé, but even foreign observers, such as the English Utilitarian John Bowring, came to study the city where the tensions of modernity seemed so sharply defined.[20]

Visitors flocked to Lyon in the 1830s because they saw it as a site where changes of universal significance to modern civilization were visible, but the city's situation also exemplified the tensions between Paris and the provinces. The Revolution of 1830 gave rise to appeals for "decentralization" of the country's cultural life, and Lyon's newspaper press was one of the few examples of a successful effort of this sort.[21] In the 1830s, the railroad and the electric telegraph had not yet broken down Lyon's relative isolation from Paris. Local events, particularly the two insurrections, played themselves out in an essentially autonomous fashion. Neither the national government nor the capital's press was able to act quickly enough to affect the outcome of these crises. Although Lyon absorbed more copies of the Paris papers on a per capita basis than any other French *département*, the local authorities and other observers hardly mentioned their influence.[22] On the other hand, this was the first period in several centuries in which the city had a print culture of its own that profoundly influenced local life. Lyon's book publishers had been important factors in the sixteenth-century wars of religion, but after 1600 Parisian competition had steadily forced them into marginal roles.[23] By the late 1820s, however, the city supported a lively and independent newspaper press that commented extensively on and participated actively in local events, as well as national ones. Parisian periodicals could not reach the city quickly and cheaply enough to compete with these local publications. Lyon's situation thus allows us to analyze the role of newspapers in a revolutionary situation with particular clarity. Not only were the local papers the only periodicals able to intervene in the unfolding of fast-moving events, but they were unquestionably the city's dominant print medium, the only one with a significant audience and a definite impact on local events. In moments of crisis, Lyon's printers turned out a few broadsheets and pamphlets, and during quieter intervals they published a few books on local affairs, but the half-dozen locally published newspapers and periodicals were far more important. The energies and tensions that were disseminated in Paris in a variety of media, including caricatures, plays, intellectual journals, books, and pamphlets, were concentrated in the press in Lyon.

The simplified nature of Lyon's print culture thus makes it easier to study than that of Paris. At the same time, however, the Lyonnais press was more extensive and diverse than that of any other French city besides the capital. During the first half of the decade of the 1830s, Lyon had newspapers representing all the major tendencies of French politics and culture. Lyonnais journalists argued for the proregime and antiregime varieties of liberalism, democratic republicanism, and Catholic and royalist legitimism. The city had "serious" newspapers and an important example of the satirical radical press exemplified in Paris by the illustrated *La Caricature*. It was home to the first solidly established working-class newspapers in France and to an outspokenly feminist periodical, and it had a cultural press that offered an alternative to other periodicals' concentration on social and political conflict. Furthermore, a diverse and extensive collection of sources, ranging from daily reports by Lyon's prefect to the autobiography of one of its leading journalists, offer a wealth of information about the motives behind these newspapers' printed content and the nature of their relations with the authorities, with their stockholders, and more broadly with the community whose tensions they represented. Lyon in the early 1830s thus provides a unique vantage point from which to examine the connection between the press and the particular configuration of forces that produced the early nineteenth century's characteristic revolutions.

The Case for Press History

Lyon's press system is both representative of the major cultural movements involved in the Revolution of 1830 and unusually well documented, but this study's concentration on newspapers is also meant to make a case for their importance in cultural studies. The current vogue for cultural history has usually left the periodical press on the sidelines. The explosion of a cultural history devoted to what the editors of a recent volume on the press call "other now canonical sites of modernity—panoramas, department stores, cinema and fashion, the boulevards and their habitual denizens," has relegated newspapers to the status of a genre taken for granted and exploited primarily as source material rather than as a subject in its own right.[24] In this study, I hope to make the case for the vital place of the press in any meaningful cultural history and for a history of the press that incorporates the insights of this new cultural history, seeing newspapers not merely as purveyors of information and ideology—although the importance of these roles should never be underestimated—but also as important sites for the construction of social and cultural identities and as an important form of literature.

Despite the vigorous questioning of traditional canons brought about by the new theoretical approaches of the past twenty years, journalistic texts still remain largely excluded from the domain of French literature, as well as that of other traditions. Denis Hollier's massive *New History of French Literature*, distinguished by its valiant effort to incorporate nontraditional perspectives into the study of the subject, includes a short chapter devoted to the debates over press freedom during the Restoration, but no substantive discussion of the characteristics of newspaper discourse.[25] This continued resistance to recognizing journalism as literature is all the more curious in that the "antiliterary" characteristics often attributed to newspaper writing are precisely those often prized by poststructuralist theorists. Newspapers by their very nature feature heterogeneous content, resist narrative closure, and exemplify the anonymous, authorless text. In nineteenth-century France in particular, the separation of newspapers from literature is unjustifiable. There was no clear separation in this period between journalism and other forms of literature, or between journalists and other writers. Newspapers incorporated in their columns almost every other literary genre. Both the national press and the Lyon press of the early 1830s published not only narratives of events conventionally classified as "news," but also parliamentary oratory, poetry, short works of fiction, argumentative essays, and letters. Nearly every French author of importance wrote extensively for the press. Journalists had no scruples about referring to their publications as "literature," and indeed often argued that journalism was in fact the characteristic literary form of the modern age, "a new power, with no model in the past, and whose influence on society has been so vast and so fecund."[26]

Two main theoretical arguments have been advanced against according journalism the status of literature: first, that the newspaper is by its nature an essentially incoherent genre, incapable of communicating meaning, and that this lack of coherence serves the hegemonic purposes of bourgeois domination; second, that journalism is an essentially corrupt form of writing, dominated by money as opposed to aesthetic values. These critiques were elaborated by nineteenth-century French authors themselves—indeed, they often had even earlier antecedents—and contemporary theoretical analyses are often largely restatements of attitudes expressed at the time. Among contemporary literary theorists, the one who has devoted the greatest effort to explicating the nature of nineteenth-century French journalistic discourse is Richard Terdiman. In the chapter of his influential work, *Discourse/Counter-Discourse*, entitled "Newspaper Culture: Institutions of Discourse, Discourse of Institutions," Terdiman gives an analysis that is partly a summary of nineteenth-century critiques and partly a late twentieth-century poststructuralist critique. Quoting Baudelaire's remark, "'I do not understand how anyone with clean hands can touch a newspaper without a shudder of disgust,'" Terdiman shows how newspapers became the target of the period's "disenchanted intelligentsia," saying, "For them the newspaper became the quintessential figure for the discourse of their middle-class enemy, the *name* for the writing against which they sought to counterpose their own."[27]

Terdiman also articulates his own critique of the newspaper, one that goes beyond the denunciations of Balzac and Baudelaire. In the first place, he maintains that the newspaper virtually defies analysis, because of its lack of coherence: "We cannot display the content of such a diffuse formation. . . . Newspapers trained their readers in the apprehension of detached, independent, reified, decontextualized 'articles,'" and in this way blocked them from seeing any larger pattern in their world. "Seen in this way, the newspaper can be understood as the first culturally influential *antiorganicist* mode of modern discursive construction. Its form *denies form*, overturns the consecrated canons of text structure and coherence which had operated in the period preceding its inception." Elaborating on this theme, he concludes that the principle behind the newspaper "is a systematic emptying of any logic of connection." The newspaper, and the larger capitalist market economy of which it is a part, "rationalize disjunction; they are organized *as disorganization*."²⁸

If it were in fact true that journalistic discourse functioned only to prevent comprehension of reality, any serious analysis of it would be bound to end in hopeless confusion. In fact, however, Terdiman's vehement critique of the newspaper is not the only possible reading of journalistic texts consistent with poststructuralist theoretical insights. Critics in both the United States and France have argued that newspapers are not vehicles of disorder, but texts that offer an intelligible representation of the world. This order of representation becomes coherent only if it is analyzed on its own terms, however, and not simply as the degraded Other of genuine literature. In the introduction to an important collection of essays, Robert Manoff and Michael Schudson have written that "journalism, like any other storytelling activity, is a form of fiction operating out of its own conventions and understandings and within its own set of sociological, ideological, and literary constraints." These conventions are not the same as those of other forms of literature, but they serve the same purpose—that of generating meaning.²⁹ In an analysis strongly informed by the critiques of such authors as Michel Foucault and Roland Barthes, as well as by the work of American sociologists of the press, such as Gaye Tuchman, the French communications specialists Maurice Mouillaud and Jean-François Tétu have shown how newspapers frame events according to a logic of their own. They agree with Terdiman that the newspaper "has, from its origin, featured a fragmented form that contrasts to the unity of the book," but not one without meaning. In fact, in modern societies, which are necessarily fragmented into different groups, the newspaper and other mass media create representations that make common meanings possible.³⁰ These representations are not easily found in any one journalistic story; they inhere, rather, in the overall plan of the newspaper, a text whose layout itself communicates an order, "a new discourse that is precisely the discourse of the newspaper,"³¹ and in the series of individual numbers that make up the larger whole of the publication. Journalistic meanings, as James W. Carey has written, "are properties of the whole, not the part; the coverage, not the account."³² The newspaper is not opaque to meaningful analysis; it must simply be read according to its own logic, rather than that of other literary genres.

A second strand of critique, frequently articulated at the time and restated by Terdiman, is that the newspaper is a degraded literary form because of the role played by money in determining its content. Honoré de Balzac dramatized this accusation in his *Lost Illusions*, where the veteran journalist Étienne Lousteau disabuses the young hero of his idealistic visions of the profession. " 'It's a dirty business, but I live by it, and so do hundreds of others. . . . At this trade—as a hired assassin of ideas, industrial, literary, and dramatic reputations—I make fifty francs a week,' " Balzac's cynical character explains, at the end of his description of how money and ambition completely dominate the newspaper business.[33] This denunciation of the corruption of the French press has roots in eighteenth-century debates about literary property and has remained a powerful theme in public discourse throughout the twentieth century; the purge of French journalists and publishers carried out after the Liberation in 1944, for example, was intended to "liberate the newspapers from money and give them a tone and a truth that will raise the public to the level of what is best in it," as Albert Camus wrote.[34] The argument depends in part on an idealization of "true" literary production, defined as writing done for its own sake, without thought of tangible reward, and on a demonization of commercial advertising. In the wake of biographies that have shown how much financial considerations preoccupied major nineteenth-century authors—Balzac, the devastating critic of the journalist-for-hire, as much as anyone—and of Pierre Bourdieu's analysis of the diverse forms of "cultural capital" that writing could earn, the notion that journalists were truly distinct from other authors because of their concern about the rewards from their work hardly seems tenable.

It is equally questionable whether the presence of advertising in the French press automatically meant, as Terdiman claims, that *"all* the space in the paper became potentially salable, potentially purchasable," and therefore devalued.[35] Economic pressures are not the only ones that affect journalists. As Manoff and Schudson argue, journalists are indeed influenced by the "political and economic structures" within which they operate, but they are also influenced by the "occupational routines of daily journalism, and [the] literary forms that journalists work with."[36] The nineteenth-century French public had little patience for journalists and publications whose opinions were visibly for sale, and the commercialization of the press was far from total, particularly in provincial markets like Lyon. There, the archival sources largely bear out local journalist Jérôme Morin's claim that "financial necessities have considerably less effect on the spirit of newspaper than you think" and that most publications were backed by political interest groups with little concern about their balance sheets.[37] Marc Martin, the leading contemporary historian of the French media in the last two centuries, has argued that the weakness of the French press was not its dependence on the market but its reluctance to seek commercial advertising, which left editors completely dependent on patrons with specific political agendas. This failure he blames partly on the very effectiveness of the denunciations that greeted experiments like Émile de Girardin's effort to create a cheap popular press by relying on advertising revenue in 1836. "The creation of this 40-franc press led the politically oriented papers representing the entire spectrum of ideologies, from the republican extreme left to the legitimists, to proclaim the impossibility of combining its traditional political function with a mercantile one. From 1840 onward, journalism had a strong suspicion of advertising."[38]

Both nineteenth-century and contemporary critiques of the press as inherently ideological or corrupt thus need to be read critically themselves. In particular, the role of newspapers in France's revolutionary crises cannot simply be subsumed under the notion of bourgeois hegemony and domination. Periodicals played an especially important part in these episodes because the nature of the medium gave them both a unique ability to respond to fast-moving events and a special claim to act as representatives of social and political groups and movements. As early as 1797, the French revolutionary journalist Pierre-Louis Roederer had offered an analysis of the reasons for the medium's special power in crisis situations that is still persuasive today. Unlike books, Roederer pointed out, newspapers do not passively wait for readers to seek them out, but actively pursue them. The newspaper "arrives in installments and always brings something of fresh interest. It arrives periodically, on a regular basis." The newspaper's impact is magnified because it reaches a whole public of readers "every day, at the same time, . . . in all classes of society, in all public places, being the almost obligatory diet of daily conversation." As a result, he claimed, "newspapers . . . not only act on a larger body of men, but they act more powerfully than any other form of writing."[39]

Modern studies of the newspaper medium have largely confirmed Roederer's intuitions about its special characteristics. The newspaper's continuing periodicity does indeed create a special kind of relationship between readers and this particular form of text. Unlike other forms of printed text, the newspaper is both fixed and changing. Its recurring features, above all its title and the distinctive pattern of text presentation that identifies a particular paper, serve both to distinguish it from other newspapers and to establish a continuity between its separate issues.[40] The newspaper thus does not have to win over its audience anew with each issue; it is already familiar to them before it arrives. At the same time, however, each issue of a newspaper is different from the one before, bringing the latest news of ongoing public events. Like Sheherazade, the narrator of *The Thousand and One Nights*, the newspaper envelops readers in a sequence of narratives that never reach closure. As a result, readers are inclined to read it promptly when it appears, creating the phenomenon of simultaneous text consumption that Roederer had already noted and that contemporary anthropologist Benedict Anderson has used as a paradigm for the process by which "imagined communities" are created. Each newspaper reader, Anderson has written, "is well aware that the ceremony he performs is being replicated simultaneously by thousands (or millions) of others of whose existence he is confident yet of whose identity he has not the slightest notion. Furthermore, this ceremony is incessantly repeated at daily or half-daily intervals throughout the calendar. What more vivid figure for the secular, historically clocked, imagined community can be imagined?"[41] Newspaper readers are thus bound together into a group, dispersed in space but united in time. "Although the readers of a newspaper don't know each other, they develop a common consciousness, a 'we-ness,'" as the German newspaper scholar Otto Groth, author of what is still the most elaborate analysis of the newspaper medium, wrote in the 1920s.[42]

Books and pamphlets could be events in their own right, as Félicité de Lamennais' sensational democratic and millenarian tract, *Paroles d'un croyant*, the first French popular bestseller of the nineteenth century, was in 1834,[43] but they could not keep up with the rapid changes that characterized revolutionary politics. Books and pamphlets also appeared as isolated phenomena, unrelated to one another; they could not convincingly represent collectivities whose existence was supposed to extend over time. The ability of periodicals to fill this role was due to their ability to interact with their readers. Readers could and did send in articles and letters, and the rising or falling number of subscriptions gave editors some tangible evidence of whether they had succeeded in evoking a favorable reaction. The publication of a newspaper also had an important symbolic dimension. Contemporaries recognized, as one Lyon journalist put it, that "newspapers are *êtres moraux*, which everyone personifies."[44] Newspapers cast themselves as the public voices of the audiences they claimed to represent. The members of all of the period's social groups were necessarily scattered and unable to speak for themselves in a unified way, but a newspaper could take their place and articulate their views. Group members might come together physically on occasion for common action, but a newspaper was always on the spot and ready to respond to any challenge. Newspapers claimed a metonymic relationship with the "imagined communities" with which they identified themselves. In their rhetorical battles with one another and with the authorities, newspapers acted out conflicts that their texts claimed were synonymous with the real social and political struggles around them.

Contemporaries recognized the special connection between the newspaper press and the revolutionary situation of the 1830s, and the period gave rise to innumerable celebrations and lamentations about the press's power, a collective metatext that both testified to and helped establish the enormous symbolic importance of these printed sheets. Charles X's ministers had done much to inflate the reputation of journalism by singling out its pernicious influence in the manifesto that announced their unsuccessful coup. "The press has . . . disseminated disorder into the most upright minds, shaken the firmest convictions, and produced in the midst of society, a confusion of principles that yields to the most sinister attempt. Thus, by anarchy of doctrines it prepares anarchy in the state," they wrote in their "Report to the King" of 25 July 1830.[45] The July Revolution seemed to prove their point, and the victors celebrated this demonstration. The chapter on "La presse parisienne," from a collective publication, the *Nouveau Tableau de Paris*, published in 1834, praised journalism as the only genre of literature that had expressed the revolutionary rupture of 1830: "The world of periodicals is the only place where serious revolutions occurred. The periodical press doesn't resemble anything seen before." Journalism had become the freest and most effective form of publication: "In the new press, outspokenness and freedom of language have replaced hypocritical equivocations and the subtleties of the *bel esprit*."[46] In the Saint-Simonian *Revue encyclopédique*, another editorialist elaborated similar ideas. Journalism was the form of literature uniquely suited to an age of mass participation in public life: "Expression of everyone's opinion, it addresses itself to everyone. It is the voice of the people; through it, the people speak, condemn, or acclaim. It is the power of the people, for it is the means by which the sovereignty of the people is manifested. . . . Journalism promotes the diffusion of enlightenment, the implantation of ideas in every member of society."[47]

The Transformation of Identities

It is precisely its characteristics of timeliness and ability to transform their readers into collectivities capable of coordinated action—characteristics already recognized in the 1830s—that make newspapers so important for understanding the cultural dimension of modern revolutions. In 1832, one of Lyon's radical papers cited the famous slogan of the *Révolutions de Paris*, a leading French revolutionary newspaper created in the summer of 1789: "Les grands ne nous paraîssent grands que parce que nous sommes à genoux. Levons-nous!" ("Those in high places only seem to tower over us because we are on our knees. Let's stand up!"), a phrase that epitomizes the importance of understanding revolutions not just as changes in institutions but as transformations in self-concept and behavior.[48] This slogan urged no specific political action, nor did it herald a redistribution of wealth. It was revolutionary because it called on readers to change their understanding of who they were, to make themselves "new men" (and, potentially, new women) by conceiving of themselves as the equals of those who had always claimed to be their superiors.[49] The newspaper's epigraph urged readers to renounce their old repertoire of social behavior, which it stigmatized in the metaphor of kneeling, and to act in a new way, summarized in the two words "stand up," which implied a rejection of passivity and an insistence on claiming all the rights previously reserved for the privileged group. In two short sentences, the *Révolutions de Paris* of 1789 thus provided readers with a script for transforming themselves from subjects to citizens, from dominated members of a hierarchically organized society to independent participants in a community of equals. The Lyon journalist who reprinted these words in 1832 understood their continuing relevance in his own time.

All revolutions necessarily involve this sort of redefinition of identities, in which passive subjects learn to recognize themselves as active citizens, long-standing institutions become discredited "old regimes," and previously revered rulers are relabeled as tyrants and despots. But how does this process come about? Classic explanations of revolutionary crises, particularly those developed in the Marxist tradition, emphasize the importance of long-term changes in group consciousness. Marx explained the Revolution of 1789 by positing the development of a self-conscious bourgeois class in France during the eighteenth century, and he explained the revolutions of the nineteenth century by referring to the growth of an urban proletariat. The actual upheavals of 1789 and 1848 were thus merely the visible signs of more fundamental social and economic changes that had already produced new group identities before they resulted in revolutions. Recent historical research has tended to question this classic paradigm. There is little convincing evidence to support the notion that France in 1789 harbored a self-conscious bourgeoisie; it appears more accurate to suggest that it took a revolutionary upheaval to produce the conditions in which such a group could define itself. Similarly, the revolutions of 1830 and 1848 were essential catalysts for the formation of a self-conscious working class—one whose most articulate members, furthermore, never resembled Marx's model of the proletariat.[50] Rather than searching the decades before major revolutions in search of well-defined groups waiting for their cues to perform in the already scripted revolutionary drama, it now appears more fruitful to conceive of revolutions as crucibles in which new social identities take form.

If one is not to embrace the utopian illusion that new identities take shape spontaneously in the heat of revolutionary action, it is necessary to identify the mechanisms by which they are in fact generated and propagated. One of these is certainly the media of mass communication, which are crucial sites for the proposal and dissemination of new definitions of identity. A "media revolution" is an essential aspect of any major revolutionary upheaval. "Media revolutions" are characterized by the sudden appearance of new media organs that claim to speak for previously disadvantaged groups and that propose new forms of collective behavior to those groups, designed to win for them the political and social recognition they deserve.[51] To the extent that these revolutionary media are successful, they become active agents in the constitution of the groups they claim to represent, catalysts around which the new social formations created in the heat of revolution take form.

In revolutionary situations, instead of providing stereotyped frames for events that are essentially variations of familiar patterns, newspapers are called on to define and label occurrences whose significance is very much in dispute. It is in fact the press that constructed nineteenth-century revolutions *as events*. By "framing" a sequence of occurrences, separating them from the larger flux of collective experience, news media give them their distinctive status. "Both in time and in space, the event seems to depend on decisions that, by assigning it arbitrary limits, make it a legitimate object of discussion," Maurice Mouillaud and Jean-François Tétu have written.[52] As William Sewell has recently pointed out, the category of "event," normally taken for granted in historical research, in fact requires as much theoretical analysis as any other analytic construct we employ. Sewell has attempted to provide this analysis, defining an event as "(1) a ramified sequence of occurrences that (2) is recognized as notable by contemporaries and that (3) results in a durable transformation of structures."[53] Although Sewell does not comment on the role of the media in shaping events, the evidence he himself uses in analyzing how his paradigm case, the storming of the Bastille, acquired its meaning comes in good part from the revolutionary press. Sewell himself comments that "symbolic interpretation is part and parcel of the historical event."[54] Some of that work of interpretation is carried out by the actors in events themselves, as Sewell shows in the case of the Bastille, but much of it occurs in the media. As Hans-Jürgen Lüsebrink and Rolf Reichardt have written in their pioneering study of the Bastille's emergence as a symbolic event, "If one understands the 'great event' sociohistorically as social knowledge constituted through news media and through processes of communication, one could say even that the storming of the Bastille and its individual aspects 'exist' for those contemporaries only because the media reported on them repeatedly and emphatically."[55]

Introduction

Revolutionary crises can be defined precisely by the fact that they alter the ordinary definitions of events that newspapers cover. Politics, normally confined to ministerial offices or parliamentary assemblies, spills over into the street. What was a riot—a violation of the legal order—one day becomes an expression of popular sovereignty and the source of a new legal order the next. Newspapers are inseparable from revolutionary crises because they are one of the principal ways in which revolutions achieve their being. "Revolution is not revolutionary in the absolute sense of the word," the literary scholar Sandy Petrey has written. "It, too, requires representational enactment."[56] The reporting and labeling in newspapers of the improbable events that constitute a revolution—the destruction of Bastilles, the panicky flight of rulers who were all-powerful only days before—is one form of this enactment. Another is the creation of new publications, particularly those whose content, through its radicalism, its utopianism, and its violence, violates the usual norms of public discourse. Revolutionary newspapers create new senses of possibility for their readers. Elaborating on events that have occurred in the "real" world, they give them new meanings and create the conditions for further departures from existing conditions.

The press has an active impact in revolutionary situations not only by virtue of what it prints, but also through the impact of changes in the "cultural field" constituted by the spectrum of papers competing for the right to define what is occurring. The concept of "cultural field," understood as "the system of objective relations between these agents or institutions and as the site of the struggle for the monopoly of the power to consecrate," comes from Pierre Bourdieu, who uses it to discuss, among other things, the role of the press in assigning cultural values to works of art and literature.[57] But it can also be applied to the struggle of rival newspapers competing for the right to "consecrate" political claims in a revolutionary situation. Just as "the meaning of a work (artistic, literary, philosophical, etc.) changes automatically with each change in the field within which it is situated for the spectator or the reader,"[58] so the significance of newspapers changes with the continual creation of new publications that is characteristic of a revolutionary situation. Thus, in France in 1789, when new papers identified with the victorious movement appeared, the officially licensed newspapers that dated from before the revolution, such as the *Gazette de France* and the *Journal de Paris*, found themselves transformed willy-nilly into counter-revolutionary symbols of a newly created "old regime." Similarly, in Lyon in after the July Days of 1830, the dominant liberal paper, the *Précurseur*, found itself outflanked on the left by more radical publications even as it lost its monopoly on the expression of "bourgeois" opinion as a result of the establishment of a new daily loyal to the July Monarchy. The *Précurseur*'s old dueling partner, the former organ of the Restoration government, went from being a proponent of established authority to being a subversive oppositional voice.

The spectacle created by the sudden appearance of large numbers of new media organs and the resulting restructuring of the journalistic "cultural field" is one of the characteristics of a revolutionary crisis. Under these circumstances, the impact of newspapers on their readership can achieve the intensity normally reserved for texts that speak more intimately and personally to their audience. The readers who found themselves in resonance with the previously unarticulated messages of Marat's *Ami du peuple* in 1789 or the workers' paper, the *Écho de la fabrique*, in Lyon in 1831 were being urged to radically redefine their notion of their own identities and to adopt new forms of behavior. The sudden appearance of these new fictive personae—the "people" embodied in Marat's paper, the "working class" represented by the *Écho de la fabrique*—forced members of the other "imagined communities" to respond by identifying with newspapers that would define new forms of collective behavior for them, forms of behavior that would meet the challenge posed by these radical publications.

The press of the 1830s had many similarities with the revolutionary press of the 1790s, but there were also important differences between these two media revolutions. In 1789, an uncensored and unregulated press was one of the most tangible signs of the breakdown of the old absolutist order, which had had no place for such a phenomenon. After 1830, an uncensored press was supposedly one of the basic foundations of the victorious regime. In 1789, newly created newspapers had been completely outside the legal order; in the 1830s, they remained incorporated within it, a fact symbolized in Lyon by the register kept by the prefect's office in which editors regularly reported the creation, mutation, and disappearance of their titles, and by the regularity with which journalists appeared in court when summoned to answer for press offenses.

Introduction

The press of the early 1830s had a Janus face; it was both a weapon for contesting the established order, and an outlet in which contestatory impulses could be expressed without necessarily threatening existing institutions.

The chapters that follow look at several aspects of the social and cultural role of the press during the revolutionary years of the early 1830s. The first traces the evolution of the journalistic cultural field, from the monotone official press of the Napoleonic years through the bipolar press of the Restoration to the increasingly diversified situation that developed after 1830. It also examines the role of the journalists who were essential to the development of this new press, and the contours of the "public space" within which newspapers were produced and read. We then turn to a close reading of the different press genres that constituted the period's journalistic system: the liberal or bourgeois press, that press shaped by and directly linked to the new society's political and economic institutions; the alternative press, which expressed an equally strong impulse within bourgeois society to escape from the constraints of power and money and to redefine gender boundaries; and the workers' press, a conscious effort to create a counter-culture fundamentally opposed to the bourgeois world. Two subsequent chapters look at different facets of the creative interaction between the press and social and political events. Such institutionalized occurrences as public banquets and political trials were examples of the ability of newspapers to construct events, and of the conflicts that these attempts generated. Press coverage of the two Lyon insurrections of the early 1830s, on the other hand, demonstrates the dilemmas an institutionalized press faced in narrating events in which it was vitally involved but which threatened to escape from its control. We then examine the consequences that followed when the task of representing the social conflicts of early bourgeois society shifted from newspapers, the medium of fluidity and immediacy, to books, with their characteristics of permanency and stability. The story of Lyon's upheavals from 1831 to 1834 became the stuff of France's first "proletarian novel," a now-forgotten text entitled *La révolte de Lyon en 1834, ou La fille du prolétaire,* and of its first classic of modern social history, the Lyon journalist Jean-Baptiste Monfalcon's influential *Histoire des insurrections de Lyon*. These two texts, both rooted in the period's journalism, foreshadowed the ways in which bourgeois Europe's class and gender conflicts would be depicted throughout the nineteenth century and long into the twentieth.

Despite the limited scope of this study, I hope to have made a persuasive case for a new understanding of several major issues. The close reading of journalistic texts offered here shows that the French Revolution of 1830, and especially its prolongation in Lyon, was indeed a catalyzing moment in the articulation of the categories that defined "bourgeois society." It also shows that this process took place essentially through the periodical press, which must therefore be at the center of any cultural history of the period. Without an understanding of the special characteristics of the newspaper medium, it is also impossible to understand the nature of modern revolutionary crises in general, or to fully appreciate the significance of the other genres of text in which they are represented. The forgotten journalists and newspapers of Lyon provide a very rich case study of the ways in which printed texts both reflected and transformed social reality and helped create the modern world.

I

Newspapers, Journalists, and Public Space

Newspapers, Journalists, and the Public Sphere After 1830

*T*he Revolution of 1830 and the media revolution that was an integral part of it opened a new stage in the history of journalism in Lyon, and in French society in general. In the wake of the July Days, both the national press system and the press in Lyon became much more diverse than they had ever been before. Not only were there more publications than ever before, but the variety of journalistic forms available to readers also expanded dramatically. A corollary of the expansion of the press was an unprecedented growth of the number of journalists. Provincial journalism in particular had been a marginal profession until 1830; during the revolutionary interval of the 1830s, local journalists became recognized as major figures in their city's affairs, a situation that gave them significant influence but also exposed them to new dangers. Finally, the new conditions created in 1830 made the question of newspaper-reading a major public issue. Questions of who had access to the press and where and how newspapers were read indicate that the definition of what we would now call the sphere or arena of public discussion was undergoing unprecedented change. At stake was nothing less than the issue of who was to participate in the formation of that public opinion which now had supposedly become the basis for political legitimacy.

Before 1830, Lyon had supported newspapers representing only two significant currents of opinion. This simplified journalistic system corresponded to a public arena dominated by a single fundamental conflict, that between supporters and opponents of the legacy of 1789. By the end of 1831, four political movements had press organs, and the number of potential political divisions in the city was steadily growing. In 1830, all of Lyon's politically significant periodicals were conventional newspapers; in the years that followed, new publications not only broadened the spectrum of political opinion but also introduced new forms of journalistic language, enlarging the range of cultural functions performed by the press. This collection of periodicals generated a substantially different image of Lyonnais public opinion than their predecessors. The issue of whether Lyon in fact possessed a public sphere worthy of representation in print—still an open question in the 1820s—was settled. New questions replaced it: Was that public arena single or multiple? Did local periodicals represent a universal public, or specific groups in conflict with one another? Could a single style of journalism represent Lyonnais public opinion, or did that task now require a diversity of approaches?

The Revolution of 1830 had thus transformed Lyon's "journalistic field," the set of publications that competed to represent the city's political and social reality. Pierre Bourdieu has introduced the concept of the "cultural field" as a form of "relational thinking" that emphasizes "the structural relations—invisible, or visible only through their effects—between social positions." "Fields" are social systems in which the actions of any one participant affect all the others. Bourdieu's concept emphasizes the fact that cultural actors, like economic actors, are in competition with one another, although the object of that competition is not simply economic profit; in a cultural field, prestige and recognition may count for more than sales. Bourdieu himself has not specifically analyzed "fields" of competing periodicals, but in fact many of his examples are constructed on the basis of evidence from the French press, and it is clear that his method can easily be applied in this domain.¹

A caricature of the Lyon journalistic scene, dating from early 1835, represents the city's periodicals standing quite literally in a field, as though posing to illustrate Bourdieu's concept ("La Revue de Lyon," Fig. 1). The design of this caricature vividly illustrates the way in which each participant's position in the press system was defined by its relationship to other publications. In the foreground, symbolic figures standing for the city's two cultural and artistic journals in 1835, the *Papillon* and the *Épingle*, square off against each other. Behind them, in the picture's middle ground, seven figures or groups represent the city's political publications. Each is accompanied by a caption deftly defining both its political orientation and the current state of its party's fortunes. At the far left, two men are identified with the legitimist *Gazette du Lyonnais*, which "works tirelessly for the Jesuits." Next to them, a figure in Roman costume and holding a sword, representing the city's main republican paper, asks, "Why wasn't I the strongest?" reflecting that movement's defeat in the April 1834 uprising. Riding on a donkey, a top-hatted bourgeois representing the proregime *Courrier de Lyon*, a paper that had preached the importance of social order even while pushing for an armed confrontation to humble the city's workers, says, "I am for subsidized, frenzied, and convulsionary moderation." Next to him, a figure in old-regime costume, identified with the city's second legitimist paper, the *Réparateur*, draws his sword and announces, "The throne or your life, miscreants!" A hog-faced bourgeois representing the *Journal du commerce* carries the makings of a good meal and explains, "Along with the cabarets, my best subscriber is the Prefecture." Next to him, two men with working-class caps on their heads stand for the city's two rival working-class papers. The *Tribune prolétaire* rides a crawfish and complains, "What the devil! My social progress goes backward instead of forward!" His competitor, the *Indicateur*, points to two columns of marchers descending the hills in the background of the picture, representing revolutionary insurgents and soldiers respectively, and tells him, "Roll up your sleeves and do it that way," presumably meaning that only violent confrontation can help the workers. Finally, an elegantly dressed man in bourgeois costume, observing the scene seated on a rock and thus not directly implicated in the relations between the others, represents the short-lived *Revue de Lyon,* which published the caricature. In the upper left corner of the picture, a circle of figures holds hands and dances, ignoring what is going on below them; they represent "the savants of the *Athénée,*" Lyon's elitist cultural group. Just above them, a funeral procession takes an unidentified publication to the "literary burying ground."[2]

This elaborate caricature was similar to many of the images published in the leading French satirical journal of the period, Charles Philipon's *La Caricature.* The Lyon artist imitated his Paris colleagues, such as Grandville, in creating an elaborate allegory, mixing human and animal figures, that displayed a scene with many of the features of Bourdieu's competitive and interconnected "fields." As in the Paris models, the picture told not one single story but many: subsets of the allegorical figures interacted with one another within the larger framework of the overall design, as in the case of the characters representing the two rival workers' newspapers, shown conducting a private dialogue that defined their position vis-à-vis each other while at the same time forming part of the overall picture of the city's press. Caricatures in which newspapers were symbolically represented in the same way as individual political figures, by human or animal personages meant to characterize them, usually in a satirical manner, were common in Philipon's journal and undoubtedly inspired the Lyon artist, who borrowed many symbols, such as the backward-moving crawfish and the oversized syringe or clyster, from well-known political designs of the period. The popularity of such caricatures indicates how clearly periodicals had come to be understood as symbolic persons, capable of representing political movements and social groupings. "La Revue de Lyon," preserved in the Bibliothèque municipale de Lyon in a beautiful, large-format colored version, is eloquent testimony to the importance the local press had assumed in the city's public life after 1830.

A similar representation of the Lyon press scene at any time from when local periodicals were first established at the end of the old regime until the late 1820s would have shown a much simpler picture. The city had rarely had more than two newspapers at any one time. This situation corresponded to a mentality, shared by old-regime administrators, revolutionary politicians, and Napoleonic and Restoration authorities alike,

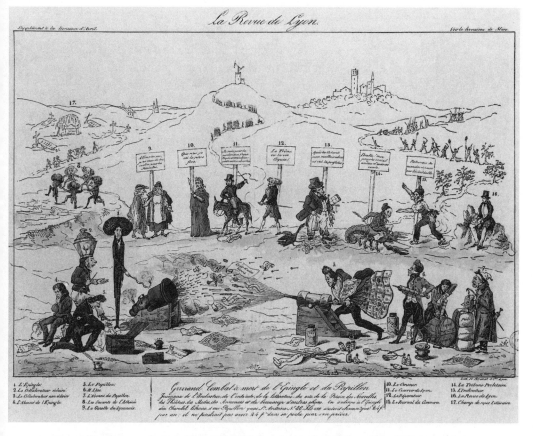

Fig. 1 "La Revue de Lyon." Published in a large, brightly colored format, this caricature portrayed Lyon's "journalistic field" in early 1835, with each periodical represented by a symbolic figure. In the foreground, two rival cultural magazines, the *Papillon* and the short-lived *Épingle*, attack each other, while the other papers, identified by captions fixing their place in the press system, look on. On the hills in the background, columns of soldiers and working-class insurgents remind the viewer of the insurrectionary atmosphere in which the city's diversified press had developed. At the upper left, an unidentified publication is taken to the "literary burying ground," a reminder of the fragility of the period's press enterprises.

that admitted the legitimacy of only one "public" opinion. A multiplicity of journalistic voices necessarily corresponded to an absence of consensus, and indeed the very existence of rival titles was bound, by the competitive nature of the journalistic medium, to create conflict. As the prefect assigned to Lyon during the First Restoration in 1814–15 wrote, "If several titles were tolerated, it is more than likely that the promoters would try to gain advantages over each other, and that, to injure each other, they would employ as a means the manifestation of contradictory sentiments about issues of politics and legislation. This could upset spirits already too inclined to become aroused and would keep alive party differences that the administration devotes all its efforts to banishing."[3]

Until the late 1820s, Lyon's press had corresponded to this model. The city's earliest periodical, an advertising sheet founded in 1748, had been joined by a literary and cultural journal in 1784.[4] Like the provincial press in other French cities, these local publications promoted local commerce and fostered the development of regional pride through articles about the area's unique history and its cultural contributions. At the same time, however, these provincial papers responded to national agendas. Particularly in the cultural arena, journals like the *Journal de Lyon* strove to demonstrate that Lyonnais readers and writers could hold their own in a cultural framework largely defined in Paris.[5] The Revolution of 1789 had contradictory effects on the development of the press in France's provincial cities, including Lyon. Increased interest in public affairs inspired the creation of new papers, but that interest was focused primarily on Parisian events, and an aggressively entrepreneurial national press claimed the largest share of this new market. In the face of the Revolution's insistence on national unity, local papers had to emphasize their contribution to integrating their readers into a larger public arena rather than their special function as organs of regional concerns.[6] "When we promised, at the time the Estates-General opened, to report all its actions, we did not foresee where this promise was going to lead us," the editor of the *Journal de Lyon* remarked at the end of 1789. "This paper has ceased to be the journal of Lyon, and become, like all the other public papers, almost exclusively the journal of the National Assembly."[7]

During some periods of the 1790s, the city briefly had two regularly appearing papers representing rival factions, but whenever one party gained firm control of local affairs, opposing papers disappeared.[8] Lyon's newspapers during the revolutionary decade served primarily to propagate national political programs defined in Paris, where rival factions worked determinedly to eliminate their journalistic opponents as soon as they seized power. For a few months in 1793, the federalist revolt against the Convention created an exceptional situation in Lyon, in which the local newspaper became an outlet for anti-Jacobin propaganda, and during the thermidorian period, the local paper reflected a counterrevolutionary backlash against the fallen Jacobin regime, but these deviations from national norms were brief.[9] In the Napoleonic period, the city reverted to the situation at the end of the old regime: it had one "political" paper, providing carefully censored news about national and international affairs, and a second paper that was devoted to advertising and local cultural articles. The Restoration government would undoubtedly have preferred to permit only a single title, but the entrenched interests of the owners of the papers authorized during the Napoleonic period prevented a fusion of the city's publications. In November 1815, after being bombarded with numerous petitions from interested parties, the minister of police finally authorized three periodicals, only one of which was permitted to report on politics.[10] Throughout the early Restoration years, public discussion about politics in Lyon was structured around reactions to the Paris political papers, rather than local publications.

In the early 1820s, there were several efforts to create local newspapers that would reflect the distinctive physiognomy of Lyonnais opinion, instead of the views dictated by the prefect or those of the Paris press. In 1820, local Ultras launched their own paper, the *Gazette universelle*, and for a few months in 1821 and 1822 a group of moderate liberals published an answer to it under a title destined to become famous later in the decade: the *Précurseur*.[11] The *Journal du commerce*, a triweekly paper devoted to business and politics, first appeared in 1823, but although it continued to publish into the 1830s, it never obtained much influence. The very notion of a local press reflecting distinctive local interests was slow to gain a foothold. In 1824, the editor of a short-lived Lyonnais periodical provided an insightful explanation for this. "I know only too well that, for several reasons having to do with the difference of location, a *provincial* journal will always be unable to present . . . the variety, the spirit, and the entertaining malice that one finds in some Paris papers," he confessed. What a local paper could offer, he claimed, was "local color, which, for the Lyonnais at least, should not be completely without importance." The problem, according to this would-be journalist, was that the Lyon public did not provide the necessary base of support for a local press. "Do I have to say that it has even happened that a *Parisian* paper has, on more than one occasion, told us about events that took place here a month earlier?" he expostulated.[12]

The second half of the 1820s saw a significant new development. In spite of legal obstacles and strong opposition from the authorities, Lyon liberals succeeded in creating a press that was more than an ephemeral phenomenon. Acting on their own initiative—the national liberal political organization, called "Aide-toi, le ciel t'aidera," was not yet in existence—the backers of the short-lived *Précurseur* of 1821–22 restarted that paper in 1826. Unlike the three-day-a-week *Journal du commerce*, the *Précurseur* was a daily and its print format was as large as that of many of the Paris papers. Its success meant that Lyon for the first time had a press organ that could truly stand comparison to the political press of the capital. The local authorities could hardly ignore the *Précurseur*'s open challenge to their control of the press and its bid to completely restructure the city's journalistic cultural field, and they therefore tried strenuously to suppress the paper, only to be foiled by local judges sympathetic to the liberal opposition. After several defeats, the authorities gave up challenging the paper's right to exist and settled for dragging it into court repeatedly for its content, although the judges continued to throw out most charges against it.[13] The *Journal du commerce* proved equally resistant to official warnings.[14] The fall of the Villèle ministry at the end of 1827 and subsequent changes in the laws governing the press appeared to stabilize the new situation. The two liberal papers complied with the new legal requirements for registration, and the authorities ceased contesting the principle of their right to publish.

The problem the Restoration authorities faced was that the consolidation of the liberal press seemed to go hand in hand with the disintegration of its proregime rival, threatening to leave the liberals a monopoly on the expression of public opinion in print. Lyon's original right-wing paper, the *Gazette universelle*, had long been under attack from all sides. Even the local prefect found it too extreme; he urged the government to encourage a rival publication.[15] By the end of 1828, the

Gazette universelle had ceased publication, replaced by the *Écho du jour*, which disappeared in its turn in January 1830, its place taken by the *Gazette de Lyon*.[16] Despite covert backing from the prefecture, Lyon's conservatives were too few and too divided among themselves to compete effectively with the liberals. Both the prefect and the Parisian authorities recognized the danger posed by the success of the *Précurseur*, "which causes such damage in Lyon, throughout the south, in Dauphiné, in Beaujolais, the Dombes, the Bresse, etc."[17] The regime's opponents had won the privilege of claiming to represent the city's public opinion, which consequently appeared as a homogeneous force hostile to the government.

The July Revolution in Lyon thus seemed to represent the conversion of the liberal opposition's dominance of public opinion into control of the government. With the liberals' triumph and the disappearance of the pro-Bourbon *Gazette du Lyon*, their newspapers became the self-proclaimed organs of a new regime and of an apparently unified public opinion. In the *Précurseur*'s own view, both the paper and the government successfully combined the principles of democracy, constitutionalism, and monarchy.[18] The *Journal du commerce*, it is true, leaned slightly to the *Précurseur*'s left, protesting against a tendency to "give a total revolution the wan appearance of simple change of ministry," but the local authorities never expressed any concern about this paper.[19] Initially, then, it seems that the July Revolution did not really unleash a transformation of the press, but rather created a press that unanimously supported the new order. At the national level, spokesmen for the new regime had no interest in encouraging a radical restructuring of the press. Parliamentary debates in October 1830 resulted in the maintenance, with only minor modifications, of the laws regulating political publications. Like the stamp taxes imposed on the British press during this period, these measures were explicitly intended to limit the number of papers, to restrain their radicalism, and to raise their price to a level where only the well-to-do could afford them. That these regulations deliberately distorted the connection between public opinion and what appeared in the press was openly acknowledged in legislative debates. When a liberal deputy made the classic argument that "journalists are sentinels posted to observe and publicize all the actions of those in power" and complained that "to weaken them by requiring sizable deposits of capital, by taxes that limit their distribution, is to act like the former government," a more conservative opponent replied by comparing newspapers to other enterprises that endangered public safety and were consequently required to provide assurances of reparation for the harm they might cause: "Great God, what a business, . . . capable, in a short interval, of spreading throughout France a lie . . . that will be avidly taken up, whereas the retraction will be ignored."[20]

The new July Monarchy nonetheless created conditions in which a more diversified press could emerge. Would-be newspaper publishers no longer needed prior authorization from the prefect, and the regime, born out of the press's successful opposition to arbitrary government, was too strongly identified with the ideal of press freedom to do away with it entirely. The reintroduction of jury trials for press offenses, suspended since 1822, promised journalists a chance to defend themselves in court if necessary. The new regime's attempt to position itself as a government of the middle also spurred the formation of new papers. As Marcel Gauchet has argued, it was under the July Monarchy that the natural dynamic of the French party system first took shape. The government's effort to stay independent of political parties while claiming the center of the political spectrum created divisions within the political currents to both sides of it, as moderate rightists and leftists sought compromises with those in power, and as more intransigent members of those groups reacted by distancing themselves still further.[21]

In Lyon, the first signs of this process sufaced in the spring of 1831. The earliest alteration of the postrevolutionary journalistic system was the appearance of the legitimist *Cri du peuple* on 7 May 1831. The situation of the last years of the Restoration was restored: the city now had a clearly antigovernment paper to go with the liberal press. In June 1831, two new papers, the *Glaneuse* and the *Sentinelle,* were established to criticize the newly created July Monarchy from the left. Although neither used the term "republican" to characterize itself, both made their dissatisfaction with the Orleanist regime unmistakable from the outset. The *Sentinelle* introduced a new political viewpoint into the Lyon press, but it used the standard printing format of the large Paris political dailies; it was a political innovation but not a stylistic one. The *Glaneuse,* on the other hand, was new in its format as well as its politics. It professed to be apolitical and presented itself as Lyon's first equivalent to the

Paris "boulevard journals," promising to concentrate on entertainment and cultural issues rather than politics. In fact, however, from its very first issues it took "the fat-bellied doctrinaires of the *juste-milieu*" as its target.[22] The *Glaneuse* delivered its message through satire, through stories presented as fiction rather than factual reporting. It thus broadened not only the range of political views represented in the press but also the language in which politics could be discussed.

The process of diversification begun by the *Cri du peuple* and continued with the founding of the *Glaneuse* and the *Sentinelle* changed the structure of Lyon's journalistic field. The *Précurseur*, which had positioned itself after July 1830 as the sole significant expression of the city's public opinion, had to redefine itself in relationship to its new competitors. Although it claimed to favor a diversified press, the *Précurseur* questioned the new papers' legitimacy as representations of Lyonnais public opinion. It stigmatized the *Cri du peuple* as proof that "the Carlists of Lyon are either not strong enough to have a regular organ giving serious expression to their doctrines, or . . . out of fear or for tactical reasons, they don't want to have a real tribune." As for the *Sentinelle*, the older paper dismissed it as both "extreme" and as non-Lyonnais: "They needed to avoid having people say, 'Who are those people? Where do they come from?'"[23] The *Sentinelle* disappeared in October 1831, but the growing divisions within Lyon's public also created a crisis within the *Précurseur* itself that was publicized at length in its own columns, which resulted in that paper becoming the representative of moderate left-wing opposition to the new government, with the *Glaneuse* to its far left, the *Journal du commerce* in between, and yet another new publication, the *Courrier de Lyon*, taking up a position to the right of the *Précurseur* as the acknowledged representative of the Orleanist government. By the beginning of 1832, the heritage of Restoration liberalism was thus divided among four different publications. The legitimist opposition was no more immune to this process. In August 1833, a new title, the *Réparateur*, offered itself as a more moderate alternative to the rhetorically intemperate *Gazette du Lyonnais*, the successor to the *Cri du peuple* of 1831.

By the time the *Réparateur* appeared, it was clear that the 1830 revolution had not only allowed a broadening of the number of positions along the political spectrum represented by conventional political newspapers but also had opened a space for the emergence of alternative forms of journalism. The more liberal atmosphere following the overthrow of the Bourbons encouraged activists to put themselves forward as the voice of groups that had previously been shut out of public debate. In Paris, several short-lived papers claiming to speak to and for workers, whose role in the July Days street fighting had been so significant, appeared in the fall of 1830,[24] and the abbé Lamennais and his followers turned to the press to promote their brand of liberal Catholicism. In Lyon, the *Glaneuse,* which first appeared in June 1831, had been the first example of this kind of alternative journalism. Although it occupied a clearly identifiable position on the spectrum of political papers, it differed sharply from papers like the *Précurseur* in tone and style. In October 1831, another form of alternative press appeared: the first workers' paper, the *Écho de la fabrique.* It was destined for a much longer career than any of the workers' papers in Paris; one or the other of its various direct successors remained in business until August 1835. The *Écho de la fabrique* was thus the first genuinely successful French newspaper that self-consciously claimed to represent workers and their interests. Eager to claim nonpolitical status and thus avoid having to post a security bond, the *Écho de la fabrique* introduced the techniques of partisan journalism to the coverage of the various conflicts between Lyon's wholesale silk merchants and silk weavers. Forbidden to discuss government policies and debates in the Chamber of Deputies, it filled its columns with discussions of political economy, such as debates about the consequences of mechanization in industry. The *Écho de la fabrique* thus extended the sphere of Lyonnais public discussion to take in the world of labor. The *Écho*'s significance considerably exceeded its limited circulation. By claiming to represent a specific social class—workers—it challenged the notion that any newspaper could speak for a public opinion representative of the whole of society. The very existence of the *Écho* thus redefined the nature of all other papers in Lyon's journalistic system, and raised the question of whether they too represented specific class interests.

The Lyon press continued to diversify in 1832 and 1833. In June 1832, another "alternative" publication was founded: the *Papillon,* a biweekly that claimed to be apolitical and promised to devote itself to culture and the arts. This paper bore the subtitle "Journal des dames" and was the first Lyonnais periodical to appeal especially to women. In October 1833, an avowedly feminist publication, the *Conseiller des femmes,* published its first number. In contrast to the *Papillon*, this weekly emphasized serious subjects and limited itself to articles actually written by women. Like the workers' press, the very existence of a "woman's journal" proclaimed the impossibility of ever reuniting all of Lyon's readers into a single, unified public. By this time, the same tendency toward ideological fission that had led to multiple left-wing and legitimist newspapers had seized the workers' press, as the *Écho des travailleurs* appeared to compete with the *Écho de la fabrique*. Government intrigue had converted the *Journal du commerce* from a left-wing publication into a pro-Orleanist paper, meaning that it now duplicated the function of the *Courrier de Lyon*. The spectrum of local papers may have been limited, compared with the Parisian press, but it was still sufficient to give public debate in the city a complexity it had never previously had. The bipolar confrontation characteristic of the late 1820s had given way to a multiplicity of viewpoints; the possibility of any one of them achieving the hegemonic status that would have allowed it to represent itself as *the* voice of public opinion had disappeared.

The last months of 1833 witnessed an even more radical widening of the public sphere, with the creation of several cheap republican pamphlet-journals intended to bring politics to a broad popular audience. The two rival republican papers, the *Précurseur* and the *Glaneuse*, each sponsored such an enterprise. These small, undated pamphlets—the absence of dates and numbers was an effort to get around the laws requiring registration for political newspapers—were designed to be sold by street vendors rather than by subscription, and their intended audience was the large portion of the population that could not afford access even to publications like the *Écho de la fabrique*. One issue of the *Précurseur du peuple* claimed that more than 15,000 copies of some of its numbers had been circulated, more than ten times the circulation of any of the regular newspapers.[25] This represented a dramatic broadening of the audience for printed political texts in Lyon.

In the hectic first months of 1834, Lyon's political press had thus grown to unprecedented size: nine significant periodical publications, together with the popular pamphlets. In the immediate aftermath of the Revolution of 1830, the city's press had appeared to represent an almost unanimous public opinion; the *Précurseur* dominated the scene, seconded by the *Journal du commerce* and opposed only by an insignificant legitimist title. By early 1834, the press was the image of a completely fractured public. Each of the country's three main political currents—proregime liberalism, Catholic and monarchist legitimism, and democratic republicanism—had its own journalistic representation. All three currents, in fact, had two titles claiming to represent their interests, and in this way the press was a living demonstration not only of the divided nature of French political opinion as a whole, but also of the tendency toward division within each of its various movements. Through the newly founded popular pamphlet press, the republican movement was in the process of dividing itself even further, into separate discourses aimed at educated and uneducated readers. The tendency toward dispersion had spread from the "bourgeois" political press to the workers' papers, where the two *Échos* battled with each other for the right to represent this new public. The *Papillon* had to compete with the *Conseiller des femmes* for women readers and with several ephemeral enterprises that offered the same kind of cultural journalism it provided.

Lyon's press had become increasingly diversified, not only in political tone and in terms of expected audience, but also in terms of format. Of the papers appearing in early 1834, five—the *Précurseur*, the *Courrier de Lyon*, the *Journal du commerce* and the two legitimist papers—were standard daily newspapers appearing on large printing sheets in three-column formats. These papers had all paid the bond required for the right to cover the full range of political topics, which was what they devoted the majority of their columns to. All featured regular editorial commentary on current events, in which their political principles were applied to specific situations as they arose. All assumed that their readers were familiar with the vocabulary they employed in their stories, and with the main ongoing issues in local, national, and international affairs. Four of Lyon's papers—the *Glaneuse*, the *Papillon*, and the two *Échos*—used a smaller, two-column format on tabloid-size paper. These papers appeared twice or three times a week. While they resembled each other in format, the *Glaneuse* and the *Papillon* differed considerably from the workers' papers in tone. In contrast to the "serious" political press, the *Glaneuse* was constantly shifting rhetorical registers, subverting both the authority of political institutions and that of standard political discourse. The tone of the workers' press was different. The *Glaneuse* attacked authority by emulating the new regime's habit—according to its critics—of saying one thing while meaning another. The *Échos* attacked bourgeois society by exemplifying a univocal moral discourse, justifying their existence by their claim to speak for the oppressed and the excluded. This moral tone linked the workers' press to the women's paper, the *Conseiller de femmes,* which also positioned itself as the voice of an outsider group with a moral claim to recognition. The ephemeral republican pamphlet press of January 1834 had some of the moralizing tone of the workers' and women's papers, but its vocabulary and rhetoric was closer to that of the serious political press. Restricted to a small format, and compelled to communicate its meaning simply and directly, such a press could not play the language games of a publication like the *Glaneuse:* it had to put its points across clearly and unequivocally.

One uniquely important reader, the prefect Adrien-Étienne Pierre de Gasparin, followed the evolution of Lyon's rapidly changing journalistic field and commented on it more closely than anyone else in these years. Other public officials—Gasparin's predecessor Bouvier-Dumolard, sacked after the November 1831 insurrection, as well as the city's mayor, the police chief, and the public prosecutor—undoubtedly did likewise, but the prefect had the special responsibility of interpreting the Lyon press to his superiors in Paris. Gasparin's almost daily letters to the minister of the interior and other authorities in Paris, the only such prefectoral correspondence from this period to have survived, allow us to follow the relationship between the government and the city's newspapers in remarkable detail. As the most visible manifestation of local political life, the newspapers were signs that Gasparin anxiously tried to decipher in order to predict outbreaks of opposition and measure the extent of support for the government's policies. Not content to simply read what appeared in the papers, the prefect sought information on the inner workings of the major titles; he knew the editors and understood the difference between the public affirmations in their articles and their private points of view. He was also well informed about the internal dissensions affecting most of the papers, and he regularly tried to profit from them to promote government policy. Arriving in the city just two weeks after the 1831 uprising, he immediately attempted to put loyal supporters of the Orleanist regime in charge of the *Précurseur,* the city's most important title, and when that effort failed he encouraged the creation of the *Courrier de Lyon,* a rival daily. In subsequent years, he tried to influence the choice of an editor for the *Écho de la fabrique,* arranged for a subsidy to change the orientation of the *Journal du commerce,* and even became a stockholder in the safely apolitical *Mosaïque de Lyon.*[26]

Active though he was in press affairs, Gasparin always realized that his power over the newspapers was limited. Gasparin's predecessor had complained that the government gave him no funds to try to buy off hostile editors; even after the 1831 insurrection had shown how unstable the situation in Lyon was, Prime Minister Casimir-Périer's admonitions to the new prefect about managing the press continued to remind him that money was scarce.[27] Gasparin quickly grasped some of the paradoxical features of the newspaper world: he informed Casimir-Périer that he had decided against a secret purchase of the *Journal du commerce* because he feared that the owner would simply use the money to start a new paper in an effort to extort even more money from him.[28] In 1833, Gasparin's effort to influence the direction of the *Écho de la fabrique* proved equally frustrating. The editor he worked to oust founded a new paper, the *Écho des travailleurs,* and his supposedly more moderate replacement steered the original workers' paper on an even more confrontational course. In May 1834, after the republican insurrection, Gasparin prepared a lengthy and detailed report on the press and the outcome of his efforts to manage it. His conclusions were discouraging. The only paper he could count on to obey his instructions unquestioningly, the *Journal du commerce,* had so few subscribers that "the results don't go very far." Each of the other papers, including the avowedly Orleanist *Courrier de Lyon,* had sufficiently strong roots in some sector of Lyon society and opinion to successfully ignore prefectoral influence. In effect, Gasparin recognized that, under the ground rules of a liberal constitution, he could not prevent dissenting voices from being heard. He hoped that the installation of a more capable team of editors at the *Courrier de Lyon* might improve the situation, but the only really effective policy he saw was to prevent the publication of hostile titles by secretly putting pressure on the city's printers.[29]

Gasparin's inability to manage the Lyonnais press demonstrated the relative autonomy of the city's journalistic field. Its conflicts had their own logic; the prefect had a clear overview of them, but his power to shape their outcome was limited. The "media revolution" set in motion by the events of July 1830 was largely out of his control. In contrast to the Parisian press revolution of the 1790s, economic forces had little to do with this development in Lyon. None of the papers in the city in the early 1830s appears to have made money for its publisher, although the papers certainly helped support the printers and in some cases provided a decent living for their editors. Of the various strategies for enlarging the press market, the one that was ultimately to have the greatest impact on the French national press—the development of profit-making periodicals supported by commercial advertising and aimed at a mass market, along the lines of Émile de Girardin's *La presse*, launched in 1836—was conspicuously absent in Lyon during these years. Although they developed within the framework of a commercial society, Lyon's new papers competed for ideological rather than economic success. Funded by wealthy patrons or, in the case of the two *Écho*s, by master silk weavers who were better off than the mass of the city's working class, these publications were willing to lose money in a bid to gain cultural capital and political influence. Their multiplication reflected a widespread recognition that, in an increasingly literate society, the printed word had become the principal instrument through which struggles for power were waged. From 1789 onward, as François Furet has argued, French politics had become a competition for power among rival discourses. The novelty of the situation after the Revolution of 1830 was the emergence of new discourses aimed not directly at the conquest of political legitimacy but at redefining the notion of the public or challenging the very notion of a unified public opinion, as Lyon's working-class and feminist press did.

The revolutionary situation of the early 1830s facilitated this process. With so much at stake in ideological terms, "movement entrepreneurs" representing not only rival political views but also such groups as workers, women, and "youth" were willing to ignore the difficulty of making a profit from their efforts in order to claim their share of public attention. The prevailing sense of crisis created a feeling of urgency and an uncertainty about outcomes that made journalists willing to risk the possibility of legal punishment. If the revolutionary rupture of 1830 permitted the diversification of the press, that process in turn intensified the revolutionary atmosphere of the years that followed. Under the Restoration, the Lyon press "field" was structured by a simple division between proregime and antiregime publications. Their antagonism suggested the possibility of a triumph of liberal principles and a change of regime, a rather simple exchange of places between winners and losers. After 1830, however, the press became a microcosm suggesting a far more complex range of possibilities. The emergence of a militant republican press meant not only the definition of a third political alternative but also the possibility of an open-ended "perpetual revolution" in which new ideological possibilities would constantly appear at the extremes of the existing political spectrum. The tendency of each political current to generate not one but several press titles created further uncertainty: no political movement seemed capable of expressing itself in a unified fashion. The appearance of several forms of alternative press was a graphic demonstration of the existence of challenges to the prevailing social order whose force had barely been suspected before 1830.

Journalists: Bourgeois on the Margins

The change in the structure of the Lyonnais press was closely related to a change in the character of the city's journalists. The new, diversified, and outspoken press required more professional and more committed journalists. In symbolic terms, *Précurseur* editor Jérôme Morin's widely publicized defiance of the police in July 1830, the most spectacular episode in the city's local version of the revolution, opened the way for an era of outspoken, individualistic newspaper writers. The post-1830 journalists became prominent personalities in the community, courted and feared. Necessarily literate and generally "bourgeois" by virtue of their social origins and style of life, the provincial journalists of Lyon were nevertheless marginal figures in the city's society. It is true that they were not stigmatized as "prostitutes," as Parisian newspaper writers of the epoch frequently were; the less commercial nature of the Lyon press and the small scale of the community, which made it difficult for writers to change their political identifications, made this kind of attack implausible.[30] But the necessarily public nature of the journalists' own activities in a society in which most people rigorously guarded their privacy, and the power of violating others' privacy that their occupation gave them, set them apart from the rest of the population.

The public that regularly read newspapers and talked about them within the confines of Lyon's bourgeois public sphere was, of course, a group whose members could read and write. But Lyon's reading public was not a community of writers. Only a minority of its members were truly comfortable expressing themselves in print. In 1826, the *Biographie contemporaine des gens de lettres de Lyon* enumerated 115 local authors, a total reached by including almost every local inhabitant who had ever published as much as a poem.[31] The question of whether Lyon, or for that matter any French city outside of Paris, could sustain a local intellectual elite was a burning one in this period. On the eve of the July Revolution, a Lyonnais writer and sometime journalist, Sébastien Kauffmann, devoted a mock-heroic poem to the effort of local authors to persuade Lyon's theaters to perform their plays, rather than only works composed in the capital, "as if talent only existed in Paris." Kauffmann's *Célestinade* delivered a mixed message, however. The author could not help admitting that most of the local playwrights' productions deserved the oblivion to which they were quickly consigned.[32] The contributors to a collective volume of essays about the city published in 1834, *Lyon vu de Fourvières*, intended to demonstrate that the city now had not only readers but capable writers. Its preface, written by Anselme Petetin, the one established Paris journalist who had migrated to Lyon after 1830, reached a pessimistic conclusion about the status of the local literati. None of them had attained the level of Paris; Lyon simply lacked the social support for real literature. "You want to have a local literature! Start by creating a local society for yourselves," Petetin advised.[33]

Both Kauffmann and Petetin acknowledged, however, that there was one category of writers who had managed to create an opening for themselves in Lyon: journalists. For Kauffmann, journalists were the essential intermediaries who could actually turn opinion in favor of the local playwrights, and Morin, then editor of the *Précurseur*, was one of the local figures who figured in his poem.[34] Petetin acknowledged that provincial newspapers had thrived after the July Revolution and had taken on a new cultural importance: "The political papers enlarged their formats to make place for a *feuilleton* [a section at the bottom of the page, often reserved for articles on cultural topics]. Every *département* center had its Janin and its Balzac, whose local celebrity eclipsed that of the great stars from Paris for ten miles around."[35] Even before the Revolution of 1830, the national press had recognized the importance of regional newspapers. According to the *National,* "The *département* press has taken the initiative on all local issues and has seconded the Paris press on general issues. Its establishment has begun the emancipation of the provinces and will continue to extend it."[36]

Journalists might not qualify as true writers in classifications based on nineteenth-century hierarchies of literary genres, but they were clearly specialists in the composition of printed texts, architects of a significant print culture. And the situation in Lyon required journalists who were willing to speak out. Their city was no sleepy provincial backwater, where a local newspaper could function by reprinting stories clipped from the Paris press. Caught in the midst of events that they knew had national significance, the editors of Lyon's papers had to make their own decisions and speak in their own voices, or recruit local contributors to do so for them. Their newspapers were a genuine form of locally produced literature, and, even if their literary talents were modest, the city's journalists were writers whose words made a real impact in the world.

The group of writers and editors who dominated the city's print culture in the years around 1830 were not the first significant journalists to have lived and worked in Lyon. At the end of the old regime, Mathon de la Cour, editor of the *Journal de Lyon,* had been a major figure in local life; during the thermidorian reaction, Alexandre Pelzin had made that paper a much-feared force in favor of purging former Jacobins.[37] The list of local writers published in 1826 named three who were identified primarily as journalists. What was new in the 1830s was that the role of journalist had become more institutionalized, and that the city's press market had grown to the point where it could support not just one or two such writers but a small-scale journalistic community. The hiring in November 1831 of the veteran Paris journalist Anselme Petetin as editor of the *Précurseur* was a symbolic watershed—for the first time, a Lyon paper succeeded in attracting a professional away from the capital. Petetin was the first local journalist to be explicitly identified as a journalist: a clerk at the Prefecture initially registered him under the vaguer term "homme de lettres," but those words were then crossed out and the word "journaliste" was substituted.[38] But Petetin, skilled though he was, did not overawe his Lyon rivals. He had to compete with a half-dozen other newspaper writers of real talent and dedication: Jean-Baptiste Monfalcon, spokesman for the government's local supporters; Adolphe

Granier, the vehement republican editor of the *Glaneuse;* Marius Chastaing, France's first important "worker's" journalist; Théodore Pitrat and Senones, the dedicated if none too skillful editors of the city's legitimist papers; and the remarkable Eugénie Niboyet, creator of the *Conseiller des femmes*. The division of a formerly unified public opinion created a situation in which there were necessarily a multiplicity of rival journalists.

One reason the journalist did not figure in the conventional hierarchy of literary figures was that he or she usually lacked the autonomy of a "true" writer. The journalist's role was tightly defined by laws and contracts and by the constraints of the profession. With a few exceptions, such as the eccentric legitimist editor-publisher Théodore Pitrat and the feminist Eugénie Niboyet, the journalists active in Lyon in the early 1830s did not create or own their own papers. All contracted to perform certain specified functions and accept certain responsibilities, in exchange for which they received monetary compensation. The contract Jérôme Morin signed with the *Précurseur* in December 1828 gives a good idea of what the obligations and rewards of the profession were. Morin committed himself to devote full time to overseeing the publication of the paper and the management of its office. He was responsible for paying a co-editor, a business manager, office boys, a bookkeeper and a handyman out of the 7,500 franc annual allotment the paper guaranteed him, so he was an administrator as well as a man of letters. In return, he was to receive 30 percent of the paper's profits up to 12,000 francs, and 10 percent of everything it earned beyond that point—undoubtedly a handsome income compared with what any of Lyon's poets or dramatists could hope to earn from their publications.[39] The republican journalist Adolphe Granier, editor of the *Glaneuse*, had a more modest contract, as befitted a paper appearing only twice a week. He was nevertheless guaranteed a minimum of 2,400 livres a year in salary, with a possible bonus of up to 600 livres if the paper made a profit.[40]

Journalists' roles were not defined only by their contracts, however. The Restoration's press legislation, taken over in this respect by the July Monarchy, required that a newspaper editor also serve as the paper's *gérant*, or legally responsible director. The *gérant* had to own a share in his newspaper—the fact that women could not legally control their own property made it impossible for Eugénie Niboyet to be the *gérant* of her own publications[41]—and he was legally responsible for the paper: if it ended up in court, he stood in the dock and suffered whatever consequences that entailed.[42] The requirement that the *gérant* be part-owner of the paper was intended to ensure that he would suffer financially as well as personally if the paper was convicted of a press-law offense. Whereas the literary author was an idealized figure, theoretically free to follow his or her muse, the journalist was a model of the legally defined bourgeois citizen, compelled to obey the laws on pain of punishment to his person and his pocketbook. These restrictions were frequently undercut by subterfuges. Many *gérants* were straw men whose only function was to go to jail if the paper lost a court case—in 1831, the *Précurseur* went even further and employed a *gérant* who turned out to have died some time before his name was removed from the paper's masthead[43]—and the paper's stockholders often provided the *gérant* with the funds needed to purchase his required share. But the restrictions of contract and law did make the journalist something less than a truly autonomous man of letters.

The early nineteenth-century journalist was not entirely at the mercy of his or her employers and the courts, however. Journalists proclaimed their moral autonomy through the institution of the published declaration of principles with which most of them began their activities. When he was invited to apply for the position of editor of the *Précurseur* in 1831, Anselme Petetin presented the stockholders' representatives both with his financial demands and with a written statement of his political principles; if the paper wanted to hire him, it was doing so in the knowledge that he would use it to promote those ideas.[44] Printed in the paper when he took over the editorship, Petetin's *credo* became a contract, not just between him and the stockholders but also between him and the paper's readers.[45] When Marius Chastaing became editor of the workers' paper, the *Écho de la fabrique,* he made a similar personal statement.[46]

The journalist's statement of principles might preserve the writer's appearance of individual autonomy, but it could not free him from the constraints imposed by the newspaper medium itself. Thanks to the memoirs of Jean-Baptiste Monfalcon, we have a vivid portrait of the life of a successful Lyon journalist in the early 1830s. Monfalcon, who was also the official physician of the large Perrache prison and the Hôtel-Dieu, Lyon's charity hospital, nonetheless put in at least seven hours a day working on the *Courrier de Lyon.* He wrote the daily editorial, the "premier-Lyon," and many of the paper's cultural and literary articles as well, and he also oversaw the selection of articles from the Paris press. Although he had a capable assistant editor, it was Monfalcon's job to work out problems with the paper's six-member administrative council.[47] Unlike the editor of a large Parisian paper, who could rely on a team of collaborators, all with their own fields of specialty, the Lyon journalist had to be ready to write on all topics. The necessity of writing every day and under the pressure of deadlines was frustrating: "One writes too fast to write well."[48] Although they were bitter political enemies, Monfalcon recognized that his rival, Anselme Petetin, had the skills journalism required: "a cutting style, a talent for the telling remark, always with something to say though sometimes wordy, and finally, a great facility for recycling the clichés of the radical opposition."[49]

More than the Paris journalist, the provincial writer lived in direct contact with the denizens of the local public sphere. "A provincial newspaper editor," Monfalcon wrote, "makes many enemies no matter how careful he is. Everyone knows him and holds him responsible for everything. He can't take a step outside his office without bumping into the people he has just written about." Rather than enabling the journalist to feel that he was the natural voice of the public, or at least of the segment of it he had chosen to speak for, however, this closeness actually became oppressive for most of the Lyon journalists. As Lyon's public became increasingly divided after the July Revolution, editors found themselves caught in violent crosscurrents. Less than a year after his defiance of the July Ordinances had made him the city's liberal hero, Jérôme Morin told his readers that he could no longer stand the strain of being at the center of their disputes: "It was easier to profess liberal doctrines in the face of the royal prosecutors and the judges of Charles X. The dangers we ran were at least compensated for by the applause of our friends. . . . Today, to satisfy half of our friends, we have to wound the other half, blot out the memories of an old struggle carried on together, and break old bonds of esteem that were often cemented by those of favor and gratitude. We admit that we don't have this kind of courage."[50]

Morin's successor Petetin fared no better. When the Paris republican journalist Armand Carrel made inquiries about the *Précurseur*'s problems in 1833, the Lyonnais republican lawyer Jules Favre explained that Petetin, despite his journalistic skills, had alienated many of the paper's supporters. "These disagreements are mostly due to his ignorance and to his contempt for our Lyonnais susceptibilities," Favre wrote.[51] On his side, Petetin complained of "the isolation in which I lived in Lyon."[52] Running afoul of their own supporters was not an experience limited to republican journalists. Being fired from their papers or quitting in exasperation because of quarrels with stockholders' committees supposedly committed to the same political views were experiences shared by almost every significant Lyon journalist of the period, including Petetin, Monfalcon, and Chastaing.

Losing a job was not the worst danger that closeness to readers posed for Lyon's journalists. One particular hazard to the profession was the risk of being challenged to duels. Although no Lyon editorialist suffered the fate of the Paris republican writer Armand Carrel, who was killed in a duel with the rival newspaperman Émile de Girardin in 1836, several were injured, including the assistant editor of the *Courrier de Lyon*. In his autobiography, Monfalcon vividly recalled the constant tension he and his wife suffered because of the repeated challenges he had to endure. Despite his keenly perceived sense of the idiocy of the procedure and his awareness that duels were sometimes deliberately provoked by political groups bent on intimidating their opponents, he and other journalists felt compelled to conform to the masculine code of honor that the postrevolutionary bourgeoisie had taken over from the aristocracy it had destroyed in 1789.[53] As Monfalcon put it, "I knew that our strength was above all a moral one; as long as I was a journalist, I had to take into account the demands of opinion." Not only did he accept challenges made to him, but in 1834 he himself challenged a legitimist editor whose paper had printed an unfavorable review of his book about the Lyon uprisings.[54] The sentiment of an obligation to duel extended even to the editors of Lyon's self-proclaimed working-class press, despite their determination to distinguish themselves from the bourgeoisie. At the beginning of 1834, in the midst of the crisis that would soon lead to the great general strike of the silk weavers and the April uprising, representatives of the two hostile papers, the *Écho de la fabrique* and the *Écho des travailleurs*, met on the field of honor, although they agreed to a reconciliation at the last moment.[55] Only Eugénie Niboyet was safe from the threat of pistols at dawn—perhaps the one incontestable advantage she derived from the gendered behavior codes of nineteenth-century France.

Duels were perhaps the most dangerous hazard of the journalistic profession, but hardly the only one. For members of the city's educated class, journalists faced an exceptionally high risk of being put on trial and imprisoned. Between 1831 and 1834, Lyon's highest court, the *cour d'assises*, heard a total of twenty-four trials involving thirteen different journalists accused of press offenses. At least twelve trials seem to have resulted in convictions.[56] Other journalists were arrested, but released without being charged, found themselves facing lesser tribunals, or, in 1835, were sent before the Cour des Pairs in Paris, which prosecuted those accused of involvement in the April 1834 insurrection. Journalists could count on their supporters to pay the fines imposed on them, but they could not always escape serving their prison sentences in person. The legitimist editor Théodore Pitrat claimed to have spent twenty-five months in prison during the seven years he ran the *Gazette du Lyonnais*.[57] Lyon's Perrache prison may have been, as an official inspector claimed, "a model for most prisons in France," but incarceration was still an unpleasant experience.[58] The proregime journalist Monfalcon, who was also the Perrache prison's official physician, made a point of trying to help colleagues, even those of opposing views, who found themselves behind its walls, but he was not always successful, as a letter from his rival Petetin indicates.[59]

Given the demands and risks of their occupation, it is not surprising that most of Lyon's journalists in the early 1830s seem to have been relatively youthful individuals with strong commitments to their ideals. Although she differed from her colleagues by virtue of her sex, Eugénie Niboyet was in many ways typical of the group. Born in 1796, she had thrown herself into the Paris Saint-Simonian movement after the 1830 revolution. Her letters leave no doubt about her dedication, her leadership abilities, and her willingness to speak out on behalf of her own ideas. When the movement's male leaders decided to rein in female activists in December 1831, she wrote an indignant protest, announcing, "I like to work with the masses because that's where I have a sense of my power!"[60] She brought these same qualities to her effort to create a woman's periodical in Lyon in 1833–34 and to the many journalistic enterprises she launched later in her life. Marius Chastaing, the principal editor of the workers' paper *Écho de la fabrique*, was about the same age as Niboyet, having been born in 1799. He had a longer record of activism, dating back to the early years of the Restoration, and, like Niboyet, he would remain an activist and journalist throughout the decades that followed.[61] A similar pattern of lifelong commitment characterized Jean-Baptiste Monfalcon, the editor of the proregime *Courrier de Lyon,* born in 1792. After his brief journalistic career, he would spend the rest of his life defending the rights of property and preaching the importance of a stable social order. The legitimist publisher and newspaper editor Théodore Pitrat, somewhat older than most of his colleagues, was also cast from the same mold. He published his first monarchist newspaper in 1819 and was still unrepentantly promoting opposition to bourgeois liberalism in 1848.[62] The moderate republican Anselme Petetin was still proclaiming the need for state intervention to aid the poor in 1849, as he had in 1832.[63]

Lyon's journalists were deeply divided over ideological issues, but they showed a certain nascent solidarity in confronting a community that often seemed indifferent or even hostile to their type of work. In his memoirs, Jean-Baptiste Monfalcon noted that only the cranky legitimist Théodore Pitrat took politics so seriously that he refused even to acknowledge his rival in the street. Monfalcon used his position as prison doctor to intervene on behalf of his republican colleagues when they found themselves incarcerated. In launching her women's journal, Eugénie Niboyet appealed privately to her male colleagues, noting her own adherence to "the courtesies that well-brought-up people owe each other" in letting them know about her project in advance, and asked them to publicize her efforts; all but Pitrat obliged, usually in friendly terms. Although her personal sympathies were on the republican side, Niboyet testified on behalf of a legitimist journalist accused of participation in the April 1834 insurrection.[64] The printer and sometime journalist Léon Boitel made his shop a rendezvous for Lyonnais intellectuals of every stripe. At one time or another, he put out journals for almost every political party in the city, and he kept up friendly personal contacts with figures ranging from Monfalcon to the republicans. His tolerant outlook and charismatic personality doubtless played an important part in promoting civility among writers of differing outlooks.[65]

Despite their personal ideological commitments, the Lyon journalists' behavior toward one another showed their acceptance of the division between public and private spheres that defined France's emerging bourgeois society. Only the legitimist editor Pitrat openly defied convention, making even his private life a battlefield in the struggle against the new order. Not really a writer, Pitrat was a genius at causing the new regime as many problems as possible. According to police reports from the early 1830s, he had refused to pay his taxes for years. When his property was seized to pay his debts, loyal friends rigged the auction, buying his furniture for a fraction of its real value and returning it to him so he could keep up his fight.[66] Rather than acknowledging a separation between his profession and his personal identity, Pitrat engaged in a kind of guerrilla theater, using his journalistic status to provoke confrontations that, in his view and that of his supporters, would expose the essential viciousness of the new regime. With the exception of Pitrat, however, the city's journalists, even those whose papers opposed the government, behaved like members of an increasingly recognized professional group, with a shared sense of their proper function in France's emerging bourgeois society, and especially its sphere of public discussion.

Print Culture and the Public Sphere in Lyon

Newspapers circulated and journalists worked in the arena in which reading intersected with politics, or, in other words, in the public sphere of bourgeois society. The concept of the public sphere comes from Jürgen Habermas's seminal work, *Strukturwandel der Oeffentlichkeit*, first published in 1962,[67] which has provided a persuasive framework for understanding the role of public opinion in modern history and has influenced most recent discussion of the subject. Habermas defines "the public" as a social formation that takes shape outside of traditional structures based on ascribed status and legal privilege. The public is a collectivity of individuals defined by their capacity to reason about their shared interests. In the eighteenth and early nineteenth centuries, the explicit markers of this reasoning capacity were education and economic independence, without which it was assumed that an individual was incapable of seeing beyond his immediate material needs; implicitly, as feminist critics of Habermas have pointed out, it was assumed both at the time and in Habermas's discussion that members of the public were men.⁶⁸ The public was therefore the totality of educated (male) owners of property.

Habermas's public is clearly related to what Marxist historians had traditionally defined as the bourgeoisie, but it differs both because Habermas's definition comprehends individuals not traditionally included in the definition of that class, such as property-owning aristocrats, and because its members are defined not just by objective criteria, such as wealth and education, but also by their participation in a specific cultural practice: the joint discussion of issues of common interest, particularly political and cultural ones. As Habermas wrote in an English summary of his book, "a portion of the public sphere comes into being in every conversation in which private individuals assemble to form a public body."⁶⁹ An important aspect of Habermas's analysis is his insistence that this process has a spatial dimension. The eighteenth-century Habermasian public manifested itself in specific locations where the conditions for free interaction between independent individuals existed, such as coffeehouses, salons, and clubs. Sites where discourse was regulated by hierarchically ordered structures of authority—Habermas specifically mentioned princely courts, but his notion would certainly apply to spaces controlled by the church, by the military, or by employers—might be loci of other forms of publicness, such as the "representational public sphere" in which absolutist rulers displayed their power, but they were not part of Habermas's "public sphere of civil society." Other areas, particularly the interior of the family home, constituted a private "sphere of intimacy" that was also excluded from the domain of the public.⁷⁰ Reading took place, of course, in many places outside of the spaces that made up the public sphere: in the home, in church, in schools and in offices. But the reading that contributed to the formation of public opinion in Habermas's sense was above all the reading of texts that prepared their readers to participate in the exchanges of opinion that took place in the sites where the Habermasian public manifests itself, and it is this subset of acts of reading that particularly concerns us here.

Habermas recognized that the development of the public and of public opinion was intimately connected with the circulation of printed texts. Unlike the city-republics of classical antiquity, whose public could be gathered in a single place, the public that emerged in the eighteenth century, being a spatially dispersed community, required media of communication to bind it together. Printed texts served both to disseminate a common fund of ideas and information and to communicate the results of public discussions. "In a large public body this kind of communication requires specific means for transmitting information and influencing those who receive it."[71] For several reasons, printed periodicals had an especially vital connection to the constitution of the public in the late eighteenth and nineteenth centuries. In the first place, periodicals held the community of the public together in time as well as in space; the fact that they were read as soon as possible after their publication meant that readers interacted with the same texts at more or less the same time, and the fact that periodicals reappeared on a regular basis meant that this experience of simultaneous reading was regularly renewed. Such coordination is inherent in the notion of public opinion, for if different members of the dispersed public had in fact been discussing different topics, the notion of their developing a shared opinion would be illusory. Second, periodicals, more than other forms of text, allowed for the interchange of views. Their readers could easily become writers. For this reason, periodicals quickly assumed the position of printed representations of public opinion that they have continued to claim for themselves down to the present.

The struggle over the spaces in which newspapers were read and discussed during the revolutionary years of the early 1830s is a particularly significant indicator of the ways in which notions of the public and the public sphere were being contested during that period, and Lyon is a particularly good location in which to observe those struggles. The city's social geography had long defined its major internal conflicts, and in the early nineteenth century that geography seemed to underline the special role of the property-owning bourgeoisie and of the silk weavers. Other forces occupied modest or marginal parts of the city. The Fourvière hill on the west bank of the Saône was a bastion of Catholic religiosity, but in the early nineteenth century the church was very much on the defensive in a city whose bourgeoisie was overwhelmingly liberal. The centralized state had not yet succeeded in imposing itself as a dominant force in the city, either. To be sure, the prefect and the garrison were key players in the period's conflicts, but both had been overwhelmed by local forces in November 1831, and it took the bloody defeat of the April 1834 insurrection to show that they could keep the city firmly in hand.

Architecturally, the *préfecture*, the main symbol of the national government's authority, was considerably less important to the city than the massive and ornate Hôtel-de-Ville. Only the construction of new forts after the two insurrections provided unmistakable visible evidence that Lyon no longer controlled its own destinies.

With the aristocracy and the church confined to marginal roles, and the national government's power in question, Lyon appeared to be dominated by two forces: the bourgeoisie and the workers. Particularly in the light of the two uprisings, both appeared to have a unity they lacked in most other French cities, or, for that matter, in most other European cities. Lyon's wealth derived to an extraordinary extent from a single industry, the manufacture of silk cloth. The silk merchants formed a bourgeois elite that was highly conscious of its own interests. The overwhelming importance of their industry to the city allowed them to dominate all other industrial interest groups, in contrast to the situation in cities such as nearby Saint-Étienne, where the entrepreneurial class was divided by conflicts of interest between textile manufacturers, mine-owners and metal-working enterprises.[72] Similarly, workers dependent on the silk trade far outnumbered those attached to any other industry.

Lyon's striking site, at the confluence of the Rhône and the Saône rivers, made it stand out from other French cities, but in many ways its urban geography was typical of early nineteenth-century French towns (Fig. 2). The wealthier residents lived in the historic center, on the west bank of the Saône at the foot of the Fourvière hill, and in the newer districts on the low-lying peninsula between the two rivers, the *presqu'île*. In a pattern characteristic of the early nineteenth century, workers dominated the urban periphery, marginal "in both a social and geographic sense: on or beyond the fringe of bourgeois society, and at or beyond the frontiers of its urban world," as historian John Merriman has written.[73] The most striking example was the northern suburb of La Croix-Rousse, on the plateau overlooking central Lyon. When the invention of the Jacquard loom in the early years of the nineteenth century drove silk weavers to seek lodgings with ceilings high enough to accommodate the new machines, entrepreneurs had covered the slopes of the steep hill leading up to the plateau with apartment buildings. By the early 1830s, the Croix-Rousse had a population of more than 16,000, almost all of them silk workers.

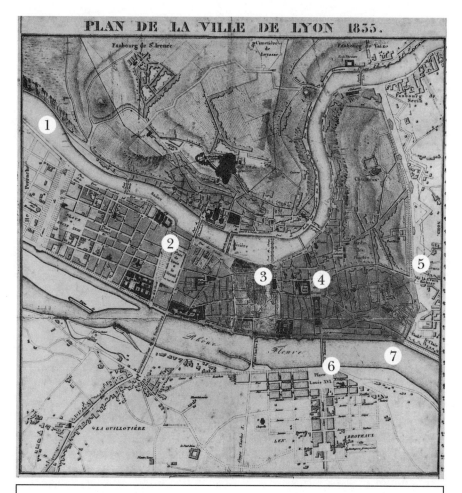

1. The Saône
2. Place Bellecour
   (Place Louis-le-Grand)
3. Place des Célestins
4. Place des Terraux
5. Croix Rousse
6. Pont Morand
7. The Rhône

Fig. 2 "Plan de la Ville de Lyon 1835." Lyon's "public sphere" was localized in the *presqu'île*, the narrow peninsula between the Saône and the Rhône. This map shows the public spaces where newspaper-reading and discussions took place, particularly the place de Louis le Grand, better known as the place Bellecour, the place des Célestins, and the place des Terraux. Completed in 1835, after the insurrections of 1831 and 1834, this map also shows the new line of fortifications separating Lyon from the working-class suburb of La Croix-Rousse on the hill at the base of the *presqu'île*.

At the foot of the Croix-Rousse hill and its concentration of workers was the *presqu'île*. The majority of the silk merchants' warehouses were located near the place des Terreaux, the open square at the foot of the hill, so that weavers could deliver their products easily. In the first half of the nineteenth century, before the Second Empire's urban reconstruction projects, the center of Lyon was a warren of narrow streets. Penned in between rivers and steep hills, the city had been forced to grow upward. The tall buildings blocked the sun and led Jean-Baptiste Monfalcon, the city's indefatigable chronicler, to deplore its lack of light and air.[74] Central Lyon had only a few public squares, such as the place des Terreaux and the place des Célestins with its theater. Near the southern edge of the built-up area was the one sizable open space within the city limits, the place Bellecour, originally the place Louis XIV, an artifact of eighteenth-century urban design whose tree-lined promenades were one of the city's main glories.

The concept of "bourgeois public sphere" or "public realm," so frequently referred to by recent historians of eighteenth-century France, often has a rather abstract quality. In Lyon in the years surrounding the Revolution of 1830, however, the term can be applied with considerable precision, both in spatial and in social terms. The public squares in the city's central district and the cafés, reading rooms, theaters, and other gathering places around them were the principal nodes of what we can designate as a bourgeois public sphere. Almost all the permanent institutions that supported the city's print culture were clustered close to these squares, particularly its booksellers, nearly all of whom had their shops on or just off one of the squares in the *presqu'île*.[75] The offices of the city's major newspapers were also close to the public squares, especially the place des Terreaux. The public sphere was defined not just by the fact that printed matter was most easily accessible in certain locations, but also by the behavior of certain members of the city's population who regularly gathered in these locations to engage in the social practices that constituted them as Lyon's political public, and the most important of these practices was the reading of newspapers and the discussion of their contents. Even at the time, the connection between public gatherings and the press struck journalists and their readers. "It seems to me that the press should be everywhere the public is, above all where it formulates its judgments—always ready to record them, sometimes to combat them," one Lyon editorialist commented in 1831.[76]

In designating certain locations in the city as its bourgeois public sphere, we are employing a modern term, but in seeking to define the specific social significance of these locations, we are following in the footsteps of Lyon's principal writers in the early 1830s. The years surrounding the July Revolution were also years when the city was literally being visualized in new ways. Older descriptions of the city, intended mostly for travelers, had emphasized its monumental public buildings and its facilities for visitors.[77] The collaborative volume issued in 1833 under the title of *Lyon vu de Fourvières*[78]—a collection of literary sketches, many of them devoted to describing the city's public spaces—inaugurated a new kind of urban literature, in which the city was conceived of as an organic whole with parts devoted to specific functions.[79] The fact that many of its chapters had first been published as articles in the city's periodicals, particularly the literary and satirical *Glaneuse*, is testimony to the importance of the press in shaping this new conception of urban space. And if the city's public squares attracted so much of these journalists' attention, it was surely because these components of Lyon's political public realm were the sites in which they and their readers performed some of their most important activities.

Several features make it possible for us to define this network of urban sites as constituting a "bourgeois public sphere." The city's streets and public squares were not, of course, closed to women, workers, or even the destitute. In his description of the place des Célestins, the newspaper printer Léon Boitel, organizer of *Lyon vu de Fourvières*, wrote, "All Lyon comes together in this circumscribed place. It is the city in all its forms, with its opulence and its misery, its silk and its rags, its innocence and its corruption."[80] But the public squares alone did not constitute a politically significant public realm; they achieved this significance only because of their connection to significant indoor spaces that were not open to all comers. These included cafés, *cabinets de lecture* or reading rooms, the foyers of Lyon's two public theaters, the premises of the city's bourgeois clubs or *cercles*, and such establishments such as the hairdresser's shop that advertised a salon where customers could find the latest papers.[81] Admission to all these locales except the *cercles*, which had closed membership,[82] was formally open to any male citizen, but in practice it required a certain amount of money, and no doubt a certain conformity to bourgeois social norms in dress and behavior. An article about the theater in the *Précurseur* in early 1831 addressed this issue directly. Its anonymous author claimed that "spectacles should, like everything else according to our *moeurs* and our laws, be open to everyone without distinction." But, he went on, "each person's liberty ends where that of others begins." The lower-class spectators in the *parterre* needed to learn not to "fill the hall with obscene catcalls and insults" that violated the "rights of the real public."[83] Even the public promenade at Bellecour, to judge by the illustration included in *Lyon vu de Fourvières*, was in practice a bourgeois preserve. Women were admitted there, as they were to the theaters and some of the *cercles*, provided they were escorted by a man; the title of an article in the feminist *Conseiller des femmes*, "The Husband at the Café, the Woman at Home,"[84] effectively communicated women's sense of exclusion from what was probably the most frequented of these social institutions. Lyon's streets and public squares, open to everyone, were in fact the periphery of the bourgeois public sphere, whose core was these closed-off spaces.

The sites included in the public sphere were defined not just by the social standing of those who frequented them but also by the specific activities carried on in their precincts. When he crossed the threshold of a café, a reading room, or a theater, the bourgeois citizen left behind both the economic arena in which he gained his wealth, and the domestic sphere in which he enjoyed its comforts. As a Lyon journalist wrote in 1835, "everyone goes to cafés, from the most important magistrate to the street performer; . . . class distinctions are more and more indistinct."[85] The spaces of the bourgeois public sphere were redoubts that emphasized the two aspects of bourgeois status highlighted in the title of Edmond Goblot's classic essay on the French bourgeoisie, *La barrière et le niveau*.[86] The various implicit and explicit entry requirements constituted barriers keeping out workers, women, and the poor, but among those admitted, a sense of equality prevailed. A *société de lecture* founded in 1827 with the support of members of the city's various learned circles provided a typical example of the mechanisms of exclusion that defined these groups: applicants had to be recommended by two members, and an annual membership cost 50 francs.[87] Though there was certainly a tendency for individuals with common professional interests to group together—the *Cercle de commerce,* for example, defined itself as a gathering place for wholesale merchants—cafés and other gathering places welcomed a mixed clientele, in which lawyers, *rentiers,* doctors, silk merchants, clerks, and military officers sat side by side. Furthermore, the public sphere was insulated from hierarchical authority structures. What one local journalist said about the difference between public balls to which anyone could buy a ticket and private affairs under the control of a patron applied also to cafés and reading rooms: "Each, having paid for himself, is under no obligation to anyone else."[88] If the prefect entered a café, he took his place as an ordinary citizen; if a priest did so, he received no special marks of respect. When individuals who did inhabit the public sphere formed highly ritualized groups, such as Masonic lodges, they imposed on themselves rules of behavior that took them out of that realm; Lyon's Masonic lodges were undoubtedly important in the conduct of the city's political life, but they were clearly separated from its public spaces. Although some Lyon journalists were Masons, we have no references to newspaper-reading and discussion at Masonic meetings.[89]

Public newspaper-reading was one of the principal activities carried on in the spaces of Lyon's public sphere. In good weather, newspaper readers could be found at the fashionable promenade in the place Bellecour, where a canvas-covered open-air *cabinet de lecture* set out its wares on a rough wooden table. "This kiosk is the devil's own lair, the home of all the newspapers. Around it are scattered all the opinions, ambulatory or stationary, seated or leaning against a tree. There, each one finds satisfaction according to his tastes and his sympathies. For the same price, each one finds out what to think, and above all what to talk about for the rest of the day," Léon Boitel wrote.[90] (See Fig. 3.) Newspaper readers also inhabited indoor spaces. Newspaper reading rooms and cafés were the most important of these. Providing its members the opportunity to read newspapers was the first goal listed in the prospectus of a "reading society" opened in 1827. Its statutes also promised to "create a central point where men for whom the taste of reading, the attraction of conversation, the charm of the *beaux arts* have become genuine needs." Members had access to the group's three rooms, one of which was set aside for conversation; to ensure the seriousness of the establishment, gambling and game-playing were strictly forbidden.[91] In addition to the reading rooms

Fig. 3 Public Newspaper-Reading as a Bourgeois Ritual. This anonymous illustration from the collaborative volume *Lyon vu de Fourvières* (1833) shows bourgeois gentlemen reading papers at the open-air kiosk in the place Bellecour, described by Léon Boitel as "the devil's own lair," around which "are scattered all the opinions, ambulatory or stationary, seated or leaning against a tree."

and cafés, newspapers were distributed on a regular basis in the city's theaters. Nineteenth-century theater-goers had not yet been socialized into concentrating on the spectacle on the stage. They traded visits to each others' loges and gossiped in the foyers, where the city's self-proclaimed cultural papers, the *Glaneuse* and the *Papillon*, were on sale.

Cafés and public reading rooms were well-known gathering places, familiar to all residents of the city. One could, for example, tell a friend to meet one at Madame Durval's *cabinet de lecture* on the place des Célestins, "the one with the caricatures pasted up on the windows."[92] Unlike the *cercles*, these were commercial enterprises that sought to draw as many customers as possible, using, among other means of enticement, advertisements in the local press. A *cabinet de lecture* at the Port Saint-Clair, near the offices of many of the silk merchants, proclaimed that it was "located in the most beautiful section of the city" and vaunted its collection of more than sixty periodicals. "Its elegance and its offerings have made it the most fashionable literary gathering."[93] Already in 1820, the busiest of the city's cafés, Calati's, drew 300 to 400 customers a day eager for news. A police commissioner reported: "The newspapers of both parties are read as soon as they arrive, they are read aloud and fifty people come to hear the *Constitutionnel*. It is in this café where all reports from both sides are exchanged, and from there they spread to the rest of the city. " A report on a second café, Targe's, gives a vivid impression of the political atmosphere of the Restoration period: in accordance with the law, which prohibited commentary on the papers, "one never hears the slightest reaction. Nevertheless, it is easy to tell what impression the speeches from the Chamber [of Deputies] have made on the readers because one can see smiles that announce approbation."[94] Reading-room etiquette had not changed much by the early 1830s. A humorous essay in a local paper depicted for readers the "living statues" arranged around the table in one such establishment. Physically immobile, they nevertheless betrayed their emotions and their political affiliations by "the lesser or greater reactions painted on their faces."[95]

The accounts we have of newspaper-reading in Lyon in the 1820s and early 1830s indicate that it was a well-organized social practice that followed a pattern familiar to wealthy and educated Europeans since the late eighteenth century. Newspaper-reading was most often done outside the home and away from places of business, in specific public or semi-public locations, and primarily by men. Those who participated in newspaper-reading thus had to make a deliberate effort to transport themselves to places where newspapers could be found. Reading the press was a ritual of public life, not an accidental occurrence or one relegated to odd moments of "free" time during the day. It was part of a larger pattern of public sociability, involving regular gatherings in the city's public places. Around the place des Terreaux, in the neighborhood of the silk-merchants' business houses, 3:00 P.M. was the customary hour for this interchange. "The offices are deserted, the cash-boxes silent, the order book closed. . . . It's the time set aside for digestion, rendez-vous, confidential conversations, and projects for the evening. . . . Enter the Café du Commerce. . . . The young men babble, gamble, swear, and talk politics. . . . [They ask,] 'What's the *Précurseur*'s line today?' "[96]

Within the cafés and *cabinets de lecture*, newspaper-reading had its own etiquette. In a reading room, one sought out the reader who currently had the title one wanted to see and saluted him politely, asking to be "after you, if you please, ready to repeat the formula to three or four others who have already put in their names."[97] Lyon's *cercles* had written rules specifying how members were to handle the newspapers their reading rooms provided.[98] Social mores also dictated polite behavior in these public places. Reviewing the city's major cafés in 1835, a journalist noted that the aggressive assertions of opinions and the leering at women that had been common a generation earlier were no longer tolerated: "A café nowadays is a large salon in which mutual respect and courtesy are automatically given and received, and habits that help spread good manners are formed. . . . Nowadays women are respected in a café as in a private home."[99]

Participation in the ritual of public newspaper-reading was a way in which the reader affirmed his social and political identity. Most descriptions of newspaper-reading in Lyon from this period describe readers of different political persuasions mixing with each other in the reading rooms, cafés, theater lobbies, and public promenades where newspapers were found. By their dress, their age, and their choice of reading matter, readers signaled to one another their political views. One newspaper sketch from 1831 portrayed a "rich manufacturer" catapulted to full citizenship by the previous year's revolution reading the *Précurseur,* at that point still a proregime journal, while a "young man with a high forehead, a lively eye" who had participated in the street demonstrations of that year called for the more vigorous *Sentinelle nationale,* and a third character, "member of every confraternity, treasurer of two parishes," read the Legitimist *Cri du peuple* and a financially strapped *rentier* made do with the *Journal du commerce.*[100] Humorous sketches of this sort were a journalistic cliché dating back to the revolutionary period, although they undoubtedly reflected a certain reality. Less frequently described but probably more common in actuality were situations in which likeminded individuals gathered to share reading material they approved of. The reading of newspapers was an essential aspect of the sociability at the *cercles* to which Maurice Agulhon has drawn attention in a wellknown study of bourgeois social practices in this period.[101] These groups, which maintained reading rooms that were open only to their members, tended to be more homogeneous in opinions than the gatherings in public places or cafés.

The social struggles in Lyon during the first half-decade of the July Monarchy were in part struggles over the boundaries of the city's public sphere, and over the question of who could participate in its informal rituals, including the reading of newspapers. Within a few months of the July Revolution, one local printer was already petitioning the prefect for permission to create "a periodical accessible to the people, whose moral and philanthropic purpose would be to instruct them on their duties and their rights." Unlike the city's main daily papers, the proposed new publication would be sold by the number as well as by subscription, thereby reducing the financial barrier to readership. Most significant is that this publication for the common people would seek its readers in places outside the circuit of the established papers' circulation; its sales points would be in the working-class suburbs, the desperately poor SaintGeorges neighborhood south of the cathedral, and in the streets where its backers proposed to set up four "itinerant newsstands that would cover the public squares."[102] Although their proposed paper apparently never appeared, the backers of this project had anticipated the issues that would be debated throughout the following years as publications aimed at readers excluded from the bourgeois public sphere sought to gain admission to it and reshape it. The planners of this proposed popular paper intended to bring working-class readers into the circuit of public discussion structured by newspapers. They explicitly sought to change the geography of newspaper-reading by extending the practice to neighborhoods that the established papers did not target, and their "moving newsstands" were intended to enlarge even the inner city's public realm, extending it from the cafés and reading rooms along the streets to the socially diversified crowds moving through them.

Most of the new publications actually created in Lyon in the early 1830s fulfilled at least some part of the unsuccessful 1830 petitioners' project. The papers avowedly aimed at workers, the *Écho de la fabrique*, created in October 1831, and its rival, the *Écho des travailleurs,* both had permanent sales points in the working-class suburbs whose significance the 1830 petition had highlighted, as well as in the main gathering points in the central city.[103] Their goal was both to create a new working-class public space in the peripheral neighborhoods where workers actually gathered, and to create a symbolic working-class presence in the traditional bourgeois public space itself. The denizens of that domain were in no doubt about the significance of this challenge. The *Écho de la fabrique* reported, in one of its early numbers, that several silk merchants had demonstratively shredded copies of its prospectus at the entry to the *Café du commerce.* "This quixotic act didn't surprise us," the paper editorialized. "These gentlemen thought they saw the rod of Aaron ready to strike them, and they were not wrong."[104]

All groups in the city understood the symbolic significance of access to the strategic sites where the discussions that defined the city's public opinion took place. The legitimist paper, the *Cri du peuple,* protested the Orleanist regime's ban on religious processions through the public squares: The exclusion of Catholics from these spaces was a striking sign of the church's exclusion from public influence.[105] In October 1831, the silk workers organized by the Society of Mutual Duty first made the general public aware of their strength by a peaceful mass march that passed through central Lyon's main public squares. Reporting on this unprecedented demonstration, the *Précurseur* commented: "It was at once order and disorder, and in the disorder itself there was a calm and self-regulation that one would have had trouble imagining in a population of workers." Because the workers had marched unarmed and in silence, the paper noted that the laws against causing public disorder had not been violated.[106] The workers had found a way to occupy the city's politically significant sites—such as the place des Terreaux in front of the Hôtel-de-

Ville, and the place Bellecour, the favorite bourgeois promenade—while remaining within the legal order. In so doing, they had demonstrated their ability to challenge the social order almost as strikingly as they would a month later, when they rose in insurrection.

Pioneered by the working-class press, the effort to expand and redefine the space of public discussion and the social composition of those admitted to participation in it was subsequently taken up by the bourgeois republican press itself. As we have seen, at the end of 1833, both the moderate republicans associated with the *Précurseur* and the more radical group identified with the *Glaneuse* began producing simplified versions of their publications, intended for a popular audience. Sold as individual pamphlets rather than as periodicals in an effort to get around the restrictions on the press, these publications were printed in large numbers and were distributed by street hawkers who wore distinctive hats to identify themselves. The *Glaneuse* announced the purpose of the republicans' campaign, "to make genuine instruction universal," and then continued: "To do this, they have to do what the government does not want to do, that is, make the sources of this instruction accessible to all. The sale in the streets, and at minimal prices, of instructive writings fills this condition."[107] The injustice of the Orleanist regime's restrictions on this would-be popular press was one of the pamphlets' main themes.[108]

Although similar efforts had been made several months earlier in Paris, the Lyon campaign for a popular republican press, and one that could be sold freely in the streets where a popular audience could have access to it, received extensive publicity in the national press. In Lyon, the mayor and the prefect blamed each other for mishandling the situation and allowing the street hawkers to create public disturbances. "These are not simple press-law violations; collisions and fights in the street are being provoked and incited," Mayor Prunelle wrote to the prefect. "In a word, they want a new REVOLUTION; they don't even avoid the word!"[109] It was the situation in Lyon, rather than the Parisian experiments, that provoked Louis-Philippe's ministers to propose legislation banning such publications and led the opposition papers to publicize the Lyonnais initiative. The *National* seized on the *Précurseur*'s claim that its popular edition had sold ten times as many copies as the regular paper to demonstrate that a popular public already existed, and to argue that the ministry's proposed law against street sales was "intended to prevent the popular press from establishing itself, and to protect a monopoly on publicity for the daily papers that address themselves to the bourgeoisie," a category in which the *National* included itself, avowing that it did not know how to address the problems of the lower classes in terms they could understand.[110] The struggle over this dramatic effort to expand the sphere of publicity led directly to an even more draconian law, the ban on associations, passed in early March 1834, which in turn sparked the April 1834 republican uprisings in Lyon and Paris.

While the working-class papers and the popular editions of the republican press sought to expand and transform Lyon's bourgeois public sphere, papers that addressed women raised the question of their exclusion from public discussion. Even before the establishment of the outspokenly feminist *Conseiller des femmes* in late 1833, the city's major cultural periodical, the *Papillon*, had drawn attention the consequences the fact that women were barred from the principal sites of bourgeois male sociability and also from the freedom of the streets, where a popular public sphere was emerging. One woman writer in the *Papillon* complained about men's habits of gathering where women could not accompany them: "Is it nothing to be kept in the dark about what they do, what they say for hours at a time? And these meetings that take them away from us so often, are frequently nothing but pretexts for their perfidy." Another article in the same paper blamed bourgeois women themselves for taking on airs of social superiority since the Revolution of 1830 and holding themselves aloof from the common people: "they have imposed on themselves the restriction of no longer appearing in places where plebeian crowds gather, although it is to the courage [of the lower classes] that they owe the overthrow of aristocratic prejudices."[111] Even in the absence of newspapers, male workers had their own gathering places, particularly the cabarets and cafés in the peripheral suburbs, but respectable women could not even dream of entering such locales. Lyon lacked the elegant salons in which upper-class Parisian women could mingle with men, and the *Conseiller*'s editor argued that this fact condemned the city to intellectual inferiority: "How can one imagine a society cut in two and so divided that one half makes fun of the other! How can one hope that the world will ever progress under such conditions?"[112]

The creation of the *Conseiller des femmes* was itself a serious effort to challenge that situation, but it was undertaken in the face of major obstacles. The paper sought to challenge the exclusion of women from the public sphere with an ambitious plan for an *Athénée des femmes*, an institution that would offer public education for women, taught for the most part by women instructors. Basic courses would teach the uneducated "the pure language of good company" and "the charm of reading," thus fitting them for participation in public discussions, while more advanced lectures would give women who already had some education a grounding in "social science . . . , political economy, education, history, literature, and moral philosophy."[113] With this preparation, they would be ready to make intelligent contributions to discussion of public issues, but whether education alone would gain them access to the venues where men formulated the city's bourgeois public opinion remained doubtful. The situation of the *Conseiller des femmes* showed that the creation of a feminist press, in the absence of social spaces in which women could translate its rhetoric into social practice, was insufficient to reshape the public sphere.

The spaces in which newspapers were read and discussed were thus critical sites for the shaping of public opinion. The liberal political regime created in 1830 intensified the competition for influence in these public arenas. The increasingly diversified newspapers published after the Revolution of 1830 were weapons with which different groups sought to dominate the public arena; they were also, in many cases, agents for efforts to change its boundaries. The periodicals published in Lyon in the early 1830s represented not only different political and social groups but also different notions of what the public sphere should be. The city's bourgeois newspapers, regardless of the nuances of their political views, were instruments for the perpetuation and reproduction of the socially limited public sphere that visibly existed in those years. The self-proclaimed workers' papers of the period and the various forms of what we can call the alternative press—satirical papers, cultural papers, and papers addressed to women—all, in different ways, acted to multiply and relocate the spaces where public discussion could take place and change the definition of who could participate in that discussion.

2

The Press, Liberal Society, and Bourgeois Identity

The Press and the Liberal Legal Order

The "press revolution" associated with the overthrow of the Bourbon monarchy gave Lyon a more diversified collection of periodicals than it had ever had before. Within the city's "journalistic field" there were several different groups of publications. Those most directly involved in the functioning of the July Monarchy's social and political system were the newspapers that paid caution money and were therefore authorized to comment directly on and to attempt to influence the country's political life. Because the July Monarchy allotted political rights on the basis of property and income, it has been characterized ever since its own day as a "bourgeois" system, and it is natural to assume that the periodicals most directly implicated in its public life constituted a "bourgeois" press. As we shall see, this term was indeed used to refer to the major political newspapers of the 1830s. The ways in which these newspapers were "bourgeois" were never simple, however, and some of them—those whose political orientations were republican or legitimist—were overtly opposed to the narrow liberalism of the new regime. Whereas the new publications created after 1830 to speak to and for groups such as women and workers trumpeted their ambition to be identified with particular social groups, the "bourgeois" press was often characterized by its denial of any such identification. When a press that was willing to adopt this label did appear, it was characterized either by its insistence that a self-conscious bourgeoisie capable of supporting a newspaper did not yet exist, and that it was the press's task to create such a group, or by its own admission that the bourgeoisie was a limited and unrepresentative group. The "bourgeois" press thus dramatized the contested and unstable nature of bourgeois identity, even in a society where the bourgeoisie was often said to rule supreme.

That French society of the years after 1830 was predominantly "bourgeois" is normally thought to go without saying. What that means, however, is notoriously difficult to define. Historian Sarah Maza has categorized the middle class as "the most obviously artificial among familiar social groupings," the one that is most difficult to define in objective terms.[1] As the persuasiveness of paradigms rooted in the Marxist tradition has diminished, social historians have become much less confident in their ability to identify such a group or to define its "interests" in any simple way. Edmond Goblot's classic analysis, revived in much recent work, stressed common values and attitudes; Jean-Pierre Chaline's exemplary case study of Rouen appropriates some of Goblot's insights, but insists on the importance of wealth as well. Adeline Daumard's empirical approach has emphasized the existence of many different subgroups with conflicting values within the bourgeoisie. Jürgen Kocka has suggested that the bourgeoisie is defined above all by its opposition to other groups with clearer identities, particularly the aristocracy and the proletariat; William Reddy has pointed to the paradoxical importance of a bourgeois sense of honor heavily shaped by aristocratic values. By introducing the issue of gender, Bonnie Smith and other women's historians have made it even more difficult to define what "bourgeois" might mean.[2] If defining the nineteenth-century bourgeoisie has come to seem increasingly problematic, it is clearly necessary to reexamine the notion of the "bourgeois" press as well.

Whereas the question of what constituted the bourgeoisie remains controversial, there is general agreement that the Revolution of 1830 gave birth to a regime whose constitution incorporated the principles of liberalism first articulated in 1789 and often said to reflect the class interests of the bourgeoisie: the legal equality of all male citizens, the protection of individual rights, particularly the right of property, and the principle of representative government. The social groups who recognized themselves in these political institutions, or at least their male members, shared certain values—a belief that rewards should be based on individual merit, a faith in the importance of formal education, a glorification of a family model with distinct gender roles.[3] Active participation in print culture was an essential element of this culture. The liberal citizen knew how to read and used this skill regularly; many male members of the group had the linguistic sophistication that comes from having mastered a second language, the Latin whose study took up so much of nineteenth-century secondary education.[4] The reading of newspapers was an essential aspect of this culture of literacy. Indeed, Benedict Anderson has argued that newspaper-reading was what defined the bourgeois class and made its members conscious of their status as a group. Through this shared experience, members of the group came "to visualize in a general way the existence of thousands and thousands like themselves. . . . For an illiterate bourgeoisie is scarcely imaginable. Thus in world-historical terms bourgeoisies were the first classes to achieve solidarities on an essentially imagined basis."[5]

Anderson's analysis suggests that the bourgeois press was bourgeois not just because it reflected some preestablished bourgeois outlook, but because it played a major role in constituting that outlook. We might consider as "bourgeois" those periodicals that inculcated in their readers, through their form, their content, and the way they were read, the values that both contemporaries and historians have labeled as bourgeois. In this respect, newspapers worked together with other forms of printed matter, particularly novels, whose role in instituting a "bourgeois" form of individual self-consciousness has been highlighted in a number of recent studies.[6] Like novels, the press often contributed to the development of bourgeois consciousness without proclaiming this as its purpose. One peculiarity of "bourgeois" newspapers is that, unlike the self-proclaimed workers' papers or the papers directed at women, they did not label themselves as representatives of a specific social group. This refusal of self-definition was itself one of the defining characteristics of the bourgeoisie: the essence of the bourgeoisie was its lack of objective definition. The bourgeoisie was the category to which all members of society could, in theory, belong, provided they purged themselves of the objectively identifiable characteristics, such as noble titles at one extreme or engagement in manual labor at the other, that implied membership in other groups. When they did reflect on what they were doing, the journalists who wrote these newspapers usually defined their roles in universalistic terms; they claimed to speak for the interests of society as a whole, rather than for those of any specific group within it. In refusing to identify themselves with any specific social interest, the journalists of the bourgeois press were being faithful to the heritage of French liberalism, going back to the Revolution of 1789. That tradition was characterized by what historian Patrice Higonnet has labeled "bourgeois universalism": an idealistic tendency to confound the interest of property-owning but nonprivileged males with that of society as a whole, and indeed with all of humanity.[7] The idealism of bourgeois universalism was genuine: both in theory and, to some extent, in practice, any adult male could qualify to participate in the community of citizens. But it was also exclusionary: those who did not or could not conform to liberal norms were *ipso facto* excluded from community participation, and their point of view was automatically stigmatized as particularistic and therefore illegitimate.

The Revolution of 1830 created conditions under which it became increasingly difficult for the liberal press to pretend that the readers for whom it spoke were a universal group and one that could remain undefined while other newspapers referred to specific social identities. In the wake of the overthrow of Charles X, a new radical rhetoric directed against "the aristocracy of wealth" assimilated the new regime's elite to a special-interest group like the titled aristocracy of the old regime. One Lyon radical paper asked in early 1832, "Why does this world slavishly prostrate itself before the *majesty* of a cashbox? Why, after having torn up the worm-eaten parchments of an elegant and debauched nobility, does it accept another nobility that has all the arrogant airs of the first, all the vices, without the brilliant exterior that partially hid them?"[8] The liberal press, which had claimed to speak for the entire community, minus a few illegitimate special interests, now faced a dilemma: Could it continue to assert that it represented "the people" or some other universal and undifferentiated category, or was it only the press of the "middle class" or some other limited grouping? The logic of liberalism required that bourgeois interests be defended as the only genuinely universal social interests, and a generation of talented thinkers constructed elaborate arguments to this effect, as Pierre Rosanvallon has shown in *Le Moment Guizot*.[9] Within the confines of a Chamber of Deputies chosen by the wealthiest 2 percent of the male population, such arguments could seem unassailable, but in the more democratic realm of the press, they stood exposed to criticism. Indeed, the experience of Lyon showed that the press market worked to generate newspapers dedicated to persuading their audience to recognize themselves as a bourgeoisie, to shed their tattered cloak of universalistic invisibility, and to recognize that they were indeed a specific group with specific interests. This "bourgeois" press thus became one of the most powerful mechanisms working to create a self-conscious bourgeoisie, ready to defend itself in a society where class conflict had come to seem ineluctable.

Whether particular newspapers explicitly defined themselves as bourgeois or not, they functioned according to rules that had been consciously designed to make newspapers one of the pillars of a property-conscious liberal society. From the earliest months of the Revolution of 1789, when the old licensing system for periodicals had broken down, observers had recognized that the legal definition of a newspaper would have a powerful effect on the dominant kind of newspapers, and that the nature of newspaper enterprises would in turn have a powerful effect on France's social and political culture. Social conservatives, like the prerevolutionary press baron Charles Panckoucke, unsuccessfully urged the National Assembly to pass laws that would have deliberately favored large-size English-style newspapers that would need to employ teams of journalists to fill their columns. Such newspapers, Panckoucke had argued, would require substantial financial resources and would consequently have to "be circumspect in order to avoid being disturbed, and run the risk of losing their capital and the money invested in them."[10]

Although Panckoucke himself managed to create a newspaper modeled after his own suggestions—the *Moniteur*, a large-format daily that became the revolutionary legislatures' "journal of record"—most of the revolutionary press in the 1790s was characterized by its low price and its deliberate attempt to adapt its language to a popular audience.[11] This press did not survive long after thermidor, however. When it imposed a stamp tax on newsprint in 1798, the Directory inaugurated a long era of legislation designed to raise the price of newspapers and thereby limit their audience to the wealthier members of the population. The Napoleonic government continued this policy, pushing newspapers to use larger paper to accommodate the *feuilleton*, a separate article at the bottom of the newspaper page. These larger papers were more expensive to produce and therefore cost more. Under the Restoration, stamp-tax rates were manipulated to favor larger formats, and the introduction of caution money, the substantial cash bond that would-be newspapers had to put up in order to gain the right to publish, perfected the system: the right to use modern society's most powerful political print medium was restricted to those who were able to make a large and financially unremunerative investment. In spite of the introduction of more efficient printing presses, which should have lowered costs of production, a subscription to a daily political newspaper, which had cost 36 francs in the early 1790s, cost 80 francs in the 1820s and early 1830s.

Given their high price, French newspapers were not items for casual purchase by all comers. Their publishers had a strong incentive not to print more copies than they could be sure of selling. A ban on the street sales of single issues of newspapers kept the papers out of the hands of those who could not afford a regular subscription. As a result, some commentators at the time used the label "bourgeois press" to refer to the standard daily newspapers. The "bourgeois press" could be defined, as "the press based on subscriptions," as the Paris republican daily the *National* suggested in early 1834.[12] This seemingly simple categorization in fact implied a whole set of political and cultural constraints. A press based on subscriptions was a press whose readers could afford to subscribe, and at the standard price of 80 francs for a year's subscription—the price not only of the Paris papers but of the Lyon *Précurseur*—that was a small minority of the population. Subscribing to a newspaper was not merely a matter of money, however. It implied a style of life that was stable and predictable enough that the subscriber could foresee having the time and the desire to have the newspaper on hand on a regular basis. Even the material form of "bourgeois" newspapers reflected certain assumptions about their readers. These papers, unlike those explicitly designed for a lower-class audience, were printed on large folio sheets of paper, the easy manipulation of which presupposed spacious surroundings. The long columns of closely packed type, unrelieved by illustrations or subtitles, were meant for experienced readers. Those readers also had to have the leisure time to concern themselves with matters that did not directly affect their daily lives, with "the affair of the moment or the unfinished intrigue," as the *National* put it, in contrast to potential readers of the popular press who were "compelled by necessity to want radical, profound, immediate changes" in underlying social conditions.[13]

The "bourgeois press" was also defined by the fact that it inhabited the same cultural space as its public. Articles in these newspapers made no concessions to their readers, who were presumed to need no special instruction to understand their content. Public figures mentioned in these newspapers' columns were identified by name or title, but readers hardly ever received any background information about them. It was assumed that if a paper made ironic reference to Guizot's or Thiers's political positions during the Restoration, readers in the early years of the July Monarchy would know what those positions had been, just as it was assumed that the reader would know the geographic locations of the major European countries and the main features of their history and institutions. It is significant that one of the few stories for which newspapers assumed their readers did need extensive background was the Lyon uprising of 1831; until that moment, the middle class had been able to treat the working classes as a nonfactor in the discussion of "public" affairs. Jean-Baptiste Monfalcon's lengthy "backgrounder" on the structure of the Lyon silk industry, written for the liberal Paris daily *Le Temps*, was reprinted in a number of papers of differing political coloration.[14] By contrast, papers aimed at groups previously excluded from the public sphere, such as workers or women, invariably exhibited a strong pedagogical bent. In Lyon, the *Écho de la fabrique* and the *Écho des travailleurs* had a regular column, "Readings for Proletarians," made up of excerpts from literary classics and designed "to awaken the imagination, to exercise the judgement, and to give workers the taste and the hunger for reading."[15] The *Conseiller des femmes* also assumed that many of its readers needed a proper education. But papers like the *Précurseur* and the *Courrier de Lyon* never spoke to their readers in such a tone.

Another dimension of the status of the press was that newspapers were enterprises defined and circumscribed by a set of laws that embodied what were normally taken to be "bourgeois" values. Like the bourgeois citizen, the paper had to have a clearly documented *état civil,* a fixed address, and a financial stake in society. "Nowadays a newspaper is both a commercial and a political enterprise, and in this dual capacity it is regulated by legislation with dual goals," a well-informed observer wrote in 1834.[16] Each paper had to register with the authorities, indicating its title, its intended periodicity, and the identity of its *gérant.* The requirement that a newspaper intending to print political articles had to post a substantial cash bond was openly designed to ensure that newspapers would have to follow the dictates of economic self-interest, and above all avoid committing offenses that would result in heavy fines. To give extra force to this constraint, the law required the *gérant* to have a financial stake in the paper and to answer physically, by risking his freedom and, in extreme cases, his life if the paper violated the law. In his famous article, "What Is an Author?" Michel Foucault argued that this insistence on defining a legal individual who could be punished was essential in the creation of the concept of authorship; there is no question that it was fundamental to the definition of newspaper enterprises in postrevolutionary France.

A legally constituted newspaper also had to be structured as a particular form of property. Newspapers were constituted internally as small bourgeois republics, in which participants had rights according to the capital they had invested and the formal provisions their papers' charters granted them. Most papers had several stockholders, and the law clearly defined their rights, which they could go to court to uphold, both against outsiders and against other stockholders or even the *gérant.* Lyon's most important paper reminded critics during one period of difficulties, "The *Précurseur* is administered according to rules; a general [stockholders'] assembly is called according to conditions specified in writing and consented to by all those involved," and a caricature of the period showed such a meeting (Fig. 4).[17] Specifically, political newspapers were usually set up as *sociétés en commandite,* a form of partnership in which, as business historian Charles Freedeman explains, "the ordinary shareholders . . . furnished capital and, provided they did not participate in the management of the enterprise, possessed limited liability," while the director—the *gérant*—assumed unlimited risks, in exchange for an almost

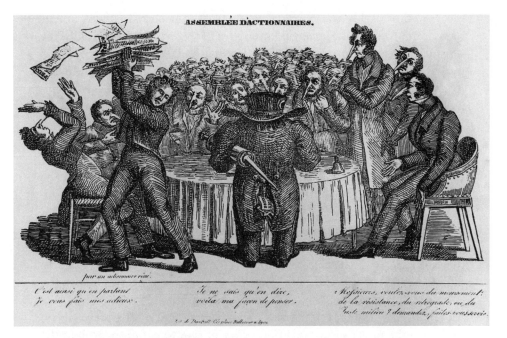

Fig. 4 "Assemblée d'Actionnaires." The squat figure of Lyon's first July Monarchy mayor, Gabriel Prunelle, addresses the shareholders of the city's dominant newspaper, the *Précurseur*, in this anonymous print signed by "a disgruntled shareholder." The man to the left of Prunelle lifts a pile of copies of the paper; the other shareholders are shown with elongated noses. The caption refers to the conflicts over the paper's political line in 1831; the final paragraph on the right reads: "Gentlemen, do you want movement, resistance, retrogression, or *juste-milieu*? Ask and you will be served."

unfettered right to run the enterprise. Newspapers and periodicals were in fact among the enterprises most likely to be constituted in this form; in 1833, half the *commandites* registered in Paris were for periodicals and other publishing ventures.[18] The *société en commandite*, which did not require government authorization, was a flexible and popular form of business organization in France all through the first two-thirds of the nineteenth century. In setting themselves up in this fashion, newspapers were integrating themselves fully into the world of legal and financial arrangements associated with the liberal order.

How thoroughly the political press was enmeshed in these norms was dramatized by the struggle for control of the principal Lyon daily, the *Précurseur*, in the summer of 1831. The battle over the paper's political orientation took the form of a classic stockholders' conflict, conducted with all the legal finesses that could be expected among a group dominated by businessmen and lawyers. Because the paper's rules specified that holders of three or more shares were still entitled to only two votes in the stockholders' assembly, both factions tried to increase the number of supporters they had in the stockholders' assembly by having holders of multiple shares sell some of them to new purchasers, who would then be able to provide additional votes for their side. Facing defeat in this contest, the progovernment faction then invoked a contract clause under which the paper could be required to preempt outside purchasers and buy its own shares if they were put up for sale. The more radical stockholders objected that this would force the *gérant*, one of their loyalists, to bankrupt either himself or the paper, and the case ended up before the *tribunal de commerce*, ostensibly as a confrontation over business practices, property rights, and money, rather than politics.[19] The prefect, who followed the paper's internal affairs closely, had to patiently explain the details to the minister of the interior in Paris, who failed to understand why it was not possible for the regime's supporters to get the paper under control.[20]

We can speak of a "bourgeois" press, then, not because the newspaper medium was inherently and inescapably, by its very nature, an instrument of bourgeois hegemony, as Richard Terdiman has asserted, or because we can demonstrate empirically that there was a distinct bourgeois class that read certain newspapers, but because newspapers had indeed been consciously constituted to help create a culture respectful of laws and property. In a broad sense, all the periodicals published during the early 1830s, even those that called themselves nonpolitical, were shaped by this "bourgeois" legal and institutional system. To the extent that there was any genuinely underground press in this period, it took the form of pamphlets rather than periodicals, and in Lyon, at least, if the prefect's reports and police records can be believed, such publications had little influence. But some newspapers can be labeled "bourgeois" in narrower senses as well. One such definition would limit the term to those papers that overtly supported the Orleanist "bourgeois monarchy." Surprisingly, this press was sparsely represented in Lyon, consisting only of the *Précurseur* and the *Journal du commerce* in the early months after the July Revolution, and the *Courrier de Lyon* after its creation in January 1832 as a response to those papers' shift to opposition, together with the *Journal du commerce* again after the prefecture began to subsidize it in 1833. For most of 1831, and particularly in the tense weeks leading up to the November insurrection, none of the city's papers supported the government. The luckless prefect Bouvier-Dumolard, sacked for his failure to prevent the uprising, complained bitterly that the ministry had left him "exposed, with no defense, to the daily attacks of five newspapers . . . , when with a little money and the help of a few good citizens, it would have been easy to change the direction of the most dangerous of them."[21]

Such a narrow, politically based definition of the bourgeois press is inadequate, however; it unjustifiably identifies the notion of "bourgeois" with support for a particular set of ministers, and raises such questions as how to classify a paper like the *Précurseur* of 1832–34, whose editors, stockholders, and readership continued to be drawn from the same social groups as those who had supported it before it broke with the Orleanist regime. Another approach to identifying the bourgeois press would be to define it in terms of the social characteristics of those who produced it, those who subscribed to it, or even those whose activities were most often described in its columns. Analyzing the press in terms of the social status of the journalists turns out to provide little information. In a broad sense, almost all the journalists active in Lyon in these years—not only the editors of the "bourgeois" papers, such as Jérôme Morin, Anselme Petetin, and Jean-Baptiste Monfalcon, but also Marius Chastaing, the failed lawyer who edited the *Écho des travailleurs,* and Eugénie Niboyet, editor of the *Conseiller des femmes*—fall into the category of the bourgeoisie. All were educated, and all appear to have led respectable bourgeois lives, with the possible exception of Théodore Pitrat, the diehard legitimist who made a weapon against the Orleanist regime out of his theatrical willingness to defy bourgeois norms by sacrificing his personal property.[22] Several journalists had risen from humble family origins, but it appears that their backgrounds did not dictate their subsequent political and social attitudes. Jean-Baptiste Monfalcon, the most consistent supporter of the July Monarchy among Lyon's leading journalists, was himself the son of a master silk weaver who had married a peasant's daughter.[23]

The journalists' own social status is thus an insufficient criterion for defining the bourgeois press. Identifying papers as "bourgeois" according to the social standing of their readers runs up against another problem, the lack of sources. No subscription lists for any of the Paris or Lyon newspapers from this period have survived. Even if they had, the large number of copies that were read in cafés, reading rooms, and other public places would create a considerable margin of uncertainty in their analysis. Furthermore, even if we had more data on subscribers, we would not necessarily be able to conclude that newspapers were shaped to satisfy their preferences. One effect of the caution-money system was that it made investment in a newspaper essentially an act of ideological commitment, rather than a calculated investment. Under these circumstances, editors had little incentive to try to maximize readership by responding to public opinion. Stockholders knew that newspapers were unlikely to be able to earn enough to return a profit. Even the *Courrier de Lyon*, which appeared to be the mouthpiece of Lyon's capitalist-oriented merchant class, was a money-losing proposition that ran through half of its initial capital of 100,000 francs before it began to cover its basic expenses.[24]

Contemporary observers' descriptions of which groups read which newspapers often seem to reveal more about their authors' hopes and fears than about the reality of the Lyonnais situation. Monfalcon, who knew the Lyon journalistic scene better than anyone else, was sure that the two *Écho*s were in fact widely read among the city's silk workers, but his references to the readership of the other papers were more moralistic than sociological: the *Glaneuse* he considered the journal for readers with "ignoble passions," and the *Journal du commerce* was for those devoted to gossip.[25] The prefect, Gasparin, made almost daily references to the local papers, but his reports on their readership were equally imprecise, relying on terms such as "the lower class."[26] The humorous descriptions of newspaper readers that appeared in several papers during these years identified each paper with a corresponding social type, but these classifications were at best highly approximate. According to the *Glaneuse*, for example, the *Précurseur*'s typical reader was "a rich manufacturer," a species too rare in Lyon at this date to have kept any newspaper in business. An impoverished *rentier* settled for the *Journal du commerce*, and a fanatical Catholic with "an air of devotion, shifty eyes, pinched lips" naturally read the legitimist *Cri du peuple*. The readers of the two outspokenly republican papers, the *Sentinelle* and the *Glaneuse* itself, were, according to this story, young, energetic, and attractive to women.[27] The empirical evidence about newspaper readership in Lyon during the early 1830s is not totally inconclusive. One would probably be safe in concluding that workers and artisans made up a larger proportion of the *Écho de la fabrique*'s readership than they did of the pro-Orleanist *Courrier de Lyon*'s. But it is also clear that no Lyon paper's readership was confined to any one social group, and that to convincingly identify certain papers as "bourgeois" strictly because their readers fell into that category would be difficult.

The lists of stockholders and backers of the different Lyon papers also provide suggestive but inconclusive evidence of their social audiences. Unfortunately, full stockholders' lists exist for only a few of the period's periodicals. As would be expected, because one had to have a certain amount of money to make such an investment, most of these lists are dominated by men of unquestionably bourgeois status. Of the *Précurseur*'s sixty-five voting stockholders in August 1832, no less than twenty were identified as "négociants." Ten were lawyers or notaries, eleven practiced other educated professions, and six held some form of public office (including three members of the Chamber of Deputies).[28] When the paper set up an organization to help raise money to pay the fines inflicted on it in 1833, supporters recruited from smaller towns throughout southeastern France were invariably a cross-section of the occupations normally classified as "bourgeois": *propriétaires*, merchants, physicians, notaries, local officials, retired army officers. Independent artisans and shopkeepers, on the other hand, were conspicuously absent.[29] The two delegations of stockholders who represented the *Courrier de Lyon* in dealings with the prefecture in 1831–32 were even more respectable: the only social categories represented among them were large-scale silk merchants, *rentiers*, and public officials, along with a single physician, which bears out Monfalcon's characterization of them in his autobiography as "merchants, *rentiers*, and property-owners."[30] The sixty-seven stockholders of Eugénie Niboyet's cultural magazine, the *Mosaïque*, founded in October 1834 after the disappearance of her *Conseiller des femmes*, were equally respectable, although considerably less business-oriented: they included twenty physicians and pharmacists (plus one veterinarian), eleven writers, artists and sculptors, two military officers, and the prefect and subprefect.[31] The disjunction between the composition of these groups of newspaper stockholders demonstrates the diversity within the middle classes themselves, a diversity reflected in the variety of "bourgeois" publications the city supported.

That the social status of the stockholders and backers of papers did have some real correlation with their orientations is borne out by the limited evidence concerning papers that proclaimed their opposition to the bourgeois order. According to an account later published by the principal editor of the *Écho de la fabrique*, the thirty-seven original stockholders behind the paper had included thirty-one master silk weavers, one master artisan outside of the *fabrique*, and five individuals whose occupations suggested bourgeois status, two of whom served as editors for the paper.[32] There is thus little question that the paper was firmly rooted in the very specific milieu of the master silk weavers, a working-class elite who were in fact small-scale business owners in their own right. Although we do not have stockholders' lists for any of the legitimist papers, a police report on a group organized to support one of the papers, the *Gazette du Lyonnais*, suggests that it drew from Lyon's small aristocratic elite. Of the nine names listed, five had noble *particules* and no indication of occupation. One lawyer and one doctor served as reminders that legitimism enjoyed some bourgeois support; the other two were women.[33] Although this paper's backers may have been more aristocratic than those of the *Précurseur* and the *Courrier de Lyon*, its readership was more varied. When the *Gazette du Lyonnais* took up a collection for the victims of the November 1831 uprising, the donors were a distinctly mixed group. They included a number of priests and vicars, but also a variety of artisans, several silk workers, some former Restoration magistrates, and a number of women whose social status was not specified. The prefect Gasparin dismissed the paper as "the journal of *curés* and *sacristies*," but its appeal evidently extended to at least some members of the lower classes. Gasparin paid more attention to the moderate legitimist paper, the *Réparateur*, which he claimed had the support of "the Carlist elite," who provided it with plenty of money.[34] In any event, however, these papers' backers were not businessmen, and the lawyers and doctors identified with the papers were a minority compared with the other groups they attracted.

The sense in which the press of the early nineteenth century can be called bourgeois is thus ambiguous. All the period's papers, even the avowedly nonpolitical ones, were embedded in the "bourgeois" order by virtue of their status as duly registered properties that could be bought and sold. The caution-money system structured political newspapers—those in which questions concerning the exercise and distribution of power were directly addressed—in such a way that they should have been compelled to defend order and property, as those in power defined those terms. In practice, however, the "bourgeois" newspaper proved to be an institution sufficiently flexible so that it could be appropriated by groups bent on challenging the existing order as well as defending it, and the bourgeoisie itself proved to be sufficiently diverse so that its press was anything but uniform. Freedom of the press was one of the principles most identified with nineteenth-century liberalism, but the press proved to have a highly ambiguous relationship with the bourgeois society whose interests liberalism supposedly represented. At the same time, however, groups that turned to the press to put their claims forward were adopting a cultural practice that served to justify and maintain the hegemony of liberal principles. Even as it opened the way for new challenges to the "bourgeois" order, the press drew contestatory groups into its framework.

The Political Press in Lyon

Although all the periodicals published in the early 1830s can be viewed as institutions rooted in a "bourgeois" legal and social order, some papers clearly fit the classic mold of the "bourgeois" political press more closely than others. On the national level, the major dailies that had opposed the Restoration government—the *Journal des débats*, the *National*, the *Temps*, the *Constitutionnel*, and the *Globe* before it became a Saint-Simonian organ—were clearly in this category. Their equivalents in Lyon were the *Précurseur*, the liberal daily founded in 1826, and the *Journal du commerce*, a triweekly that had appeared since 1823. When the Revolution of 1830 occurred, these papers seemed almost literally to have taken power;

the disappearance of the paper that had supported Charles X's government left the two liberal papers with a complete monopoly on the expression of public opinion. This situation was short-lived: by the end of 1831, as we have seen, the city's journalistic cultural field had become more diversified than ever before. The picture of a liberal press challenged from the outside by legitimists, republicans, and a working-class press is only part of the story of what happened to these papers after 1830, however. Equally significant is the fact that the liberal press itself became increasingly divided and fractured. In Paris, this process reflected many kinds of conflicts: ideological differences, personal rivalries, and economic competition even among papers with similar political leanings. In Lyon, however, the evolution of the political press after 1830 starkly reflected a fundamental conflict over the very nature of liberalism and the issue of whether a bourgeois class with a distinct identity existed. The fate of the city's dominant daily paper, the *Précurseur*, is almost too perfect a parable of the contradictory impulses contained in the bourgeois breast. The battle for control of that paper, and the subsequent rivalry between it and the *Courrier de Lyon*—a new title founded by the losing faction in its stockholders' assembly—dramatized the fact that bourgeois identity was not fixed in the early 1830s; it also demonstrated the vital role of the press in articulating the conflict between the bourgeoisie seen as a universal category representing the general interests of society, and the bourgeoisie seen as a distinct class whose interests were necessarily in conflict with those of other social groups. The inability of either paper to eliminate or discredit the other, and the ambiguous relations that both maintained with a government often supposed to have been devoted to the narrow class rule of the bourgeoisie, demonstrate the impossibility of imposing a single representation of the bourgeoisie.

The *Précurseur*, Lyon's best-known newspaper in the early 1830s, was the central actor in the creation of Lyon's public sphere during the Restoration, and in its reshaping after 1830. The *Journal du commerce* had predated it, but the smaller paper had never won much of an audience, in part because it was the property of a single owner, Antoine-Louis-Christophe Galois, who seems to have been something of an outsider in the city's life. The *Précurseur*, by contrast, had numerous stockholders, drawn from the city's business, professional, and political elites. In the last years of the Restoration, the *Précurseur* had become the public symbol of Lyonnais liberalism, and in July 1830 it was, in a most literal sense, the place where the liberal public made its stand against the Bourbon regime's effort to silence its opposition. The *Précurseur*'s history epitomizes the development of the liberal bourgeois press in France as a whole during this era, as it evolved from defending the "principles of 1789" in the name of a universalized and socially undefined public to expressing the conflicting values of a bourgeois class that was, for the first time, achieving real self-consciousness.

Liberal opponents of the Restoration in Lyon had founded a paper called the *Précurseur* in 1821–22, but the press laws imposed by Villèle's newly installed Ultra ministry in 1822 soon put it out of business, leaving the conservative *Gazette universelle* as the city's only political newspaper.[35] In 1826, the liberals returned to the fray, getting around the legal ban on the founding of new titles by buying the rights to the defunct earlier version of the paper.[36] The paper's backers were a cross-section of Lyon's bourgeois elite. A list from the end of 1828 shows that the paper's sixty-eight shares were divided among thirty-four owners, including seven *négociants,* probably silk merchants, eight medical professionals, six lawyers and notaries, and five landowners. The investors included several future members of the city's political elite under the July Monarchy: Prunelle, mayor and deputy; Jean-Claude Fulchiron, another deputy who would support the Orleanist regime; and Jean Couderc, one of its opponents.[37] Unlike the three-day-a-week *Journal du commerce,* the *Précurseur* was a daily and its print format was as large as that of many of the Paris papers. Its success meant that Lyon for the first time had a press organ that could truly stand comparison with the major papers of the capital. The *Précurseur*'s first number began with the most systematic justification of the principle of a local press published in Lyon up to that time. "Lyon, France's second city on the basis of its size and its population, the first in industry and commerce, the most outspokenly attached to constitutional government, has long felt the need of a journal that would contribute to the national spirit, protect citizens against abuses of power and magistrates against unjust attacks, censure error and arbitrariness in the acts of the administration, and serve as a tribune for the discussion of social interests," it began. The paper thus appealed to Lyon's sense of its own importance, and, by asserting that Lyon was the "most outspokenly attached to constitutional government," claimed that the metropolis at the junction of the Rhône and the Saône had a distinctive political coloration of its own, more advanced than that of Paris.[38] The theme of Lyon's autonomy and uniqueness echoed again in the first issue's comprehensive "Exposition of the *Précurseur*'s doctrines." "It would be erroneous to think that [Lyon's] inhabitants wait for the Paris papers to form an opinion on the events that constantly modify the state of society. There is a foolproof thermometer: its industry, its commerce. A center of activity, Lyon is less under outside influence than it is an influence on the outside." Lyon had a unique social character that determined its political outlook and made it different from—and in some ways superior to—Paris. In effect, the paper claimed that liberal concerns were more clearly expressed in Lyon than in the capital: "Its mores are simple, shaped by the habit of work. The dominant political opinions there are openly constitutional and liberal, because commerce and industry require independence and legal protection. The Lyonnais aren't concerned about the often decried centralization in Paris; fashion, intrigue, the hunt for jobs, and servility are not the way to get ahead with them."[39]

The *Précurseur* argued that there was a Lyonnais version of public opinion that was more in tune with economic realities and more moral than that coming out of Paris, but it also justified its project with a broad argument about the importance of print culture. As journalists had ever since the middle of the eighteenth century, the *Précurseur*'s editorialists asserted that newspapers were the modern world's substitute for the civic forum of classical antiquity. Thanks to the discovery of printing, "the public forum is the entire world. The tribune is everywhere. The audience is everyone who can read." The printing press had launched an era of progress, from the Protestant Reformation through the development of modern science and philosophy to the Revolution of 1789. The establishment of a press free to represent Lyon's civil society would thus be part of the march of human progress.[40] The founders of the *Précurseur* thus staked out claims that made their paper a symbolic site of central importance in Lyon's political public sphere. An attack on the paper, and on freedom of the press more broadly, was presented in advance as an attack on the progress of human civilization and on the rights proclaimed by the French Revolution. It was also presented as a symbolic attack on the city of Lyon, whose public opinion the paper claimed to represent, both against the Restoration government and against the overweening predominance of Paris. Finally, an attack on the paper would be an attack on commerce and industry, whose interests the paper claimed to embody.

Of course, the words in the *Précurseur*'s prospectus were mere rhetoric. But by 1830 they had become part of the living reality of Lyonnais life because the paper had succeeded in connecting itself to a significant audience. Jean-Baptiste Monfalcon, a well-informed source, later claimed that the *Précurseur* had more than a thousand subscribers on the eve of the 1830 revolution; there is no doubt that it quickly eclipsed the eight other local periodicals appearing on the eve of the July Revolution, six of which were nonpolitical enterprises.[41] Government officials and journalistic rivals acknowledged its importance. From 1827 onward, the paper was the Lyon voice of the organized liberal movement, "Aide-toi, le ciel t'aidera," printing its electoral propaganda and celebrating its candidates' victories. Whereas the conservative *Gazette universelle* "is not read in any public place, and is found only in three or four cafés in Lyon," according to one rival editor, the *Précurseur* was widely available.[42]

The paper's positive definitions of its intended audience were vague and general: it would represent "l'esprit national," the "citizens," the "interests of society," "men" (in the sense of humanity in general). The only groups it specifically excluded from its sympathies were "a king who would try to be absolute, ministers who refuse to admit any real limits on their power . . . , an aristocracy of pride and pretensions. . . , a Christian clergy who were neither tolerant, nor liberal"—in other words, the catalog of enemies defined by the liberal revolutionaries of 1789, with whom the paper specifically identified itself.[43] Over the next few years, the paper's content gave a few more clues as to its intended audience. A series of articles on "the education of the well-to-do ("la classe aisée") in late 1828, for example, identified that group as "this numerous and powerful class whose examples and words give direction to a nation, testify to its character, and define its moral standards, particularly in France under the constitutional regime, where no upper class dominates it and none separates it from the lower class." This group was not selfish: "The more enlightened the well-to-do class is, the more enlightenment it spreads among the people."[44] The paper's own definition of its mission was virtually identical with that which it attributed to the "well-to-do class," and it is clear that the two were closely identified.

In its form, the *Précurseur* was typical of the journalistic culture of the postrevolutionary era. Initially published in a modest two-column format, the paper expanded to the four-page, three-column folio size of the Paris dailies at the beginning of 1828 and, like them, opened its columns to advertising. The body of the paper was news, primarily reports on national politics from Paris, coverage of major foreign events, and a limited amount of reporting on local events. The largest single "story" in the paper was normally a summary of the latest debates in the Chamber of Deputies, provided by a reporting service in Paris that served a number of provincial papers. Advertising was confined to a modest place at the bottom of page four. What differentiated the paper, and the rest of the nineteenth-century French press, from the eighteenth-century news gazettes was the presence of a regular leader article on the first page. This article indicated that the paper's goal was not just to inform readers but also to shape their opinions. Its function was to comment on the news in the body of the paper. The lead article did in print what the readers in the city's cafés and *cabinets de lecture* did orally, transmuting news into ideologically motivated opinion.

The July Revolution in Lyon appeared to mark the point where the liberal opposition's dominance of public opinion was converted into control of the government. In a very literal sense, Lyon's July Revolution was a confrontation between the regime and the *Précurseur*, whose office served as the opposition's headquarters during the crisis. From the moment the first rumors of a ministerial coup reached Lyon on 28 July, the paper outlined a strategy of opposition for its readers, urging officials to ignore illegal orders, and ordinary citizens to withhold their taxes. For its part, the paper proclaimed, it would appeal to the courts against any unconstitutional limitation on its freedom, and even if only one judge in France had the courage to resist the government, "from that asylum, [the press] would spread throughout France the word that is so powerful when it serves to transmit the truth."[45] As the news of the attempted coup was confirmed and the first reports of popular resistance in the capital arrived, the paper's editor, Jérôme Morin, publicized his defiance of the police commissioner sent to seal his presses, announcing, "In this time of crisis, to defend one's civic rights is to defend those of the country." Although the city's other two papers had yielded to the authorities, Morin promised that the paper would continue to appear "as long as it is not prevented by *force majeure*," even if he had to enlist supporters to copy it by hand.[46] When a police raid on his regular printer's shop prevented publication, Morin found another printer and pointed out to readers the typographical evidence of the "wounds" the paper had suffered, which limited it to three pages.[47] While Morin was raising the banner of resistance in print, the *Précurseur*'s editorial office served as the meeting place for the liberal emergency committee that took over the city and succeeded in negotiating with the military commander to prevent an armed clash between the troops and the citizens. By 1 August, the *Précurseur* could report the victory of the liberal movement in Paris, and on the following day it could assure readers that Lyon too was calm.

The heroism of Morin and the *Précurseur*, narrated in the paper itself and documented in the archival sources, became Lyon's local revolutionary myth, its provincial version of the legend of the July Days in Paris, repeated in a number of pamphlets and finally incorporated in Jean-Baptiste Monfalcon's *Histoire des insurrections de Lyon*, which served for more than a century as the definitive account of the event.[48] With the liberals' triumph, their newspaper became the self-proclaimed organ of a new regime and of an apparently unified public opinion; the pro-Bourbon *Gazette du Lyon* disappeared, and the *Journal du commerce* had to apologize for having contributed nothing to the liberal victory.[49] In the *Précurseur*'s own view, both the paper and the new government successfully combined the principles of democracy, constitutionalism, and monarchy.[50] It appeared, therefore, that the July Revolution did not truly unleash a transformation of the press, but rather created a unanimous press in support of the new liberal order. True, the change of regime did pave the way for a physical transformation of the *Précurseur:* like most of the Paris dailies, the paper responded to the greater freedom of discussion after the July Revolution by enlarging its pages and introducing for the first time a *feuilleton*. The *Journal du commerce* also enlarged its pages.

As the new regime consolidated itself, the *Précurseur* defined the victory of July 1830 as the triumph of "democracy" and stated: "The dogma of popular sovereignty has replaced the dogma of divine right."[51] But who were the people? The new circumstances resulting from the revolution forced the *Précurseur* to redefine its notion of the public for which it intended to speak. Instead of vague references to "democracy," "citizens," and "public opinion," it had to take a position on such issues as qualifications for voting. Initially, the paper supported a regime based on property, but not the elitist "bourgeois monarchy" that the July Revolution had actually installed in power. "Extend political influence to the entire middle class, [and] you will have force and right [on your side]," proclaimed an editorial borrowed from a Paris journal and printed in January 1831 with the *Précurseur*'s endorsement. "I say force, because this intermediate mass will dominate both the working classes who are dependent on it for their work, and who through work itself are constantly rising into it, and the upper classes, whose intelligence it equals and whose number it surpasses. I say right, because it is this class that really represents all the social interests the legislator should satisfy, and it is in this class that the intellectual and moral force of opinion takes shape."[52] The logic of this argument was the same as that used by such government ideologues as Guizot to justify the limitation of political rights to the country's wealthiest taxpayers; the difference was that the anonymous columnist employed a much broader definition of the property-owning middle class.

Even before the silk workers' insurrection of November 1831 and Anselme Petetin's appointment as editor, however, the *Précurseur* had experienced an internal crisis that was one of the first visible signs of a broader split in liberal attitudes resulting from the July Revolution. From the end of 1830, the paper had begun to print articles suggesting a critical attitude toward some of the new government's policies. It highlighted the problems of unemployment and poverty in Lyon, although the solutions it proposed were essentially limited to appeals to charity.[53] The paper also urged a more forceful foreign policy than the government was willing to adopt, calling for active intervention on behalf of revolutionary movements elsewhere in Europe.[54] It publicized the speeches given by the first Saint-Simonian "missionaries" to visit Lyon, although it carefully distanced itself from the new principles they were putting forth.[55]

By the middle of 1831, the division between the Lyonnais supporters of Casimir-Périer's "party of resistance" and the proponents of the rival "party of movement" had made the *Précurseur*'s effort to pose as the representative of a unified liberal opinion untenable. Changes in Lyon's journalistic field threatened the paper's claim to represent the forces of progress. The satirical *Glaneuse* and a new political daily, the *Sentinelle*, both began publication in June; if the *Précurseur* could initially afford to ignore the former, a paper ostensibly aimed at theater-goers, it could not overlook the latter, which explicitly claimed for itself the role of spokesperson for the "party of movement"[56] and thus challenged the *Précurseur* to define itself. In an unusual front-page article, Morin, the paper's editor, told readers that he had hoped to bring the rival factions in the city together, so that "a single organ could satisfy the needs of liberalism," but that he was finding it impossible to do so. "To understand the conflicting pressures we have had to deal with, one has to have been at the head of a journalistic enterprise since last July, and have put up with these internal and external conflicts in which old friends lean on a poor journalist from opposite directions and leave him no choice but to stay immobilized between two opposing forces." As a way out, he suggested a poll of readers.[57] Two days later, Morin returned to the subject with a long disquisition on what he saw as the paper's proper political role. He contrasted the Paris political press, where each title overtly aspired to shape opinion in conformity with its ideological preferences, to the situation in Lyon. "The *Précurseur*, published in a great provincial city, . . . where it is so widely read that the opinion of its subscribers can be taken to be the opinion of the city, is in this respect in a very unusual situation." If he found his opinions to be different from those of the community, "it would not necessarily be a reason to abandon our convictions, but it would certainly be a reason to make a serious effort to understand the reasons for this difference . . . to ask ourselves why so many enlightened, conscientious, and liberal men have abandoned the opinions that we used to defend together." Although he did not spell out the problems at issue, Morin concluded that it was no longer possible for him to accommodate the conflicting views of Lyon's now-divided liberals. "Today, to satisfy half of our friends, we have to offend the other half, forget the memories of an old battle fought in common, and break off long-lasting relationships."[58]

By publicizing the *Précurseur*'s inability to provide a representation of public opinion that would be acceptable to all of its readership, Morin made the growing division among the participants in Lyon's public sphere a public issue in its own right. He was forced into this because the assembly of the *Précurseur* stockholders had become the locus of a struggle between two groups representing two very different notions of what policy the city's leading newspaper should follow. Morin's articles and his subsequent resignation made it obvious that there was no longer any such thing as a unified public opinion in Lyon, even among the prosperous and educated male elite that had supported the liberal movement before July 1830. Behind the scenes, the conflict took the form of a battle over naming a replacement for Morin. Supporters of the Casimir-Périer ministry's conservative policies backed the paper's associate editor, Jean-Baptiste Monfalcon, but a majority of the stockholders "wanted to go further, even to the point of a republic" (as Monfalcon later recalled), and blocked his appointment. From the paper's offices, the dispute spilled over into the city's public arena; Monfalcon's opponents even circulated a satirical song about him.[59] This public struggle for control of Lyon's dominant newspaper revealed that the very definition of bourgeois interests was now a subject for public debate, with little prospect of a generally acceptable resolution.

After an unlucky experience with a first replacement for Morin, who died shortly after accepting the job, the more radical stockholders fixed on a young but experienced Paris newsman, Anselme Petetin. The extent to which Petetin would change the *Précurseur*'s orientation was not immediately clear. He initially appeared to accept the primacy of propertied interests and condemned the idea of universal manhood suffrage, "because we value intelligence above numbers."[60] Although he disavowed any love for republicanism, universal suffrage, or violence, however, he expressed sentiments that were bound to produce some disquiet among Lyon's more property-conscious bourgeois. The statement of principles he published when he took over the paper promised "to introduce political economy into politics and to do something for the material interests of the masses, who have been neglected for so long because of the vanity of castes. The frightening increase of the population of proletarians in all European countries, the workers' uprisings in all manufacturing cities, are sufficient signs for those who know how to use their eyes." The answer was "a healthy organization of industry and commerce" and the recognition that "there is only one force in the world, intelligence; one privilege, merit; one aristocracy, talent made productive through work."[61] Although this proposition contained nothing explicitly subversive of bourgeois interests, it made no reference to wealth or property as criteria for political rights.

The silk weavers' uprising of November 1831, the appearance of rival publications that claimed to represent workers, on the one hand, and the true interests of the property-owning classes, on the other, and the first legal measures against the paper all helped drive it to a more radical position. Petetin, who had arrived in Lyon and taken editorial control of the paper only three days before the fighting broke out, had had no chance to take a position on the issues dividing the weavers and the silk merchants before the insurrection. The *Précurseur*'s editorials on the crisis in the weeks before he arrived had been sharply divided, a sign of the underlying split among its backers. One series of articles had expressed alarm about the weavers' demands for a *tarif* or government-imposed price for the weavers' products, and had objected to the peaceful mass demonstrations the weavers had begun to stage as a form of intimidation.[62] When the prefect promoted negotiations between the weavers and the merchants, the paper alternated between accepting the outcome as a way of defusing tensions and denouncing it as a first step down a slippery slope that would inevitably undermine property rights.[63] These articles were balanced by others emphasizing the workers' suffering, stressing that their protests had remained within the bounds of legality and insisting that "the question has an aspect even more serious than the interests of industry, I mean the aspect of humanity, of justice, of true liberalism." The worker should not be treated as "a machine whose purpose is merely to work a loom. As a man, as a citizen, he has rights equal to ours. Teach him how to use them in an orderly fashion, so that society, continually and completely ruled by just laws, will never be left at the mercy of acts of force."[64]

As these contradictory voices in the *Précurseur*'s editorial columns, and similarly divided statements in the pages of the city's other liberal paper, the *Journal du commerce*,[65] demonstrate, the looming confrontation between workers and merchants made middle-class journalists and readers painfully aware of the contradictions in the liberal tradition they represented. Both the editorialists whose emphasis on entrepreneurial freedom and property rights led them to oppose a *tarif* and any government involvement in economic bargaining, and their rivals, who stressed the natural rights of workers and the practical importance of showing them that they could defend their interests by legal means, could cite the spirit of the Declaration of the Rights of Man and Citizen to justify themselves. This crisis was exacerbated by the appearance in the midst of the *tarif* negotiations of a new periodical claiming to speak for the city's workers, the *Écho de la fabrique*. This paper's rhetoric challenged the invisibility of the bourgeoisie—by explicitly calling the merchants a selfish special-interest group—and the universality of its values; it also challenged the monopoly that bourgeois papers and journalists had taken for granted. The mere existence of such a paper required other publications to define their own identity more openly. The appearance a month after the November 1831 insurrection of the *Courrier de Lyon*, a daily newspaper claiming to speak for "the interests of property, of commerce, of industry, of the wealthy classes," thus appeared as an almost inevitable development, even though the new journal's prospectus added that those interests were also the interests of the "working classes."[66]

For the *Précurseur*, the appearance of these two new rivals and the drastic reconfiguration of Lyon's journalistic field that they represented was a dramatic challenge. The backers of the *Courrier de Lyon*, who included the minority stockholders of the *Précurseur* and several of that paper's former staffers, made a concerted effort to destroy the older paper. Petetin claimed that an "office of slanders" had spread rumors against him, that the conservative stockholders were attempting to paralyze the *Précurseur*'s operations, that its subscription registers had been stolen, and that the new paper's backers were telling advertisers and subscribers that the older one was on the verge of bankruptcy. To establish his own legitimacy, he took the highly unusual step of publishing the documents exchanged between him and the paper's oversight committee at the time of his hiring in October 1831, demonstrating both that he had a valid contract and that he had committed himself to a balanced policy of urging better conditions for workers but opposing an extension of political rights to them. He concluded with a ringing promise that "the *Précurseur* will not die . . . because it is not a tool of a party, a daily petition for honors and positions, but an independent organ of the country's needs."[67] Although these polemics were directed against the paper's "bourgeois" rival, the *Courrier de Lyon*, the universalist assertion that the

Précurseur spoke for the entire country was also a defensive response to the *Écho de la fabrique*.

The *Précurseur*'s effort to avoid being labeled as either bourgeois or antibourgeois was complicated by its increasing antagonism toward the Orleanist regime. As soon as royal troops had reentered the city after the November insurrection, the paper had been seized for having "incited the revolt, approved the revolt, and justified in advance all violations of property rights," a preposterous charge that seems to have been quietly dropped. Gasparin, newly installed as prefect, met with Petetin and reassured Paris that "he is no anarchist and he has promised me to stick for some time to a moderate opposition without bitterness."[68] The authorities became increasingly irritated by the paper's oppositional tendencies, however, and on 3 June 1832 Petetin reported that he had been indicted again, this time for having "incited hatred and contempt for the king's government and having offended the person of the king." Two days later, the funeral of General Lamarque in Paris set off an abortive republican uprising whose brutal repression led the *Précurseur* to openly declare its opposition: "The government of July, so radiant at its birth because of the hopes with which we surrounded it, is today red with French blood."[69] This provoked a series of prosecutions, to which the paper responded by making its republican sympathies increasingly clear.

Whereas the *Précurseur*'s political coloration became more sharply defined after June 1832, its relationship to bourgeois interests remained ambiguous. In the immediate aftermath of the Lamarque insurrection, Petetin conceded that the paper had "lost some support among the bourgeoisie. We freely admit that it now has fewer friends there than during the Restoration, but, on the other hand, how many new supporters it has found among the people!"[70] Defending the paper in court two months later, the deputy Louis-Marie de Cormenin asked whether "it would be fair if the people and the young had no representation of their opinions, while commerce and the bourgeoisie had theirs"—in effect conceding that the *Précurseur* no longer spoke for the latter groups.[71] Later in the year, Petetin denounced municipal theater subsidies that supported an institution that appealed only to "this little bourgeois aristocracy that dominates our city as it dominates the state."[72] At the same time, however, the paper continued to claim a connection with the middle classes. "How is it . . . that the *Précurseur* and the part of the bourgeoisie that it represents have taken on republican opinions?" Petetin asked in March 1833, and he answered his own question by insisting that it was the result of press articles that had convinced workers to abandon violence.[73] The press had thus created a union between people and bourgeoisie, reestablishing the possibility of a universalist public opinion.

By this point, the *Précurseur* explicitly rejected all arguments for basing political rights on property rights. "It is . . . the *man* and not the *thing* that must be represented," Petetin now wrote. He recognized that the deputies chosen by urban workers would be heavily outnumbered by those representing "rural property," but he claimed that experience in Lyon had shown the political maturity of the masses, and that a bicameral legislative system would ensure adequate protection of property rights.[74] In private, Petetin defended this radical position against the best-known Paris republican journalist, Armand Carrel of the *National*, who believed that one house of a bicameral republican legislature should still be chosen exclusively by property owners.[75] The *Précurseur*, once the recognized organ of Lyon's middle classes, had now become the advocate of a system of government in which that class would be decisively outvoted by peasants and workers. Even in its democratic incarnation, however, the *Précurseur* remained conscious of the fact that it was not a people's paper. Launching its pamphlet-size popular edition in January 1834, the paper conceded that "the newspapers, because of their price, are the press of the rich," and asked "whether the people cannot now have its own press. . . ."[76] As Petetin explained, the logic of democracy required the creation of a press that could reach the entire population, which publications like the *Précurseur* could not. The "popular party . . . demanded political rights for the majority, and the cost of newspapers, driven up by the stamp tax and postage, made it impossible to give that majority the simplest notions of politics. Nevertheless, civilized ideas must always precede and lead to the perfection of political institutions."[77] In Paris, the *National*, the paper closest in outlook to the *Précurseur*, explicitly underlined the fact that the creation of these new popular papers made the larger and more expensive dailies a "bourgeois press";[78] ironically, it was the democratic papers, with their universalistic ideology, that now became the first newspapers to openly embrace this label and to insist on the limits of their own audience.

The *Précurseur* had chosen to emphasize the idealistic, egalitarian aspect of the liberal tradition; while acknowledging the existence of a bourgeoisie, the paper denied that this group's interests were necessarily distinct from those of the mass of the population. On the left, the paper was outflanked in 1832 and 1833 by the more outspoken *Glaneuse*, the Lyon voice of the Société des droits de l'homme, the period's most important republican movement. Paris republican periodicals, such as Philipon's *La Caricature*, recognized the *Precurseur* and the *Glaneuse* as the movement's two representatives in Lyon (see Fig. 5). Although the two papers rarely quarreled with each other—the *Glaneuse* even published a poem praising Anselme Petetin for having "sacrificed, without hesitation, tranquillity, happiness, joy, and leisure" to the cause of liberty—the smaller paper went further than the *Précurseur* in distancing itself from any identification with middle-class interests and liberal principles. Although it admitted that perfect equality among all people was impossible, the *Catéchisme républicain* that the paper published in April 1833 emphasized the importance of approaching this ideal as closely as possible, and it called the right of property an unfortunate legacy of the past that had to be maintained "to avoid any chaos . . . but without forgetting, nevertheless, that it was originally established through usurpation, and to the detriment of the majority."[79] Despite outspoken sympathy for the workers, the *Glaneuse* also refused to adopt the class-conscious rhetoric of the *Écho de la fabrique*. The *Glaneuse* clearly looked forward to a society in which distinct social classes would disappear—a utopia that neither the more cautious *Précurseur* nor the workers' papers ever endorsed. Later, Marxist critics would stigmatize the *Glaneuse*'s type of radical democracy as "bourgeois republicanism." In the context of the early 1830s, such radicalism was an effort to escape the confines of liberalism altogether by denying any contradiction between liberty and equality and refusing any self-identification with a specific social group, above all the bourgeoisie.

For the *Précurseur*, the *Glaneuse*'s challenge from the left was less significant than the competition it faced from the right. The *Glaneuse* clearly separated itself from the concerns with parliamentary politics and business affairs that characterized the *Précurseur;* readers who identified with it had decided to abandon the framework of bourgeois liberalism altogether. The rival daily paper founded in January 1832, the *Courrier de Lyon*, was a different matter: it claimed to be the true rep-

Fig. 5 "Grande Croisade Contre la Liberté." Grandville's drawing from *La Caricature* shows the reputation Lyon's republican papers had gained by early 1834. Figures representing supporters of the July Monarchy prepare to storm fortresses labeled with the titles of leading republican press organs. The names of the *Précurseur* and the more outspoken *Glaneuse* are inscribed on the cliff, just to the left of the lamppost representing *La Caricature* itself.

resentative of the interests and ideas that the *Précurseur* had always tried to speak for. The *Courrier* was backed by the prefect and funded by a group that included a number of stockholders from the losing faction among the investors in the *Précurseur*, those who had favored making Jean-Baptiste Monfalcon rather than Anselme Petetin the paper's editor.[80] Forced to choose between the two positions it had previously occupied in Lyon's journalistic field—that of voice of specifically bourgeois interests and that of progressive critic of the regime—the *Précurseur* had opted for the latter role and left an opening for a more conservative rival. The result of this rivalry, as the two papers' unsuccessful competitor the *Journal du commerce* reported, was a highly public division within the city's public sphere. "The political press is divided into the two utterly opposed camps of the *Courrier de Lyon* and the *Précurseur*, and this has resulted in a crowd of people who have divided themselves, either in the name of the principle of order or in the name of the principle of liberty."[81]

The *leitmotif* of the *Courrier de Lyon*'s editorials was concern for the right of property, "an institution that is the fundamental guarantee of the social order."[82] The paper paid homage to liberalism's universalist tradition by insisting that the interests of property owners, and particularly of the silk merchants whose situation in Lyon was so precarious, were in reality identical to those of the workers they employed. Articles with titles like "Workers' Coalitions Make Trade Impossible and Worsen the Condition of the Working Classes" reiterated this basic argument.[83] Although the *Courrier de Lyon* claimed that the interests of bourgeois and workers were fundamentally the same, it was equally convinced that protection of those interests required a political system in which direct representation was limited to a qualified elite, which it defined as "the middle class." This group the paper defined as "the collection of all social superiorities; . . . superiorities that are not permanent or hereditary, that have no character of privilege . . . , that anyone can attain by his own efforts," and continued: "Defining the middle class practically amounts to answering the question of who should direct society." The paper resolutely refused to tackle the question the *Précurseur* had raised soon after the July Revolution: Did the Orleanist regime's narrow definition of voting rights in fact exclude much of the supposed "middle class"? But it was firm in rejecting the calls for democracy that its rival had endorsed by 1833: "If universal suffrage is admitted, the middle class disappears completely from the political scene."[84]

Between the two of them, the *Précurseur* and the *Courrier de Lyon* thus divided up the heritage of univeralist liberalism. The *Précurseur* retained the universalist element, whereas the *Courrier de Lyon* accepted the mission of articulating the interests of a "middle class" that had previously refused to acknowledge its distinctive status as one social group opposed to others. Indeed, the middle class's lack of organization in the face of the new menace from below was one of the *Courrier de Lyon*'s main editorial themes. "What effective resistance to unjustified demands can they put up as long as their class consists of units without any ground to stand on and without any mutual relations?" the paper asked in 1833. It raised the specter of "workers, now united in a single family," directing "their mass against isolated men," and called on bourgeois *fabricants* to adopt the organizational tactics of their opponents, even though—true to its insistence that there was no real ground for conflict between the two classes—the paper continued to insist that an employers' group would not be "a system of attack against the working class."[85]

In contrast to the *Précurseur*, the *Courrier de Lyon* thus evoked a distinct and identifiable bourgeoisie, defined essentially in terms of its economic interests. Often, the paper gave so much emphasis to the specific problems of the silk trade that it appeared to be the voice of that single industry. Coming just two months after the creation of the *Écho de la fabrique*, the first self-proclaimed workers' paper, the *Courrier*'s very establishment appeared to constitute symbolic recognition that the bourgeoisie was no longer a universal class capable of representing all real social interests. The violent polemical exchanges between these two papers were a verbal enactment of class warfare. But the *Courrier de Lyon* knew that its journalistic battle was not confined to opposing this new form of proletarian journalism. Whereas the *Écho de la fabrique*'s claim to represent workers' interests was largely uncontested, even when a second paper, the *Écho des travailleurs*, began to claim the same role, the *Précurseur* continued to challenge the *Courrier de Lyon*'s effort to inspire a genuine bourgeois self-consciousness. The *Précurseur* maintained both that it represented "a portion of the bourgeoisie" and that it spoke for the rights of all citizens.[86] In both respects, it destabilized the *Courrier de Lyon*'s efforts to put itself forward as *the* symbol of bourgeois opinion.

That paper's editor himself recognized that the *Précurseur*, like his own paper, reached an audience confined to the "middle classes."[87]

Even with the assistance of the government's regular harassment of its rival, the best the *Courrier de Lyon* could achieve was a prolonged stalemate in the battle to represent bourgeois opinion. The two papers seem to have had approximately the same number of subscribers, roughly 700 to 800 apiece.[88] The *Courrier*'s own editor acknowledged this failure to displace the *Précurseur* and blamed it on the peculiarities of his paper's place in the city's journalistic system. "It was criticized for not being sufficiently outspoken and sufficiently provocative," Jean-Baptiste Monfalcon wrote. He justified himself on the grounds that "in order to attract readers, a newspaper that defends something needs three times as much talent as a journal that attacks; opposition is the lifeblood of a newspaper." The fact that his paper had been "obliged, by its position and its principles, to abstain from personal remarks, scandalous anecdotes, and hostile articles against public officials" had been a further handicap.[89] The continued rivalry between the two papers was in fact ineluctable, because they represented two fundamental aspects of middle-class identity. Although the *Précurseur* refused to be identified with any specific class, its sponsorship of the *Précurseur du peuple* had been a clear recognition that it was, in fact, a paper for a wealthy and educated readership. A part of that readership was always responsive to the idealistic universalism espoused by the paper. Even the *Courrier de Lyon* retained elements of that universalism. Its argument that a self-conscious bourgeois class, aware of what separated it from the masses, was necessary, was coupled to an insistence that the interests of that bourgeois class were in fact the universal interests of society, that the prosperity of the bourgeoisie would ensure the prosperity of the workers.

The "bourgeois" character of the liberal and democratic press was underlined by the critiques it endured from the legitimist publications—the only other papers in Lyon that paid caution money, used the large three-column format, gave the same emphasis as the *Précurseur* and the *Courrier de Lyon* to the Chamber debates, and had the same division between news and editorials. The *Cri du peuple* and its successor, the *Gazette du Lyonnais*, the two titles published by the veteran Bourbon loyalist Théodore Pitrat during the early 1830s, were distinguished by their attempt to appeal to all conceivable social groups other than the bourgeoisie, which Pitrat dismissed as "a respectable class that ceases to be so only when it wants to rise above its station."[90] Both papers openly associated themselves with the Catholic Church, the only Lyon publications to defy the liberals' Voltairean prejudices during this period. The *Cri du peuple* protested the ban on public processions on religious holidays, and the *Réparateur* publicized the efforts the city's Catholic schools were making to continue their work without the public subventions they had received during the Restoration.[91]

Pitrat's paper went beyond defending the church and called for the restoration of all pre-revolutionary social distinctions: "By destroying the difference of ranks, we have established differences according to wealth. As a result of having abolished the titled, decorated, privileged nobility, we now have no simplicity in our customs. . . . We have the *high and mighty lordships*, messieurs the bankers; their immediate vassals, the great landowners and the rich merchants; finally, this arrogant and proud petty nobility who happen to pay from two hundred to five hundred francs in taxes per year."[92] The *Réparateur*, the more moderate legitimist paper founded in 1833, avoided such direct assaults on the new bourgeois elite but joined Pitrat in taking the side of the city's workers against the middle classes. Addressing itself directly "to the workers of Lyon" just prior to the November 1831 insurrection, the *Cri du peuple* told them: "With a soul saddened by your sufferings, the royalists dream of ways to end them, and on the day when they know what you need, you will see them fly to your aid."[93] The *Réparateur* justified the formation of workers' associations as being "nothing but the need to defend oneself collectively, when individuals are not [defended] by the institutions," and the *Gazette du Lyonnais* blamed the rising tensions that preceded the April 1834 elections entirely on the authorities, claiming that the city's workers were only responding to provocations.[94] The legitimist papers thus joined the workers' paper in challenging the claim of the liberal press to represent general, as opposed to particular, interests.

The divisions among Lyon's liberals symbolized by the rivalry between the *Précurseur* and the *Courrier de Lyon* created difficulties for the local representatives of a national government supposedly devoted to bourgeois interests. The journalists of the liberal press sometimes questioned that devotion, fearing that the authorities might try to position themselves as mediators between classes, even at the expense of property rights, or that the government's overriding concern to put down rebellion might lead to a sacrifice of local bourgeois interests. These fears were not entirely unfounded. Under the Empire and the Restoration, prefects had overridden silk merchants' objections and imposed *tarifs* demanded by the workers on several occasions.⁹⁵ When the prefect Bouvier-Dumolard, following that policy, urged the silk merchants to accept the *tarif* in November 1831, one editorialist in the *Précurseur* warned against "the willingness of authority to make concessions, born from an honorable sentiment, but untimely."⁹⁶ Long after his dismissal, memories of Bouvier-Dumolard's policies inspired an extraordinary outburst in the *Courrier de Lyon* by an editorialist who described the unfortunate prefect's conduct as "this sort of copulation of legitimate authority with the authorities of revolt."⁹⁷ On the other hand, the army's repression of the April 1834 uprising raised the question of the government's willingness to destroy private property in the course of maintaining public order. "If the government makes monstrous laws that stir up violent passions, and lead to revolts, should the property-owners of Lyon have to pay the bill?" the *Précurseur* asked, reflecting a concern that was not limited to its left-wing supporters; the proregime journalist Jean-Baptiste Monfalcon also complained about the refusal of the Chamber of Deputies to vote funds for the city's reconstruction.⁹⁸

For its part, the government found that managing the liberal press was a difficult art. From the July Revolution to the November 1831 uprising, the newspapers seem to have been left to themselves; Bouvier-Dumolard, as we have seen, complained that when he requested secret funds to influence them, he was told that such efforts were useless, since new opposition papers would just take the place of those bought off.⁹⁹ His successor Gasparin, dispatched to Lyon after the November 1831 uprising, took more aggressive steps to create a pro-government press. He initially hoped to "get hold of the *Précurseur* for the benefit of sound ideas," but he quickly realized that he would have to accept the paper's independence and work instead to help create a second press organ, which became the *Courrier*.¹⁰⁰ Not surprisingly, Gasparin consistently supported the *Courrier*, intervening with the government to settle problems about its caution money and using it as a conduit for putting out information.¹⁰¹ But the fact that the newspaper's place in Lyon's "journalistic field" compelled it to make continual attacks on other papers, particularly the

Précurseur and the *Écho de la fabrique,* meant that its tone was frequently at odds with the prefect's desire to keep the city's political temperature as cool as possible. In the lengthy survey of the press he wrote just after the April 1834 insurrection, Gasparin complained: "The *Courrier de Lyon* usually supports the government's interests, but it is not a dependable organ. . . . It has the defect of regularly injecting irritating words into discussions and to prevent [*sic*] the rapprochement of classes."[102] The *Courrier*'s editor, for his part, complained that the authorities had never given the paper the respect or support it deserved.[103] Ironically, the *Précurseur* often served the interests of order better than the *Courrier*. Petetin's desire to distinguish his paper from the more radical *Glaneuse,* and to keep himself out of trouble with the authorities, made him a consistent advocate of keeping opposition within legal bounds and of policies that would permit the "rapprochement of classes" that Gasparin sought. When he was put on trial after the April 1834 uprising, Petetin had no trouble assembling a large dossier of articles showing that the paper had consistently "urged the workers and the silk merchants to [make] mutual concessions" throughout the tense months leading up to the fighting. Monfalcon, the editor of the *Courrier de Lyon,* confirmed this claim, and also Petetin's testimony that he had visited the prefect to warn him of impending danger just before the revolt.[104]

Rather than providing a coherent representation of a self-conscious and self-confident bourgeois class in harmony with the government of the "bourgeois monarchy," the liberal political press of the early 1830s thus dramatized the conflicts and contradictions that beset the new ruling elite. Instead of giving its readers clear guidelines for action in the public arena, rival papers offered them contradictory choices. Furthermore, the newspaper that most vocally supported the "bourgeois monarchy" did so in such a way that the government's local representative often found it more troublesome than its republican rival. A close reading of the period's press shows how difficult it was to construct a stable notion of bourgeois or middle-class identity, even at a moment when that class had supposedly achieved a hegemonic position. It is tempting to see the *Précurseur*'s classless idealism as an earlier, more primitive version of bourgeois identity, and the more outspoken and combative attitude of the *Courrier de Lyon* as evidence that the Revolution of 1830 generated, for the first time, a distinctive bourgeois self-consciousness. In fact, however, the *Courrier de Lyon* did not displace the *Précurseur;* the two papers continued to square off against each other all through the July Monarchy. The *Précurseur* ceased to appear at the end of 1834, but it was immediately replaced by a new publication, the *Censeur,* which took over both its subscribers and its political positions.[105] The *Censeur* and the *Courrier de Lyon* continued to articulate their rival versions of liberal politics and bourgeois identity throughout the July Monarchy, until February 1848, when, in a repetition of the scenario of July 1830, the *Censeur* filled the political vacuum left by the collapse of another monarchy. The Orleanist prefect quite literally turned the government of the city over to a committee composed of the *Censeur*'s editors, while the *Courrier de Lyon* resigned itself to the disappearance of the regime it had supported to the end.[106] Subsequent nineteenth-century regimes—the Second Empire and the Third Republic—exhibited both elitist and democratic characteristics, and their "bourgeois" press continued to be deeply divided. Newspapers, supposedly one of the pillars of a hegemonic bourgeoisie, were in fact one of the most public demonstrations of the deep fissures that were one of bourgeois society's most notable characteristics.

3

Reshaping Journalistic Discourse

The
Alternative Press
in Lyon

Alternative Voices

Like all "press revolutions," the transformation that followed from the events of 1830 was not merely a matter of an increase in the number of political viewpoints with journalistic representation in Lyon. It also involved the creation of what can be termed an alternative press: a group of publications that introduced journalistic formulas not previously employed in the city and opened up for public discussion new domains that the conventional political press described in the previous chapter had neglected. Until 1830, the Lyon press had been essentially devoted to politics, legal affairs, and business, described in the language of those who were directly involved in them. The radicalism of the liberal papers under the Restoration—the *Précurseur* and the *Journal du commerce*—had consisted in their opposition to the government, and not in any effort to redefine the nature or scope of political discourse. Lyon readers were, of course, familiar with Parisian periodicals that did not fit the conventional mold of the serious political press, although few such journals had managed to establish themselves, even at the national level, before 1830. There had been occasional efforts during the 1820s to create such publications in Lyon, but none of these titles had lasted more than a few issues; local cultural life was not sufficiently dynamic to support them, and local press entrepreneurs could not find the necessary backing to sustain their efforts. After the July Revolution, however, there appeared publications that were new not just because of the ideology they represented but also because of the forms of journalistic discourse they employed. Less directly tied to ongoing political events and to the language used by the actors in them, these periodicals challenged the notion that politics and commerce were the only objects worthy of serious public discussion, and that the public language of the *pays légal* and the business community was the only appropriate discourse for such interchange. These alternative publications modeled new ways of participating in public life and created possibilities for "imagined communities" that had not previously participated in it.

The series of newspapers created to represent the city's workers, examined in the next chapter, was perhaps the most striking example of this expansion of the role of the press, but these papers were not the only important journalistic innovations of the early 1830s. Several other periodicals founded in the early 1830s played particularly important roles in this alteration of the topography the public sphere. Three of them were particularly important, because they lasted long enough to have a real impact: the *Glaneuse*, a republican satirical journal that appeared from mid-1831 until the insurrection of April 1834; the *Papillon*, a biweekly published from mid-1832 until early 1835, devoted to literature and the arts but also originally advertised as a *Journal des dames;* and the *Conseiller des femmes*, an explicitly feminist publication founded in October 1833 that lasted until mid-1834. The impulses that led to the publication of these three journals also inspired a number of shorter-lived periodicals: the *Furet de Lyon*, a paper similar to the *Glaneuse*, in the spring of 1832; *Asmodée*, a republican journal in verse, in early 1833; and *L'Épingle*, a cultural publication similar to the *Papillon*, in 1835. All these papers shared some important characteristics that separated them from the conventional political newspapers and from the workers' publications as well. By declaring themselves "nonpolitical," all took advantage of the fact that the law exempted such periodicals from the onerous caution-money requirement that weighed so heavily on conventional political papers. Although this law only required that publications taking advantage of it eschew direct comment on politics, these publications all distanced themselves equally from the field of work, business, and commerce: they carried no stock market reports, made no claim to represent the interests of Lyon's economy (as the *Précurseur*, the *Journal du commerce*, and the workers' press did), and printed little or no advertising. In an early issue, the *Glaneuse* announced that the public "that fills our business offices, our stores, our factories . . . , that sells shawls, dresses, discounts bills of exchange, speculates on stocks, weaves wool," was not the audience it sought. The people who carried out these activities needed to be attracted to cultural institutions like the theater, where they could be taught other values. "We want the merchant to experience emotions besides those of the accounting office, the rich man to also recognize the price of our literary glories. . . . We don't want visitors to think they have changed countries when they leave Paris to come to our commercial city."[1]

Although they thus distanced themselves from the two characteristics that defined the citizen in the July Monarchy—participation in constitutionally authorized political debate and engagement in the marketplace—the *Glaneuse*, the *Papillon*, the *Conseiller des femmes*, and their ephemeral imitators must nevertheless be considered parts of the city's bourgeois culture. It is true that the *Glaneuse* and the *Conseiller* both printed some articles that explicitly addressed the problems of the urban working classes, but neither made any attempt to express a worker's point of view, as did the *Écho de la fabrique* and the other self-proclaimed working-class publications, nor did they share the insistence of the workers' papers that productive toil implied a right to make one's views heard. The topics that preoccupied the workers' papers—measuring workers' actual incomes, achieving reforms in the procedures of the *conseil des prud'hommes*—never figured in the *Glaneuse*'s articles; its policy was to concentrate on the issues of republican government and press freedom as the unique panaceas for all social ills. The *Conseiller des femmes*, whose editor had been a Saint-Simonian militant and had absorbed that movement's concern for the poor, did publish articles denouncing the low pay and sexual exploitation of working women, but the majority of its contributions dealt with the problems of women who were not obliged to work and who therefore had the leisure time to educate themselves and to contrast their situation with that of bourgeois men. Even an article calling for common elementary schools for women of all social classes justified its demand with the argument that working-class girls would assimilate "the manners of the well-brought-up young lady" and that "a general fusion wouldn't be dangerous as long as the teachers showed their pupils the advantages of their own situations."[2] The *Papillon*, avowedly dedicated to art and literature, was unmistakably reading for the leisure classes and rarely mentioned social problems.

The social space that these three papers claimed for themselves was thus contained within the larger bourgeois social sphere, unlike that of the workers' press, but the alternative press claimed imagined territory that was not coterminous with the domain of the conventional political press. In the various literary representations of newspaper-reading in Lyon published in the early 1830s, the *Glaneuse* was normally mentioned alongside the liberal and legitimist papers as a publication that would have been found in cafés, but the paper itself was emphatic about staking out a special place for itself, above all as reading matter for patrons of the city's two theaters. The *Papillon*, too, stressed that it was especially designed for that particular audience.[3] It was also the only paper that covered social events, like balls.[4] The paper thus identified itself with public gatherings devoted to entertainment, as opposed to those where questions of money or power would normally be discussed. Alone among the city's periodicals, the *Conseiller des femmes* did not suggest a concrete image of the conditions under which it might be read. This was the logical consequence of its feminist critique of bourgeois society; one of the paper's main complaints was that there were no public spaces where women could gather freely. Confined by social convention to the interiors of their homes, bourgeois women lived an entirely private life. By not imagining social situations in which the journal might be read, the *Conseiller* implicitly underlined one of the main aspects of women's life that it wanted to challenge.

Imagining alternative social spaces in which they might be read was not the only way in which Lyon's alternative publications set themselves apart from the other periodicals of the period. The way in which these journals imagined their readers, and sometimes their contributors, was another important method of emphasizing their special characteristics. To a large extent, all three projected themselves as expressions of the same social and cultural milieu. Unlike the workers' press, all were unmistakably products of bourgeois society. Their language was that of an educated public, at home in theaters, salons, and cafés. They might, and often did, concern themselves with the problems of the poor, but they did so from the position of outsiders offering philanthropy, education, or political leadership, rather than from the insider's position claimed by the *Écho de la fabrique* and its journalistic progeny. All also defined themselves to a large extent as publications by and for the young. Looking back on the paper's first few months of publication, the *Glaneuse*'s main editor, Adolphe Granier, claimed: "I made an appeal to the young writers of this city, they heard my voice, from that moment the success of the *Glaneuse* was ensured."[5] The *Papillon* filled its columns with tales of unrequited love whose protagonists were often in their teens and identified itself with the young heroes of the romantic movement in literature, music, and art. A typical article, "An Artist at Home," imagined the spartan domicile of a painter and his companion, "a pale young woman with long ebony eyelashes," as a classic utopia of youth, one that the middle-aged could hardly dream of sharing. Their "happiness was perfect: love, painting, poetry, music, that was their whole life."[6] The *Conseiller des femmes*, too, concerned itself essentially with the problems of young women striving for education, marriage, and possibly careers.

The idealized image of their readers that these publications projected corresponded to a certain degree to the reality of their overlapping pool of contributors. Many were indeed young themselves, and just starting lengthy careers, like Eugénie Niboyet, editor of the *Conseiller des femmes*, who was born in 1796 and was still be publishing in the 1870s; and the printer and cultural entrepreneur Léon Boitel, born in 1806, who contributed to both the *Glaneuse* and the *Papillon* and also served as printer of the *Papillon* and the *Conseiller des femmes*.[7] Although the success of the *Glaneuse* and the *Papillon* can be attributed in part to the fact that they had editors with professional journalistic experience, and although the growth of the entire sector of the alternative press owed a great deal to the dedication of Boitel, who served at various times as director and printer of many of these titles in the early 1830s and who then founded and kept alive for over a decade Lyon's first serious local historical and cultural journal, the *Revue du Lyonnais*, these alternative papers were mostly the work of enthusiastic amateurs. They were unabashedly Lyonnais in their orientation, determined to demonstrate that their city could produce and support interesting publications without relying on the capital.

Papers of the young, the titles that made up Lyon's alternative press also distinguished themselves by the way they used gender as a way of defining the "otherness" that set them apart from the conventional press. This was most obvious in the case of the *Conseiller des femmes*, which was by definition a paper by and for women, a group that, the journal claimed, had no place in the regular press. In its prospectus, the *Conseiller* complained that every other social group and every other cultural domain had its own publications, but that "women have found few outlets to represent them."[8] Well before the *Conseiller* appeared, however, the *Papillon* had already claimed for itself the status of a *Journal des dames*, and those words appeared as a prominent part of its subtitle for the first ten months of its existence. The *Papillon* promised to rigorously eschew all political discussions and to devote itself exclusively to literature and the arts. In so doing, it would constitute itself explicitly as a women's publication, for women had, according to the paper's editor, a special relationship to culture. Male readers might indeed need political reading matter, but "women need literary journals, [the only journals] that are truly in accord with their tastes and habits, in perfect harmony with their soul." Lyon had the reputation of being a hostile environment for such concerns, but the *Papillon* was confident that the city contained "many cultivated women, in no way pedants, who understand Victor Hugo, Rossini, or Horace Vernet perfectly. . . . It is close to these flowers that our *Butterfly* will take pleasure in fluttering."[9]

If the *Conseiller des femmes* used gender in a serious sense to define its *raison d'être* and if the *Papillon* used it to define the boundaries of its chosen aesthetic domain, the *Glaneuse* seized on the issue in a more complicated and metaphorical way. The paper did offer special thanks to women readers, who, it claimed, had played an important part in its success.[10] In its pages, however, gendered images served to subvert not the prevailing definition of gender roles, as in the *Conseiller*, but the values and institutions of the constitutional monarchy. In its descriptions of itself and its ideal readers, the *Glaneuse* sought to claim a special kind of masculine identity. With its appeal to youth, the *Glaneuse* positioned itself as reading matter for dashing young men with sex appeal. When the *Glaneuse* imagined its typical reader, it described "an elegant young man, . . . an attractive youth, spirited and lazy, he writes verse, courts the ladies, . . . loves his country and champagne, and is filled with the same animation when he speaks of the Polish revolt and of his mistress."[11] By implication, the reader of the *Glaneuse* had no interest in the mundane virtues of order and routine that were so often hailed as the basis of success in bourgeois society. The paper claimed some of the same seductiveness it attributed to its ideal male reader; an early issue attributed a good part of its success to women, and thanked them.[12]

But the editors of the *Glaneuse* also regularly played on the feminine associations of their paper's title. Throughout the first half of its short life, the paper appeared on pink paper, and when the Lyon prosecutor charged it with printing political propaganda, the paper indignantly protested against this refusal to accept its fictive persona: "Can you see me as a goddess of liberty? In place of the pretty, blonde hair that covers my head, they're giving me a Phrygian bonnet." The Phrygian bonnet, or liberty cap, was of course the headgear of the militant female allegorical figure associated with the radical republican movement.[13] Later, this bantering tone changed to one of melodrama as the paper used its fictive gender status to make the authorities' assaults sound particularly repulsive. "The *Glaneuse* is a well-behaved young lady who has never sold her favors, her pink dress is the only wealth she has, and when that dress has been stained by contact with police spies, gendarmes, court officers . . . , when you have torn off the last remnant, the poor girl will give up the ghost," one editorial announced.[14] In using this gendered imagery, the paper both stigmatized its opponents and transcended the literal speech of everyday political debate. By the powerful emotional responses they evoked, the paper's gender images trumped ordinary arguments rooted in law or practical considerations—the arguments familiar to readers of the conventional press.

In general, then, the journals that composed the alternative press attempted to portray themselves as reading matter for an alternative public sphere, one that was no less bourgeois than the world of businessmen and voters who read the conventional press in their cafés, but one in which a different set of values held sway. In the world of the alternative press, readers were concerned with questions of truth, justice, and beauty, rather than with practical considerations. By claiming to discuss higher or more fundamental issues than the daily political papers, these journals, even if they were not founded with the intention of promoting opposition to the society around them, all conveyed that they were at least potential critics of the emphasis on power and money that defined the concerns of the conventional press. Ironically, this implicitly critical position was to some extent forced on these publications by the legal system's distinction between political and nonpolitical periodicals. The special requirements imposed on political newspapers were onerous, but they were also an acknowledgment that the content of those papers really mattered. Nonpolitical papers thus faced the challenge of demonstrating that their content did indeed serve some public purpose that justified asking readers to spend their time and money on them. They sometimes did this by arguing that the topics they treated were more significant than the constantly changing and quickly forgotten issues that preoccupied their caution-money-paying colleagues: this was the sense of the *Papillon*'s glorification of the life of the artist as indifferent to material things. More often, however, nonpolitical publications sought disguised ways to convey ideological messages.

In order to carry out this strategy without forfeiting their nonpolitical status, these periodicals had to comment on politics indirectly, through language that did not immediately or unequivocally reveal its purpose. They were thus driven to avoid the language of political debate and courtroom proceedings and to employ forms of discourse borrowed from literature, the theater, and other "nonpolitical" domains. The conventional political papers might occasionally print a piece of poetry in their *feuilleton*, but it was only the alternative journals that used verse to convey their most significant messages. Thus the *Glaneuse* proclaimed its open adherence to the republican cause by printing the emotional stanzas of local republican César Bertholon entitled "Pourquoi je suis républicain" long before that paper presented a reasoned justification of republican ideology, and the *Papillon*'s most forceful statement in favor of the rights of women was a long poem, "A la femme," by Sophie Grangé (see

Appendix 1).[15] In 1833, another Lyon writer founded a political paper written entirely in verse, *Asmodée*, in hope that this tactic would enable him to avoid the caution-money requirement.[16] Such papers destabilized the boundaries between serious political language and language used for nonpolitical purposes, introducing an element of indeterminacy into public discussion that made it no longer clear which statements were serious and which were meant to be taken as amusing, exaggerated, or false. Indeed, the *Glaneuse* responded to one court conviction by announcing that, because it had been prosecuted for publishing the truth, it would now accommodate the authorities by publishing nothing but lies. It then published a list of statements made by the regime's defenders: "Lie number 6: France is happy at home and respected abroad. . . . Lie number 9: We live in the best of republics."[17] Other nonpolitical papers altered the notion of political discourse by treating subjects normally excluded from the public realm with the seriousness usually reserved for the matters conventionally defined as matters of public discourse. The treatment of women's issues in the *Papillon*, a journal whose very physical appearance (each issue appeared on paper of a different pastel shade) was meant to emphasize its devotion to the frivolous topics of art and entertainment, and in the *Conseiller des femmes* was a way of challenging the definition of which topics were in fact fit for serious public discussion.

Despite these important similarities, there were also significant differences between the three major papers that, between them, best represented what can be called the alternative press in the wake of the July Revolution. The *Glaneuse*, despite its ostensibly nonpolitical character, was an unmistakably political enterprise; in fact, several of whose editors were founding members of the Lyon section of the Société des droits de l'homme, the period's most important republican movement.[18] When the nonpolitical merchant and diarist Joseph Bergier happened to dine with some of its editors, he came home and noted, "It was a republican meal. . . . Much discussion of politics."[19] The *Glaneuse*'s innovative use of literary forms not normally classified as political stemmed in large part from its effort to evade the caution-money requirement and protect itself against government repression, rather than from any real commitment to broadening the notion of political discourse. The *Papillon*, on the other hand, was a journal with little discernible political orientation, even though it shared a number of contributors and a number of literary characteristics with the *Glaneuse*. Its content was in harmony with its professed aspiration to limit itself to literary experimentation and aesthetic criticism, although it occasionally allowed itself a brief comment celebrating the acquittal of one or the other or the local republican journalists. In his review of the city's press, Gasparin, the prefect, categorized it as "a literary journal with republican tendencies," but he never bothered to take any action against it.[20] Many of the journal's female contributors also published in the *Conseiller des femmes*, which did have a clear ideological purpose—the transformation of the status of women—albeit one divorced from the conventional political debates of the period. If politics meant overt discussion of the forms of government and the proceedings of the national legislature, then the *Conseiller des femmes* was in fact as apolitical as the *Papillon*, but Eugénie Niboyet's journal was obviously aimed at promoting significant social change, whereas the *Papillon*, despite its openness to outspokenly feminist articles, was not. The alternative press of the early July Monarchy was thus a diverse set of publications. Even when these periodicals carried similar materials, they framed them in ways that conveyed very different meanings.

Satire and Republicanism

After the July Revolution, the *Glaneuse* was the first such endeavor on the local scene. When its first number appeared, on 16 June 1831, well before the November 1831 silk workers' uprising that was to transform the city's atmosphere, the Lyon press market was just beginning its diversification. The two liberal papers, the *Précurseur* and the *Journal du commerce*, were both still supporting the new regime, and the legitimist party had just regained journalistic representation with the appearance of Théodore Pitrat's *Cri du peuple*. The first numbers of the *Glaneuse* illustrated the fluidity of the alternative press: they gave little indication of its future political role. The paper initially appeared to be aimed at creating a niche for cultural discussion. Its early numbers were filled almost entirely with theater reviews, short stories, and amusing anecdotes. The paper's special feature, according to a notice to potential advertisers, was that it would be sold every evening in the foyers of Lyon's two theaters, and would therefore serve as a privileged means of communicating with the members of the city's population who had the time and the desire to escape from the workaday world of commerce and the politicized discussions of the cafés.[21]

In reality, the paper's founders undoubtedly had political purposes in mind from the start, even if their early issues approached politics only obliquely. Among the short satirical items published in the very first number was one entitled "What is the aristocracy?" with this response: "In 1600, it was a castle; in [17]89, a place in the king's carriage; in 1800, it was a sabre; in 1816, a parchment; in 1831, a 1,000-franc banknote.— What is the aristocracy good for?—To keep the thrones shiny-new; it is the shoe-polish of monarchy."[22] The critical tone of this seemingly offhand comment was obvious, especially in 1831, when the role of the Chamber of Peers in the new constitution was a major issue of contention, although the political intention behind this journalistic jab is difficult to define precisely. Is it an open challenge to the Orleanist monarchy, or an endorsement of the views of its more "advanced" supporters as opposed to those who still clung to more conservative views? This subversive-sounding jest was balanced, to some extent, by an item elsewhere in the paper casting Louis-Philippe in a good light, by reminding readers that he had sung "La Marseillaise" during the Restoration years, when that anthem of the original revolution had been banned. If there had been any doubt in readers' minds about the paper's intent to mix entertainment and politics, it should have been dispelled within a few months. Articles in July 1831 clearly challenged the justice of limiting the right to vote to the wealthy, criticized the notion of monarchy, exalted the memory of the Napoleonic era, denounced religious obscurantism, belittled the minister Casimir-Périer, and called for an alternative celebration of the first anniversary of the July Revolution. The final issue of the month announced the official consecration of the paper's subversive role: its editor had been summoned to court.

As important as the paper's emergence as the city's organ of extreme left-wing opinion was the tone in which it expressed itself. The *Glaneuse* drew freely on a repertoire of satirical formulas, many of them dating back several centuries. The paper's first direct criticism of the electoral system, for example, was embedded in an article in the *poissarde* style, in which two working-class characters, "MM. Marteau, locksmith, and

Rabot, furniture-maker," mangled the language in the course of pointing out the obvious injustice of their situation: "I'll explicate you that, I will. When someone has to get killed, they need us, so we're the *people*. But when it comes to naming the deputies, they don't need us any more; then the *people* is the erectors [*sic*]." The same two characters reappeared a few issues later to define the meaning of the word "politics." One told the other: "Under Louis XV, it was a cotillion; under the Republic, the people; under the Empire, a cannon; under the Restoration, the English, the Austrians, Prussians, and the Cossacks; under Charles X, the Jesuits and a holy-water sprinkler; in July 1830, the paving stones," adding that he had recently seen supporters of the new monarchy keeping a close eye on the paving stones. "If the stones change their place, they know they could well change theirs too."[23] These articles revived time-honored linguistic techniques for representing lower-class speech, including orthography that indicated the mispronunciation of words and the employment of plural verb forms with singular subjects ("J'les voyons marcher la tête baissée . . ."), although Rabot and Marteau were considerably toned-down versions of their revolutionary predecessor from the 1790s, the *Père Duchesne*.[24] They used no scatalogy and suggested no violence, and when a violent insurrection did break out in Lyon later in 1831, Rabot and Marteau disappeared entirely from the *Glaneuse*'s columns.

Along with articles in the *poissarde* style, parody was another time-tested genre freely employed in the *Glaneuse*. When the paper was indicted for the first time, it announced the fact in the form of a mock funeral announcement: "You are invited to the place St.-Jean next Tuesday, 2 August, to attend the funeral and burial of Mademoiselle la *Glaneuse*, legitimate and adult daughter of the very honorable and very powerful Lady Liberty. . . . The royal prosecutor will conduct the memorial service."[25] To protest against the repeated fines it had suffered by early 1833, the paper issued a "current account statement" of what it owed "the enterprise of *Thing and Company*" in Paris, where republican-minded satirists had made the word *chose* ("thing") a synonym for the king.[26] Another of the paper's favorite strategies was to present a political message in the guise of a story, or *conte*. One such article, printed in July 1831, described the scene at the Tuileries palace in July 1830, after Charles X had fled. The people found themselves in possession of an old piece of furniture they didn't know what to do with, "something called a throne." They would have liked to make a present of it to a "young man, grandson, by adoption, of a temperamental old soldier, great artisan of thrones, who, when the spirit seized him, demolished them and put them back together in the twinkling of an eye," but their neighbors would not allow it. "Then a man came along who told them: I'm a good fellow, pay me a good income and I'll keep the peace with the neighbors, since I'm their humble servant."[27] It required no great intelligence to recognize the allusions to Napoleon and Louis-Philippe in this story, but its wording was sufficiently allegorical and the label *"conte"* sufficiently powerful to forestall successful prosecution. The government was not so tolerant in October 1832, when the paper published "The New Tom Thumb, a tale translated from the Arabic," the story of "a great country far away! far away! far away! called *the Isle of Dupes,"* ruled over by "an avaricious, hypocritical, and evil prince, who had broad shoulders, long hands, and crooked fingers, and a pear-shaped head, ornamented with an enormous false hair-piece," whose chief minister was "a *tiers* of a man who had been labeled *the Tom Thumb of the century.*"[28] These references to Louis-Philippe, conventionally caricatured in the form of a pear, and to his diminutive minister Adolphe Thiers, were too obvious for the prosecutor to pass over, but the form in which they were cast presented him with insurmountable problems. The paper's defense lawyer began by remarking that the citizens of the imaginary country described in the article "are very happy, very satisfied with their king, that they furthermore enjoy paying so that he can have a very large personal revenue.... From this I conclude from the start that we cannot be referring to France." The ruler described in the article was "avaricious.... Is it possible that Louis-Philippe could by any chance be described as avaricious?" With the prosecutor confined to the unhappy role of straight man, the lawyer finally arrived at the question of Louis-Philippe's head: " 'a pear-shaped head, ornamented with an enormous false hair-piece!' Let's see, is there a bust of Louis-Philippe on hand?" he demanded, inspiring, according to one reporter, "unanimous and repeated laughter from the audience."[29]

Readers certainly recognized the peculiarity of the *Glaneuse*'s use of language, and even those who sympathized with the paper's political views were sometimes made uneasy by the ambiguities of its satirical tone. On one occasion when the *Glaneuse*'s editor wound up in court, an editorialist for the *Conseiller des femmes* told her readers that she had attended the trial to support the "daring writer." She had been disappointed, however, when she heard his lawyer interpret the incriminated text in such a way that "the expressions . . . , decomposed, broken up, twisted from their natural sense, interpreted in a strained and unnatural way, were given a completely innocent meaning. . . . We were struck by the dangerous flexibility of a language used in such a way as to blow hot and cold on every idea."[30] The *Glaneuse*'s mixture of satire, romantic rhetoric, and political propaganda challenged the primacy of reasoned political discourse exemplified by the conventional political press and by the women's and workers' papers that modeled themselves on it. It called into question even the basic assumption that words have intelligible meanings, and threatened to turn journalism into a pure language game. From the regime's point of view, the reluctance of juries to convict the paper's editor for his satirical articles showed that the public was a willing accomplice in this subversion of public language.

The *Glaneuse*'s editorialists were highly conscious of what they were doing when they printed satirical articles. In response to one of the repeated prosecutions brought against the paper, its principal editor, Granier, openly defended his strategy. His paper, he insisted, was a literary journal, not a political one. It exemplified "a literature born of the July Revolution, a literature of sarcasm, a literature inspired by patriotism and that arms us with the lash of satire, to be applied to political ambitions and nullities. . . . The literature I have defined, this literature of allusions, of epigrams, can invade the political realm without violating the press laws, because . . . to speak of politics in the sense of the law is to discuss the government's actions."[31] The *Glaneuse* identified its strategy with that of like-minded Parisian journalists and above all with the embattled Charles Philipon, native of Lyon and founder of *La Caricature*, for whose benefit the *Glaneuse* took up a collection in March 1832. "Irony is the only weapon left to us," it proclaimed.[32]

Richard Terdiman, in his analysis of subversive counterdiscourse in nineteenth-century French literature, has interpreted the ostensibly nonpolitical satirical press of the years after the passage of the repressive September Laws in 1835 as a largely defensive reaction to the apparently impregnable power of the Orleanist and bourgeois order.[33] Beleaguered as it was, the *Glaneuse*, like the Parisian *Caricature*, nevertheless projected a more aggressive air. At least until the defeat of the republican insurrections of April 1834, the July Monarchy hardly appeared invulnerable. Conscious of their growing support among the urban lower classes, and doubtless misled by their clear preponderance in the press market, republican journalists were confident that history was on their side. And, despite its editor's protestations, irony was certainly not the only weapon in the *Glaneuse*'s arsenal. The November 1831 insurrection caused a dramatic transformation of the paper's journalistic personality. Instead of satire, the *Glaneuse* veered toward the language of melodrama: "As we write, killing is going on in the streets. Blood covers the pavement!!!" The paper claimed it had no choice but to switch from veiled and indirect language to unequivocal statement of its views: "In the face of such events, it would be cowardice to be silent. . . . Our sympathies, let us say it straight out . . . are for the most numerous and poorest class."[34] Granier, the paper's editor, had briefly been pushed into a leading role in the evanescent provisional government that took over the Hotel-de-Ville for a few hours during the insurrection. Whether the experience truly radicalized him or merely led him to express his sentiments openly, the paper's treatment of social issues subsequently became much more direct. At his trial, Granier said he had felt compelled to speak out, "and our words had to be inspired by our sympathies for this unhappy class whose rights have not been recognized. We pleaded the cause of the workers; we measured the depth of the abyss opened by despair and hunger."[35]

The paper soon proclaimed that only a republican regime could improve the lot of the common people. Even before the Paris republican insurrection of June 1832, which led many advanced liberals to state their republicanism, the *Glaneuse* had published items such as César Bertholon's poem, "Why I Am a Republican," which condemned the July Monarchy for putting financial interests ahead of support for freedom in other nations.[36] In April 1833, the paper printed a "Republican Catechism," which represented the program of Lyon's semiclandestine republican movement. While promising that "the coming republic is not that of '93" but rather "that of an enlightened nation, ready for its liberty," the paper came out firmly for a long list of civil and social rights, the greatest possible degree of equality, and a spirit of fraternity and mutual aid among all citizens."[37] By this time, the *Glaneuse* had abandoned its distinctive pink paper, and as it became steadily more enmeshed in court battles and direct militancy on behalf of the republican movement, it became essentially a political paper, the organ of Lyon's more militant republicans, as opposed to the *Précurseur*, which represented a more moderate tendency. The lighthearted tone, the ambiguous language, and the mixture of political and cultural articles that had distinguished it at the outset disappeared. The *Glaneuse* ceased to be a true representative of an alternative press and became instead the leftmost end of the spectrum of Lyon's political press.

A Press for Women

This evolution from alternative forms of journalism to political radicalism was not an inevitable process, however. Even before the *Glaneuse* had clearly abandoned its claim to represent the city's cultural life and to address readers who wanted something other than political and business news, the *Papillon* had appeared and taken over many of the genres the other paper was in the process of giving up. Several of the less political contributors to the *Glaneuse*, such as Léon Boitel, migrated to the new title. Whereas the *Glaneuse* claimed to use literary language to discuss politics, the *Papillon* was sincere in its promise to stick to nonpolitical topics, and its contributors eschewed satire and parody. Its content was primarily reviews of theater productions, art exhibitions, and other cultural manifestations, together with very short stories, which rarely ran more than two or three columns of type, and word-sketches of different sectors of Lyon, many of which were later reprinted in *Lyon vu de Fourvières*. The favorite subject of its literary contributors was the joys and miseries of love; before long, they had run through every conceivable variation on the theme of the remorse of the unfaithful lover, and if some of their efforts produced comic effects—one story ended when the abandoned swain put a pistol to his head with such effect that "a bloody brain plopped at the feet" of his *inamorata*, who promptly went mad—they were unintentional.[38]

By mid-1832, when the *Papillon* first appeared, the obvious positions in Lyon's journalistic field had all been occupied. Introducing his creation, the editor of the city's newest publication anticipated hearing protests of "Yet another journal!" The biweekly *Papillon* justified its appearance by distancing itself from the existing Lyonnais press on two grounds. In the first place, it argued that the city's press, devoted to political and commercial issues, gave no attention to the aesthetic realm. All the other papers were under the spell of politics, "this tenth muse of our time, a brusque and bloody muse who always starts out pure and reasonable, but, by the force of things, gradually reveals herself as an intoxicated man." The *Papillon* would therefore shun politics, not because it could not afford a caution-money payment, but because politics was an inherently degraded activity. For all the dubious quality of its literary contributions, the journal articulated an early version of the nineteenth century's defense of art for art's sake, as a refuge from the realms of power and money.

The second justification the *Papillon* offered for its appearance underlined even more strongly its desire to present an alternative to the conventional press. Its subtitle was *Journal des dames*, and, although the paper's prospectus, as we have seen, linked women to the arts with the claim that the realm of culture was especially suited to the female psyche, the *Papillon* has a genuine claim to be regarded as the first Lyonnais periodical to take women's issues seriously. Its role in this respect requires some modification of the traditional emphasis on the part played by the Saint-Simonian movement in the development of early French feminism. In the early 1830s, that movement's radical critique of marriage and the family did raise fundamental questions about gender roles. For a brief period, the male leaders of the Saint-Simonians also took the dramatic step of giving women places in the movement's internal hierarchy, although the group's charismatic chief, Prosper Enfantin, soon changed his mind on this point, driving some of the Saint-Simonian women in Paris to establish a breakaway group, one of whose initiatives was to publish *La Femme libre*, the first feminist journal edited exclusively by women.[39]

Saint-Simonian "missionaries" appeared in Lyon in 1831, and in 1833 a former Saint-Simonian, Eugénie Niboyet, launched a feminist publication that, like the Parisian *Femme libre*, was written entirely by women. Nevertheless, it was not only the early utopian socialists whose activities opened new possibilities for women in the press. Even before 1830, local journalistic entrepreneurs had seen women readers as a potentially important audience. In 1824, the prospectus for a short-lived cultural magazine had promised that it would be on elegant paper, because "a journal whose nature dictates that it will sometimes be skimmed by women cannot have too agreeable an appearance."[40] No local attempt to exploit the market of women readers had succeeded before 1830, however, and the new papers created after the July Revolution initially made no special appeal to women.

This association of women with an apolitical aesthetic realm was, of course, a version of a familiar argument in which the sexes were assigned to separate but supposedly complementary spheres, and in which the idealization of women as more delicate and refined than men served to exclude them from the arenas where power and money were distributed. The *Papillon* did everything it could to emphasize its adherence to conventional notions of feminine gentility, even printing its numbers on pastel-tinted paper. An early number expatiated on the appropriateness of women's attendance at the theater, where they were truly in their proper place, holding men in their thrall thanks to their special sensitivity to emotion: "They are queens of the loges they occupy, queens of the circle around them. . . . Their social subordination disappears before the productions of genius that elevate the soul and give to their sex in imagination the supremacy that it ought to have in the social order."[41] Nevertheless, the paper did provide women with a genuine opening to speak for themselves. Although its editors throughout its existence were men, the paper appealed to female contributors to offer it their writings, "treasures that modesty too often confines to their elegant albums."[42] A number of women in Lyon took the paper at its word, and the *Papillon*, for its part, responded by publishing a number of articles that sharply criticized the treatment of women in French society.

From the early weeks of its existence, the *Papillon* gave space in its columns to female contributors who contested its own definition of their proper role. A "Mlle. R." protested the assertion that women had no interest in political or social issues. Criticizing the editors' promise to be "atheists in politics," she wrote: "We don't want to be atheists in politics; we have a soul that needs to love our country, and that has no more fear than yours of having to sacrifice itself for it. . . . Yes, I have a need to suffer in contemplating all these creatures weighted down by difficult and unrewarding toil."[43] A three-part article entitled "On Literature Among

Women" argued eloquently on behalf of women's right to publish on serious subjects. "Women are permitted to sacrifice their domestic obligations to the fashions of the world and its amusements, but any serious study is labelled pedantism."[44] A young woman, Sophie Grangé, published the most eloquent denunciation of sexual inequality to appear in the Lyon press during these years. Her poem, "To Woman" (see Appendix 1 for full text) pointed out that families celebrated the birth of children regardless of their sex, and that women and men had an equal share in "the air above, eternal heritage [*l'air qui vole, éternel héritage*]." As they grew, however, girls discovered that society did not treat them equally. They should not be fooled by men's apparent devotion to them, Grangé insisted:

> Oh! Don't tell me that your gilded chains
> Are light: I know that you've been draped
> With silk and with flowers;
> I know that, with humiliating insipidity,
> They talk of love, tell you lies . . .

Her conclusion was unequivocal:

> Raise yourselves up, my sisters; it is time for us to erase
> The imprint of the hot iron that burned our face.
> It is time that our brows,
> Finally purged of infamous offerings,
> Show themselves tall and proud,
> And look down with scorn on Lacedemonia.
>
> What! For us always ribbons and dolls?
> For the man power, thunder and swords? . . .
> It had to be thus, that man belittled us:
> But to keep us always under his rule!
> God has not willed it so![45]

To call the *Papillon* a feminist journal would be excessive; the feminist articles it published were regularly offset by others that relayed the most conventional stereotypes of women's roles. The same publication that carried Sophie Grangé's poem also gave space to another young woman poet who was fully satisfied with her life: "I want nothing but to love, and above all to be loved. / To be woman in this world, what more could one want?" and to a jocular article entitled "Does One Have to Beat One's Mistress?" whose author, a regular contributor, maintained that occasional blows were the surest proof of real love. It also printed extravagant idealizations of womanhood, such as "Woman must be an angel, and her breath the breath of heaven."[46] The paper dropped its subtitle, "Journal des dames," in April 1833, and the number of women contributors diminished over time. Articles about the "woman question" never disappeared from its columns, but by mid-1834 the main contributors on the topic were men, and their sentiments were increasingly misogynist. "Now woman throws off the yoke, wants freedom, independence, and, using her destructive leveling device, reduces man to something she can control," one columnist complained, while another denounced divorce in ringing terms.[47] Even in this more conservative phase, which followed the repression of the April 1834 insurrection, the paper continued to offer a variety of views. In response to Charles Nodier's warning that women who engaged in politics would be abandoning all hope of love, a local woman writer could still expostulate that he had misunderstood their complaints about "the kind of helotism to which the organization of society condemns us."[48] The *Papillon* was thus a significant forum in which women could express themselves in public and seek to redefine their place in society, even if it was not a genuinely feminist journal.

Important as the *Papillon* was in giving literate Lyonnais women a chance to discuss their situation, that journal remained solidly anchored in the male-dominated world of the Lyon press. The *Papillon* reached out to women as an essential part of its market, and found in women's issues a topic that allowed it to affirm its difference from the city's other papers, but men—above all the editor, Eugène de Lamarlière—set the terms under which women could participate in the enterprise. In the cultural paper's columns, the feminist point of view was simply one position among others, frequently challenged or ignored. The creation, toward the end of 1833, of an explicitly feminist journal, the *Conseiller des femmes*, created and edited by a woman and intended to publish only articles by women, represented a significant new effort to reconfigure the city's print culture. It cannot be asserted that the *Conseiller* was launched in explicit opposition to the *Papillon;* the *Conseiller*'s editor, Eugénie Niboyet, sent Lamerlière a courteously worded request to announce her enterprise in the columns of the older publication, a request that he generously honored.[49] The *Conseiller des femmes,* a weekly that was more of a magazine than a newspaper, lasted only ten months. Short-lived though it was, the *Conseiller des femmes* nevertheless succeeded in posing essential questions about the place of women in public discussion and public affairs.

The *Conseiller des femmes* was the first women's journal published outside the capital. The paper's editor, Eugénie Niboyet, born in Montpellier in 1796, had been a leader of one of the Saint-Simonian movement's Paris sections in 1831; like a number of other women initially attracted by the movement's feminist sympathies, she had quit to protest its leaders' refusal to give women a real role in setting its direction. Her creation of the *Conseiller des femmes* was the start of a long career of campaigns for philanthropic causes. Her later writings included books and periodicals promoting the education of the blind, prison reform, the abolition of the death penalty, and pacifism, and in 1848 she helped edit *La Voix des femmes,* a women's paper in Paris.[50] Although Niboyet was the guiding spirit behind the publication, it was by no means a one-woman operation. A number of other Lyonnais women, including Louise Maignaud and Jane Dubuisson, signed important articles with their full names, while others, such as Aline M., ventured into print but guarded their anonymity. The paper also had several contributors who appear to have resided in Paris, most notably the very outspoken Sophie Ulliac Trémadeure, a regular writer for Fanny Richomme's *Journal des femmes.*[51] Whereas the pressures of regular publication pushed Lyon's other papers toward reliance on professional editors, the *Conseiller* was the product of a group of amateurs brought together by a common moral passion.

That passion was, of course, the right of women to participate in public debate and to use the press to demand their rights. The paper's prospectus justified the venture on the grounds that the supposedly universal public of the press had ignored a significant social group. "For the past three years," it announced, "the periodical press has held up to our eyes its prism of a thousand colors, in which are faithfully reproduced the opinions and systems of a society marching rapidly toward a more perfect civilization.... However, ... in this great movement of the press, in this energetic agitation of humanity, . . . women have found few organs to represent them." Paris had a women's literary journal and a paper aimed at providing pleasurable entertainment for women. These examples showed that "the spark is there, the female portion of humanity is waking from its long lethargy." But Lyon needed something more: "We have conceived the project of founding in Lyon, a populous city where most women are in workshops and factories, a practical journal aimed at improving their condition regardless of their social position."[52]

Like the *Écho de la fabrique*, the *Conseiller des femmes* was no simple "voice" of its presumed readership. In one sense, it was closer to women than the *Écho de la fabrique* was to workers: with a single exception, every contribution it printed in its two years of existence was written by a woman. But the paper's contributors were themselves conscious of the difficulties involved in speaking for all women. Influenced by the Saint-Simonian movement, Eugénie Niboyet, the journal's remarkable editor, wanted above all to improve the lot of the poor woman: "It is she, above all other women, who suffers the greatest dangers, being the creature our social institutions treat the worst."[53] Niboyet knew that to achieve this goal, she would have to "plead her cause to women from the privileged class, so that they will put aside distinctions of rank and fortune and give the love and care needed by this other needy class of society." Her goal was to promote a sense of unity among all women. "Woman will never truly be strong until she is . . . the friend of her sex."[54]

The *Conseiller des femmes* had to work not only to get women to think of themselves as a group with common interests but also to get them to conceive of themselves as oppressed. Precisely because women, as women, had no tradition of collective action, the paper used remarkably forceful language—indeed, despite its respectful references to religion and its contributors' assurances that they aimed to create nothing more than a "pacific coalition," one might well claim that it was the most incendiary periodical published in Lyon in the early 1830s. The cause of women was "too just not to triumph in the very near future," one contributor wrote, sure that she would see "the feudal towers of the Bastille fall before a people armed for the conquest of its rights."[55] Men were "with respect to women, what an absolute monarch is to the people subjected to his law. You have muzzled her intelligence, to make her submit to your oppression without too much murmuring."[56] In existing society, women were in the condition of helots, but if they protested, men "think they see bold hands raised to steal their scepter" and react violently.[57]

Despite the force of this rhetoric, the contributors to the *Conseiller des femmes* knew that they faced an uphill struggle to create a social space in which their publication could actually have any impact. The problem was quite simply, as the title of a short story in the paper put it, that "the husband is at the café, the wife is at home."[58] Although bourgeois women, unlike silk workers, lived in the center of the city, they could not enter most of the spaces where men read their newspapers and discussed their contents, for those spaces—with the exception of such mixed gathering-places as the theater and the promenade de Bellecour— were part of an exclusively male world. Feminist contributors to the *Papillon* had already raised complaints about this situation. The most extensive article it published in its first year was a three-part series, "On Women and Literature," protesting the treatment of educated women. Men, the anonymous author charged, were more willing to pardon women "who forget their duties" than those "who attract attention by remarkable talents." It would be more reasonable to give women a serious education, so that they could hold their own in conversation with men and thereby contribute to intellectual progress. This essayist was especially concerned about the fate of the woman writer, who was, she claimed, an anomalous figure, rejected by society, and often even by other women. The female author was not allowed to participate in the normal interchange of the public sphere. She could not convert her celebrity into a significant public position, and she could not respond to criticisms by revealing the details of her private life: "For a woman to defend herself is one more fault, to justify herself, a new scandal. Women know that there is something pure and delicate in their nature that is injured even by the public's curiosity."[59] The *Conseiller des femmes* repeated many of the same protests. Especially in the provinces, contributor Louise Maignaud complained, "one cannot speak of a female author without her appearing pretentious and pedantic, as if it was not as simple to put one's dreams and one's thoughts on an innocent sheet of paper as to speak of them in the middle of a salon or in a private conversation."[60] The *Conseiller des femmes* sought to challenge the exclusion of women from public discussion with its ambitious plan for an *Athénée des femmes*, which we have mentioned in discussing Lyon's public sphere. Over time, the *Athénée* would have created the public for a serious women's press and a space in which it might have been read, but, like the paper itself, it was doomed to a short existence.

Whereas the *Papillon*'s women contributors limited themselves for the most part to calling for a place for women in public discussion and in the aesthetic realm, the journalists of the *Conseiller des femmes* ranged more widely. Ulliac Trémadeure, in a strongly worded article "On Women in General and on Their Real Emancipation," denounced the discrimination built into the Napoleonic Code. Although they contributed to the state by paying taxes and rearing children to become citizens, women were excluded from all political and administrative positions, and their testimony was inadmissible in filling out public documents, even birth certificates. They were denied all legal rights over children and family property: a husband could "ruin his wife and children as he sees fit," but a wife had to put up with his outrages even if she brought legal action against him. Ulliac Trémadeure denied that women wanted the complete legal autonomy that adult men had, but she insisted that they should have meaningful rights within the family.[61] In another article, the same author protested the idea that, for women, "marriage is the destination one must arrive at, no matter what." Whether they eventually married or not, women needed to be prepared to make a living, and Ulliac Trémadeure argued that, for middle-class women, schoolteaching was an especially attractive and appropriate career, in part because it allowed for the exercise of the intelligence without the hazards of being a published writer.[62] Jane Dubuisson, a local contributor, denounced the exploitation of working women, who had to grant favors to male clerks and supervisors to get jobs, or curry favor with the *maîtresses d'atelier* in establishments that used female labor.[63]

Although the contributors to the *Conseiller des femmes* minced no words in condemning injustices against women, they were careful to state that they were not seeking full equality with men. Ulliac Trémadeure, in her articles on women's emancipation, took issue with men who, in the tradition of Condorcet, argued that there were no significant differences between the sexes. "Sensible men know, as we do, that there cannot be complete parity between he who is destined for tasks that require physical or intellectual force and she whom nature has made to devote all her time to the care of infants and the aged," she wrote. "Let us be women, entirely women, and we will in fact be superior [to men]."[64] Eugénie Niboyet, the paper's editor, also insisted that women had special moral qualities that made them different from men. Women did not want to play men's roles; they simply wanted freedom to develop their own abilities and to make their contribution to "the rising movement that characterizes our civilization."[65] None of the articles in the *Conseiller des femmes* suggested that women should enjoy political rights—which, in the early 1830s, were still limited to a small minority of men—or even that they should have access to male-dominated professions.

This apparent moderation has led Michèle Riot-Sarcey, the leading French historian of early nineteenth-century feminism, to claim that Niboyet's paper represented a retreat from the more radical positions of the Paris Saint-Simonians a few years earlier: "In effect, Niboyet's journal reinforces the system of values that structures social relations in the name of a political morality invoked by the liberals, reelaborated by the church, put together by the moralists, redefined by the republicans."[66] It is true that Lyon's feminist journalists employed what Claire Moses has classified as a "difference" argument on behalf of women's rights, arguing for "sex-differentiated" policies.[67] Furthermore, although they protested the conditions of marriage under French law and insisted on the need for women to be able to earn a living on their own, the Lyonnais feminist journalists never endorsed the sexual radicalism of the Saint-Simonians. The *Écho des travailleurs* specifically praised the paper because "one doesn't find in it any of the things that justifiably cause shock in Père Enfantin's doctrines," and concluded that "it should be the guide of mothers of families."[68]

For its time, however, the *Conseiller des femmes* was a highly radical initiative. It posited the right of women not only to participate in public discussions in the press—a right several women had already exercised in the pages of the *Papillon*—but also to do so on their own terms. Limiting its pages exclusively to women contributors, the *Conseiller des femmes* created a truly "alternative" press and a symbolic space, however modest, within which women governed themselves and from which men were barred. Remarkably, although the creation of the *Conseiller des femmes* had radical implications for the notion of the public sphere, Lyon's other journalists generally gave the project a favorable reception. This was in part because Niboyet carefully courted the protection of her male colleagues. "I think I would be forgetting how well-brought-up people should treat each other," she wrote to the editor of the *Papillon*, "if, before publishing my journal, I didn't try to gain the goodwill and respect of our city's journalists! . . . You will surely give me some credit for the efforts I am making to raise women up to the level of their century?"[69] The generous praise heaped on her project in other Lyon papers suggests that she must have written similarly ingratiating letters to their editors. The *Écho de la fabrique* welcomed an initiative that it saw as parallel to its own: "It is good that, by raising up a tribune from which she can make her voice heard, WOMAN takes her place in the arena where a multitude of serious questions to which she can no longer remain indifferent are disputed."[70] The rival workers' paper, the *Écho des travailleurs*, was no less positive, but so was the deadly enemy of these papers, the bourgeois *Courrier de Lyon*, which said that the *Conseiller*'s first issues "have completely fulfilled expectations and are written with great talent."[71] Reviewing the responses to her publication after its first three months, Eugénie Niboyet noted that the only criticism had come from the legitimist *Réparateur*, which had blamed her for not explicitly endorsing Catholicism.[72] This benevolent acceptance implied that the *Conseiller des femmes*, alone among the new press titles launched in Lyon after 1830, was not explicitly infringing on the territory of any of the city's other periodicals. Even the *Papillon*—which, it is true, had quietly abandoned its original subtitle, "Journal des dames," six months before the *Conseiller* appeared—praised Niboyet's initiative.[73] The *Conseiller des femmes* was so different from the rest of the press that other publications did not see it as a threat.

Alone among the new forms of press created in Lyon during the early 1830s, the *Conseiller des femmes* failed to achieve the honor of being publicly prosecuted for attacking the regime. Its rhetoric was sometimes revolutionary, but there was no feminist insurrection analogous to the workers' uprising of 1831 to lend dramatic confirmation to its claims for importance. Eugénie Niboyet did hail the collective action of a group of nuns who had asserted their right to admit a woman to their order without the explicit approval of their bishop. "All the emancipatory thought of the century is in this resistance by women accustomed to obey higher authority without questioning; what one alone would not have dared to do, the group has done," Niboyet wrote.⁷⁴ But this was a rare instance of women taking collective action. Women journalists were unable to emulate male journalists by holding public banquets and rituals that would underline the importance of their initiatives. In fact, as Niboyet pointed out, the laws that prevented women from being the legal owners of any business enterprise required her to have a male director for her paper, and thus deprived her of the ability to use the legal system to defend her paper's rights.⁷⁵ In the conditions of the early 1830s, women's patterns of sociability did not lend themselves to representation and restructuring through the periodical press. Despite its limited impact, however, the very attempt to create a women's journal served to challenge the boundaries of the liberal public sphere.

In different ways, then, the alternative press organs that appeared in Lyon in the aftermath of the July Revolution all extended the boundaries of the city's public sphere. The *Glaneuse*, particularly in its early phase as a satirical journal, challenged the hegemony of the rational political language in which bourgeois political debate was carried on. Its mixture of literary forms blurred the boundary between fiction, the supposed domain of art, and the realistic representation of the world to which the press was supposed to be devoted. More than any other Lyon paper of the period, the *Glaneuse* drew the ire of the authorities, for its activities threatened to subvert both the institutions of the constitutional monarchy and the regime's control of language and political discourse. The *Papillon* also mixed literature and more conventional journalistic language, but in a less subversive manner. Its effort to define a nonpolitical public realm of aesthetic discussion offered an alternative to the concentration on issues of power and money in the conventional newspapers, but suggested that the two realms could coexist: aesthetics as the realm of women, the young, and those with imagination; politics and commerce as the realm of men, the middle-aged, and those unable to see beyond practical and concrete concerns. Throughout the nineteenth century, French poets and artists were to demonstrate that this apparently apolitical realm could in fact generate a critical counter-discourse directed against bourgeois society, but the result of this critique was alienation, rather than active and critical engagement with the dominant system. Finally, the *Conseiller des femmes*' feminist critique of gender relations called for recognizing the right of women to participate fully in the public sphere, at least to the extent of seeing their words published, and for a women's version of the public sphere, one in which women's words did not need the imprimatur of a male editor.

The *Conseiller des femmes* faced an even more difficult challenge than the other alternative publications in its effort to make a lasting mark on the Lyon press market. Even with the ostensible endorsement of most of the city's other periodicals, Eugénie Niboyet's creation could not garner enough subscribers to make itself financially viable. Basically, its difficulty stemmed from its attempt to represent an "imagined community" whose potential members were probably largely indifferent to its existence, if they were aware of it at all. This is not to say that women had no forms of community and no ways of exercising influence in the society of the early 1830s, or that all of them were necessarily hostile to the feminist concerns of the *Conseiller des femmes*. It is rather to suggest that the medium of the periodical press was not well adapted to their situation. The *Conseiller des femmes* deliberately turned its back on fashion reporting, the one genre of journalism that had already proved its ability to attract a feminine audience. Eugénie Niboyet and her collaborators refused to define women as consumers, the function that bourgeois society was most willing to concede to them, and one that some historians, notably Bonnie Smith, have argued did in fact offer bourgeois women an important outlet for self-expression.[76] The *Conseiller des femmes*, even as it created an exclusively female form of public space, sought to offer women the cultural equivalent of the male-dominated press: an ideological organ, divorced from the marketplace and the domestic sphere. Because of the gendered social structure the paper denounced, however, women in the early 1830s did not possess the social space in which such a periodical could put down solid roots. Unable to control their own property, they could not subsidize such an enterprise; unable to gather for public discussions, they could not generate the social interchanges that would have prolonged its influence. The number of contributors to the *Conseiller des femmes* demonstrates that a significant group of women who did recognize themselves in Eugénie Niboyet's enterprise and were willing to support it. Unlike the pioneers of the labor press, however, these determined and outspoken women were not able to find the audience that would have enabled their project to thrive. Their failure shows that the print medium's ability to transform the social reality around it was limited.

Together with the appearance of the labor press, the alternative press convincingly demonstrated that the July Revolution had set in motion a transformation of the public sphere. These publications extended discussion and debate into new territories, literally and figuratively. Their very existence highlighted the limits and boundaries of the supposedly universal public sphere defined by the conventional political press. Their development had ambiguous implications, however. The regime itself chose to read the *Glaneuse,* with its policy of provocation, as a form of alternative discourse that could not be tolerated; it was forced out of business at the time of the April 1834 insurrection, and no journal featuring either explicit republican propaganda or political satire reappeared in Lyon until the Revolution of 1848. The *Papillon,* on the other hand, even though it ceased publication in 1835, represented an assimilable form of alternative journalism. The bourgeoisie defined itself in part by its appreciation for culture, and a press devoted to the subject was in some senses a necessary aspect of the public realm. Parisian journals devoted to literature and the arts flourished throughout the July Monarchy. The difficulty, as Lyon's journalists repeatedly recognized, was to establish a place for such a publication outside the capital. "There is a great obstacle to the literary emancipation of Lyon," one of them wrote in 1835, "and this obstacle is produced not by a lack of writers but by the absence of people who encourage writing." The city's business classes "are too concerned with material matters and too calculating to imagine that our city might acquire some glory from cultivating the liberal arts."[77] Fifteen years later, Jean-Baptiste Monfalcon, former editor of the arch-bourgeois *Courrier de Lyon,* lamented with Léon Boitel, the indefatigable promoter of cultural journalism from the early 1830s, the fact that "since 1815, there have been seventeen attempts to publish a review, and seventeen failures." A provincial publication either had to attempt to compete with the Parisians or explicitly limit itself to local cultural life and thereby condemn itself to be "deathly dull." The provincial environment did not tolerate serious cultural criticism: "One is always either flattered by one's friends or torn down by the envious; in either case, not the slightest impartiality and not the slightest credit." In any event, the Lyonnais public simply was not interested in culture.[78]

Monfalcon's gloomy conclusions certainly matched the reality of the early 1830s. The three titles discussed here were the most long-lived of the various "alternative" publications launched during these years. Several others had also been launched and disappeared virtually unnoticed.[79] Although the repression following the insurrection of April 1834 did not specifically target them, the cultural climate was distinctly unfavorable to even the most anodyne forms of nonconformity. When Léon Boitel, the young printer who was also the center of all efforts to create an autonomous local cultural life during these years, tried to rescue something of the spirit of the *Conseiller des femmes* and the *Papillon* by merging the two titles in October 1834 to form the *Mosaïque lyonnaise*, he had to excuse its appearance "at a moment when the press seems to have fallen into discredit, disgraced as it is by its own children." The new journal attempted to occupy the terrain of its two predecessors, promising articles on all the arts and "enough editorial variety so that the ladies will always find something to charm them in reading it." At the same time, however, the publication differentiated itself from the now discredited formulas of the other alternative papers by promising to devote equal attention to the sciences, a strategy intended both to identify it with solidly masculine concerns and to associate it with a form of intellectual activity acceptable to the July Monarchy. The *Mosaïque* also published a list of its stockholders in one of its first issues. All men, they represented a cross-section of the city's educated bourgeois elite, and one chosen deliberately from all competing political camps; César Bertholon, the prominent bourgeois republican, rubbed shoulders on the list with the prefect Gasparin.[80] Eugénie Niboyet edited the first few issues of the *Mosaïque*, becoming for a few weeks the first woman editor of a Lyonnais journal largely written by men, but her name soon disappeared from the masthead and the journal itself folded after three months, giving way to a resurrection of the *Papillon* in its original format, which disappeared in turn seven months later, as Boitel turned his attention to the more serious and scholarly *Revue du Lyonnais*, the only "alternative" Lyonnais periodical founded in the early 1830s to survive beyond the end of 1835.[81] The fate of these periodicals demonstrates that even those that appeared least political and least critical of the bourgeois social order could not survive once the revolutionary interregnum opened in 1830 had definitely ended and the space of public debate had been brought back under effective control.

4

Echoes of the Working Classes

Publicity for the Working Classes

Of the various developments involving the Lyon press in the early 1830s, none was more significant than the creation of newspapers claiming both to represent and to speak to the city's working class. A similar paper had appeared in England in 1825, but aside from a few short-lived papers issued in Paris in the fall of 1830, these were the first such publications in France.[1] The series of Lyonnais workers' papers begun in 1831 included the first French periodicals to use the words *travailleur* and *prolétaire* in their titles, and the series itself—the *Écho de la fabrique* (1831–34), the *Écho des travailleurs* (1833–34), the *Tribune prolétaire* (1834–35), the *Indicateur* (1834–35), the *Écho de la fabrique de 1841* and the *Tribune lyonnaise* (1845–51)[2]—endured, with interruptions, for some twenty years. More explicitly than the periodicals that made up the alternative press discussed in the previous chapter, these papers challenged the bases of the bourgeois public sphere and set in motion a fundamental realignment of Lyon's journalistic field.

From the outset, both promoters and opponents of these newspapers recognized their appearance as a major historical novelty. They represented the "union of journalism and industry," as a contributor to one of these papers put it, using the word "industry" to indicate the world of work.[3] The combination would "mark an epoch in the history of the proletariat" by giving it direct access to the power of the printed word, but it would "also serve to give journalism the rank it deserves" by linking it directly to the common people.[4] At a time when Karl Marx was a mere thirteen years old, this press announced itself as the representative of the interests of a "working class" distinct both from the owners of property and from the generalized *peuple* that the radical newspapers of the French revolutionary period had claimed to speak for. "Society . . . is now divided into two camps: that of the WORKERS or *proletarians*, [and] that of the idlers or *nonproductive*. We have planted our flag in the first camp," the *Écho des travailleurs* announced in 1833.[5] Lyon became the laboratory in which the new possibilities and the new dilemmas resulting from this effort to marry the press and the workers' movement first became clear.

In claiming to represent workers, the *Écho de la fabrique* and its successors eventually found themselves trying to create a new social reality: a unified, self-conscious working class. Initially, the creators of the *Écho de la fabrique*, the first of these papers, had not set out to do this. The original prospectus for the paper called it a "journal of the *chefs d'atelier* and silk workers."[6] In Lyon, the term *chef d'atelier* referred specifically to master silk weavers, skilled artisans with their own workshops who contracted with the *fabricants* or silk merchants to weave raw silk into cloth. Economically dependent on the *fabricants*, the *chefs d'atelier* were often as impoverished as ordinary workers, but, as small business owners with their own employees, they were not ordinary proletarians. Although the specific reference to the *chefs d'atelier* was dropped when the paper appeared, the use of the word *fabrique*, the traditional term for the silk trade, in the title of the paper indicated that its concern was still focused on one specific group, rather than the entire working-class population. As the paper's second editor later put it, "the *Écho* . . . in its origins was created to be only the organ of the silk workers."[7] Within six months, however, the paper had moved beyond this conception of its mission, rooted in the corporatist institutions of the past. Articles in May 1832 spoke of a "proletarian class" whose emancipation had begun with Jesus' proclamation of the moral equality of all men and called for "a universal union among workers" that would break down not only the barriers between trades but those between nations as well.[8] The paper's second prospectus, issued in September 1832, boldly proclaimed: "We will be the paper of the whole PROLETARIAN CASTE. Rally to us, artisans of every profession, *industriels* of every sort, *ouvriers travailleurs* all over France." Anticipating the *Communist Manifesto* by some sixteen years, the *Écho de la fabrique* called for a "universal association of workers" that would be the basis of "the holy alliance of peoples."[9]

In casting its net so widely, the *Écho de la fabrique* was going beyond the social reality of the day. As its editors well knew, French workers were divided into a multiplicity of groupings. Skilled artisans were divided by trade, and even within a single trade, competing *compagnonnages* or journeymen's brotherhoods engaged each other in bitter, often violent hostilities. Artisans, with their traditions of corporate and *compagnonnage* organization, had little in common with the mass of unskilled workers or with the workers in the emerging sector of factory production, and male workers showed no solidarity with women. To make workers imagine their community in a new way, as a unified group, the press had to openly challenge their existing self-conceptions. In 1833, a fight between rival *compagnonnage* members drove the *Écho de la fabrique* to ask, "How long will the workers of all trades close their eyes to their true interests? When will they understand that these internal divisions only serve to rivet more strongly the links of the chain in which egoism and selfishness bind them?"[10] By reprinting from Lyon's feminist paper, the *Conseiller des femmes,* Louise Maignaud's article on the special exploitation suffered by women workers and several other pieces, the *Écho de la fabrique,* although edited exclusively by men, raised questions about an even more fundamental division in the working class.[11]

To justify this challenge to traditional conceptions, the writers for the working-class press had to demonstrate that there was in fact a bond uniting the whole of their intended audience. Their boldest efforts in this direction opened new perspectives for future socialist theorists, but they also revealed the ambiguities concealed in the radical-sounding formulas of these newspapers, ambiguities rooted in the effort to transpose the universalist rhetoric of the French Revolution to a new social key. The opening numbers of Lyon's second workers' paper, the *Écho des travailleurs*, offered the most thoroughgoing attempt to spell out the logic of working-class unity. By its title, this paper, as well as its successor, the *Tribune prolétaire*, broke with the corporatist heritage that still influenced the *Écho de la fabrique*. *Travailleur* and *prolétaire* were general terms, applicable to members of all trades, and indeed the Lyon papers were among the earliest examples of the use of these words in what was to become their modern sense.[12] What the members of those groups had in common was that they earned their living by the sweat of their brows, as shown by their hands, "blackened by noble and hard work."[13] The opposite of the *travailleur* was the *oisif*, the "nonproducer," and contemporary society was divided between these two groups. But the *Écho des travailleurs* employed a broad definition of work. "Whoever makes use . . . of the physical and intellectual faculties with which he has been more or less abundantly provided by nature is a *travailleur*," the paper explained. Its concern was to include in this category those who worked with their minds, "the artisans of thought," as well as those who did manual labor. Both forms of work constituted productive use of natural faculties, and so both groups should recognize their natural unity.[14]

This assertion of the natural affinity—indeed, the identity—between workers and intellectuals was destined for a long career in the Marxist tradition, but it ran counter to the lessons of everyday experience in early nineteenth-century France. The need to make the argument was evidence that the "union of journalism and industry" required workers to recognize themselves in the words of newspapers that, by their very nature, could not be composed by ordinary manual laborers. The story of the Lyon workers' press is the story of one of the first experiments to determine whether the printed sheets of the press could serve as representatives for a phenomenon—the "working class"—whose inability to speak directly for itself in print often seemed to be one of its defining characteristics.

The first of the Lyon workers' papers began publication at the end of October 1831, three weeks before the silk weavers' insurrection of November. The *Écho de la fabrique* was an eight-page weekly priced at 11 francs for a year's subscription—considerably cheaper than the city's standard daily newspapers. The paper's initiators knew their audience: they had arranged to sell the papers, not only in the familiar confines of Lyon's central city where the conventional political papers were distributed, but also in the areas where workers lived and gathered in their time off: the Café Orssière on the place de la Croix-Rousse, the central square of the working-class suburb on the hill above Lyon's peninsula, two cabarets in the suburb of La Guillotière, on the eastern bank of the Rhône, where workers commonly went on their days off, and the Café des Trophées in the equally working-class suburb of Vaise, northwest of the city.[15] The paper thus made a bid to establish a new geography of public discourse on the margins of the city, away from the customary centers of newspaper-reading and discussion in the urban center.[16] Whether or not workers actually purchased and read the papers offered in these locations, the mere listing of these sales points demonstrated the *Écho*'s determination to identify itself with a public different from that found in the center of the city.

From the outset, the *Écho de la fabrique* presented itself as a paper for workers, and particularly for the silk weavers who made up nearly half of Lyon's working-class population. The paper's initial prospectus explained its purpose and identified its intended audience: "Up to now defenseless against the intrigues of commerce, subject to the brutalities and the flagrant injustices committed by certain merchants . . . , the poor workers have chosen publicity as a defensive arm to protect their rights."[17] A second prospectus, published in the paper eleven months later, repeated for the benefit of workers the claims for the power of the press familiar from the liberal papers: "The press is nowadays a power greater than any physical force, and we promise to make it your protecting shield. In this way, a universal association of workers will take shape, which will give them the power to successfully resist the egoism, the selfishness, the tyranny of the idle."[18] Rhetorically, the *Écho de la fabrique* left no doubt of its intention to represent workers as a distinct social group.

The *Écho de la fabrique* put itself forward not merely as a paper for workers, but as a paper *by* workers. Four years later, its principal editor claimed that it had been "the first organ of the proletarian cause, a journal born in the workshop, living in the workshop."[19] The inspiration for its creation came from one master silk weaver or *chef d'atelier*, Joachim

Falconnet, a well-known figure in the silk weavers' community during the 1830s and 1840s. He served as the paper's first *gérant,* or legally responsible director, and contributed articles to it and to some of its successors, while also serving at times as one of the worker representatives on the silk trade's *conseil de prud'hommes* (arbitration council).[20] In a broader sense, the paper was clearly a product of the movement toward collective action among the Lyonnais silk weavers that manifested itself in 1831 in strikes and demonstrations aimed at compelling the silk merchants to pay fixed prices—a *tarif*—for their work. The organization behind this movement was the famous Société des Mutuellistes, the Society of Mutual Duty, the mutual-aid society formed in 1828 by a group of master silk weavers. It was this group that had organized the campaign for the *tarif* that had led Lyon's prefect to order negotiations between the weavers and the merchants.[21]

But the assertions that the paper represented the workers, and that it was linked to the *mutuelliste* movement, are more problematic than the coincidence between the founding of the paper and the first public manifestations of the *tarif* movement would make it appear. In the first place, the master silk weavers themselves were not, strictly speaking, proletarians. They were small, independent subcontractors firmly subordinated to the wealthy wholesale silk merchants of Lyon but at the same time heads of family enterprises, owners of their looms, and often small-scale employers in their own right; in some cases, they left the entire manual labor of actual weaving to their journeymen and apprentices. "An *atelier,* or workroom, is, in fact, a little kingdom governed by a chief, in which four or five gradations of society frequently exist," the English Utilitarian John Bowring wrote after a visit to Lyon in the early 1830s. "The master weaver has no other lien on the *compagnon* than that of mutual agreement."[22] In 1833, one of Lyon's "bourgeois" papers triggered a flood of letters to the *Écho de la fabrique* when it charged that the *chefs d'atelier* were themselves parasites who "feed and fatten on the sweat and toil of the workers." The responses printed in the *Écho* made it clear that the *chefs d'atelier* were primarily intermediaries between the *fabricants* who provided the silk and the *compagnons* who in effect rented looms to work on from the master silk weavers.[23]

Among the differences John Bowring observed between the master weavers and the rest of the working population was that most of the former could read and write, while the majority of the latter could not. The master weavers could envisage holding their own in a print-oriented culture; the journeymen weavers and the other, less skilled, members of the workforce were much less ready to do so. Visiting a café frequented by *chefs d'atelier* in the spring of 1835, the bourgeois expert on the working classes Villermé commented, "One would have thought them, by their dress and by their conduct . . . , well-to-do bourgeois."[24] Although the master weavers, the *chefs d'atelier*, were able to mobilize support from much of the rest of the working population in November 1831, there were clear differences of outlook between this group and others, such as the journeymen silk weavers who worked for the masters and who formed their own association, the Société des Ferrandiniers, in late 1831. At a banquet celebrating the first anniversary of the *Écho de la fabrique* in 1832, the lone representative of the journeymen invited to propose a toast drank to "the union of the *chefs d'atelier* and the workers" and called on both groups to remember that "they have a single goal, a single dream, that of making a living through their work, so they should remember that union creates strength."[25] His words reminded the audience that all silk weavers might have common interests, but they also underlined the fact that the weavers did not form a homogeneous group.

The ambiguities resulting from the unique status of the *chefs d'atelier* surfaced repeatedly in the early 1830s, both in the press and in other publications. In 1833, when the master silk-weavers' Society of Mutual Duty decided to take over direct control of the *Écho de la fabrique*, its former editor, Marius Chastaing, founded his rival paper, the *Écho des travailleurs*, and denounced their narrow conception of working-class interests.[26] One reader of the new paper, who signed himself "a silkmaker and non-*mutuelliste*," praised Chastaing's decision: "The *Écho de la fabrique*, which claims to be the journal of all proletarians, . . . is nothing but the *Écho du mutuellisme*," the organ of a group that did not even represent the majority of the master silk weavers.[27] Jean-Baptiste Monfalcon, editor of the aggressively bourgeois *Courrier de Lyon*, was quick to recognize the significance of the split between the two working-class papers, which symbolized the instability of working-class identity. "How can one imagine that harmony and unity of views could last for long in an association where there are so many prejudices, so much ignorance of the interests of industry, so many corrupt passions, and such great elements of disorder?" he asked.[28] By 1835, the *mutuellistes'* own paper would write: "The master weaver [*chef d'atelier*] is always, by virtue of his position, placed between two perils, the merchant and the worker. The first . . . takes advantage of his misery; the second . . . , when prosperous conditions give him a chance to make up his losses, makes himself the agent of his ruin."[29] To the extent that the *Écho de la fabrique* and its successor paper, *L'Indicateur*, spoke for the organized *chefs d'atelier*, it cannot be maintained that the papers voiced "the" interests of the Lyon silk weavers, much less those of the entire Lyon working class. Letters to the editor printed in the papers came from articulate master silk weavers and frequently from silk merchants denouncing the paper's editorial positions, but not from journeymen weavers or from the many women who made up much of the work force. The papers did attract some ordinary workers, as the published list of donations collected to pay off a fine inflicted on one of them showed,[30] and they were more successful than a short-lived journal edited by a self-proclaimed "proletarian" who had no roots in the city's working-class milieu.[31] The culture they reflected, however, was that of the wealthiest and most educated members of the working class.

To further complicate the question of its relationship to the working class, the *Écho de la fabrique* employed professional editors of nonproletarian origin throughout most of its existence. Falconnet, the *chef d'atelier* who had founded the paper, had recruited one Antoine Vidal, a patriotic poet and schoolteacher, as the paper's original editor. Vidal, in turn, brought in an old friend of his, a veteran like himself of local campaigns against the Restoration: Marius Chastaing. Vidal died before the *Écho de la fabrique* had reached its first anniversary, but Chastaing went on to become the most dedicated exponent of "working-class" journalism in Lyon, keeping the tradition begun by the *Écho de la fabrique* alive for twenty years. There was no question about either man's utter dedication to the cause of the working class, as they understood it. Chastaing devoted twenty years of his life to editing "proletarian" newspapers, beginning in May 1832 when he took charge of the *Écho de la fabrique* and continuing in the *Écho des travailleurs*, the *Tribune prolétaire*, the *Union des travailleurs*, and the *Tribune lyonnaise, Revue politique, sociale, industrielle, scientifique et littéraire des prolétaires*, whereas the master silk weavers' organization gave up its efforts after the passage of the repressive press law of September 1835. But neither man could be considered a "proletarian" in the sense in which the term was normally employed at the time.

Jacques Rancière has made historians acutely aware of the way in which the decision to become a writer set even intellectuals of unquestionable proletarian extraction apart from their fellows.[32] At his death in August 1832, Vidal, originally from the Cévennes, was memorialized primarily as a man of letters, "the Lyonnais Béranger."[33] At the start of his journalistic career, Marius Chastaing called himself a "proletarian" and urged workers to regard him and his associates as "brothers, who have taken up the pen only to defend a class, unfortunate but noble in its conduct and its virtue,"[34] but he was never a manual worker. Chastaing was a native of Lyon, born shortly before Napoleon's coup of 1799 and thus, in his own words, "republican by birth." The son of a *huissier* or court clerk, he had studied law, although he never obtained a diploma; he ran an "Office for Commerce and Legal Claims," offering to represent clients in all kinds of civil proceedings and to broker investments and real-estate purchases.[35] He had published his first republican political pamphlet against the Restoration in 1815 and had joined a conspiratorial group in that regime's early years, but had then become "a man of thought, and not of action."[36] In the 1820s, both he and Vidal wrote on behalf of such "progressive" causes as the system of *éducation mutuelle*, a form of popular education in which older students taught the younger ones.[37] In the 1830s, he belonged to a Masonic lodge whose other members were a cross-section of the city's petty bourgeoisie, with a sprinkling of independent artisans.[38] Chastaing was thus a bourgeois *déclassé* who voluntarily adopted the workers' cause. By 1848, when he ran unsuccessfully for the Constituent Assembly, he had come to see himself as "located at the border of these two classes, that is, fortunate enough not to belong exclusively to either of them."[39]

If the backers of the original *Écho de la fabrique* were thus something other than ordinary proletarians, and if the leading proponent of the workers' press was a sympathetic outsider, the very decision to create a newspaper represented a radical departure from traditional strategies for defending the group interests of workers and of small subordinate producers such as the master silk weavers. Publicity, particularly printed publicity, was no part of any of these traditions. Before 1789, Lyon silk weavers had tried to defend their interests within the corporatist structures of the *fabrique*, a term that referred to the collectivity of artisan weavers, master weavers, and merchants. When struggles within the community of the *fabrique* became too acute to be resolved, the weavers appealed directly to municipal and royal authority, via petitions, or compelled outside intervention by collective violence, as they had in the insurrections of 1744 and 1786. The Revolution of 1789 had dissolved the old corporate structure of the *fabrique*, and the Le Chapelier law of 1791 outlawed workers' organizations. In the Napoleonic period, the creation of a *conseil de prud'hommes* provided a new forum for the regulation of intraindustry disputes, but one whose proceedings were not publicized and one weighted toward the interests of the wealthy merchants.

Unable to organize publicly, workers in the Napoleonic and Restoration periods turned to the model of the *compagnonnages*, the secretive journeymen's brotherhoods whose origins went back to the seventeenth century.[40] The Lyon master silk weavers had no *compagnonnage* tradition, but their Society of Mutual Duty borrowed the *compagnonnages'* practices of secrecy, of limiting membership to participants in a single trade,[41] and of insistence on proof of members' moral character as well as their economic status. An editorialist for the *Écho des travailleurs* stated the issue clearly after that paper's break with the *mutuellistes*. "Our sympathy does not go to secret associations," he wrote. The Society of Mutual Duty actually enrolled only a small fraction of the master silk weavers, in part because of its rumored membership requirements. "If a system of exclusion exists in this organization, if the admission of *chefs d'atelier* depends on hidden conditions, this society is wrong."[42]

The establishment of the *Écho de la fabrique* in 1831 had thus constituted a major innovation in the strategies employed to defend the silk weavers' interests, and it appears to have been controversial at the outset. The only extensive account of the paper's founding should be approached cautiously, because it was published some years later by the dissidents who had quit that paper to launch the rival *Écho des travailleurs*, but it has a ring of authenticity. According to this account, which appeared in the *Tribune prolétaire* in 1835, the decision to found the paper in October 1831 was made independently of the Society of Mutual Duty, and, although a majority of the original stockholders in the paper (twenty-nine out of thirty-seven) were master silk weavers, only ten are known to have also been *mutuellistes* at the time.⁴³ The *Tribune prolétaire* also published the list of the paper's other stockholders, who included a *rentier*, a physician, a merchant, and several members of other trades.⁴⁴

Not only does it appear that the *Écho de la fabrique* was conceived independently of the official structure of the Society of Mutual Duty, but a latent conflict between the strategies of the two became manifest when the Society ousted the paper's editors in 1833. "Its editors thought it best to defend the interests of the working class solely by means of the press, without depending on anything other than their conscience and public opinion," the *Tribune prolétaire* recalled in 1835.⁴⁵ The leaders of the *mutuelliste* movement certainly learned to appreciate the value of the press, but they never accepted the claim that it could truly replace structured organization within the working-class community. When repressive legislation broke up their pyramid of lodges and their pattern of regular meetings after the insurrection of April 1834, they started a new paper to replace the *Écho de la fabrique*, but announced, "We say it clearly, we are continuing the work of *mutuellisme* by other means."⁴⁶ For them, collective organization remained the true method of representing workers' interests; the newspaper was a secondary instrument.

Defending the workers' cause in public and in print caused a considerable cultural shift, however. In Charles Tilly's terms, it meant that key elements of a time-tested "popular repertoire of contention" were abandoned and that new ways of promoting workers' interests were adopted.⁴⁷ Secrecy had been the traditional weapon of groups representing workers and artisans. It had enabled them to settle their internal disputes without the knowledge of their adversaries and to present a unified face to the outside world. The traditional weapon of last resort for the working classes—collective violence—derived much of its power from its unpredictability, and that unpredictability owed much to the fact that, by their very nature, these sudden outbreaks were not the outgrowth of previous public discussion. The adoption of a strategy of publicity through the press meant that workers' organizations would henceforth broadcast their intentions not only to their own supporters but to their adversaries as well. The *Écho de la fabrique* and its numerous progeny were intended for a working-class audience, but their authors were always conscious that once they published their papers, anyone could read them. The

Écho de la fabrique boasted that it was read in Lyon's Café Corti, "the temple of the *juste-milieu*," whose habitués "are happy to know what is going on in the enemy camp." But its editors understood that this involved risks: they imagined bourgeois readers poring over its columns and saying to themselves, "Perhaps one could find something subject to prosecution in it."[48] They carried on a regular series of hostile exchanges with the *Courrier de Lyon,* the city's Orleanist daily paper, and they periodically cited mentions of the *Écho,* favorable and unfavorable, in the Paris press. The *Écho*'s editors may not have known that one of Adolphe Thiers's secretaries had written to the prefect Gasparin in 1833 to say that "the minister wants to receive the newspapers that serve as guides for the workers. Aren't they the *Écho de la fabrique* and the *Précurseur? * Certainly the former," but they had to consider the possibility, and they surely knew that the prefect himself followed the paper closely.[49] They quickly learned that their paper's public existence was a two-edged sword that could sometimes be turned against the group they hoped to serve, as it was when some merchants announced that they would not buy silk from weavers who subscribed to the paper.[50]

The payoff for the creation of working-class newspapers was, of course, that the workers' cause gained a foothold in the wider arena of public debate. The *Courrier de Lyon* might refer to the *Écho de la fabrique* as "a poisoned tool in the service of the lowest passions, and the scourge of the silk industry," but even such intemperate denunciations amounted to a recognition that a press outside the bounds of bourgeois discourse had established itself and could not be ignored.[51] In his *Histoire des insurrections de Lyon,* which served as the definitive history of the city's upheavals until the mid-twentieth century, Jean-Baptiste Monfalcon devoted more space to the *Écho de la fabrique* than to any of the city's other newspapers, even though he claimed that the city's merchant class had exaggerated its importance.[52] Lyon's left-wing press treated the *Écho* as a significant ally. The *Précurseur* reprinted several articles from it, referred to it as "a publication consecrated uniquely to the defense of the working population's interests," and gave sympathetic coverage to its legal troubles.[53] From the outset, the *Écho*'s notoriety extended to Paris. When the paper first appeared, the Saint-Simonian *Globe* reprinted part of its prospectus, calling it "very commendable for the most part."[54] The liberal *Le Temps,* on the other hand, blamed the *Écho* for the April 1834 uprising and asked, "Will it still exercise its influence on the most ignorant and most impassioned workers of Europe in order to prepare them for a new insurrection?"[55] Indeed, thanks to the English Utilitarian John Bowring, the paper claimed that it "enjoyed the favor, which no other French newspaper has obtained, of being translated and reprinted in London."[56]

At the same time as the *Écho de la fabrique* gave Lyon's workers a symbolic voice heard even across the Channel, it also gave them a new kind of power within Lyon's own public sphere. In its pages, workers were no longer an anonymous crowd. Some of them—primarily literate *chefs d'atelier* but on at least one occasion even a self-proclaimed "female proletarian"[57]—could see their own words in print. The paper not only gave workers a chance to see their names published, but also proposed that they seize the power to name themselves in the most literal fashion. In 1832, it conducted an "open competition for the adoption of a general term to designate the class of silk workers in a simple, comprehensive, and euphonious manner." The aim was to replace the epithet *canut*, which the editorialist claimed had become "an insult, though I don't know why." The suggested alternatives—which included *textoricarien*, *tissutier*, and *bombitissorien*—indicated that the competitors, many of them *chefs d'atelier*, had some knowledge of Latin and a determination to give their class a label that had nothing to do with working-class argot.[58]

The existence of the *Écho de la fabrique* not only gave workers a chance to name themselves; it also allowed them to name their enemies. The *Écho de la fabrique* could, and did, single out by name specific *fabricants* whom it accused of abusing their workers. Solid members of Lyon's bourgeoisie found themselves exposed to a kind of public personal attack they had never previously experienced. When two of its targets brought a libel suit against the paper, its lawyer eloquently defended what its opponents saw as an outrageous violation of privacy. "When a citizen's actions have public effects, when his interactions with other citizens have social consequences, they no longer belong to his private life, they are part of his public life, which belongs to the domain of the press."[59] The existence of the *Écho* thus redefined the boundary between public and private, making the unequal relations between employers and workers a matter for publicity. The paper's existence also raised questions about the relationship between workers and the government. When the prefect, Gasparin, tried to induce its editors to publish an anonymous letter, they responded by insisting that he would have to sign it; in other words, they wanted him to recognize the *Écho* as a valid interlocutor, like the bourgeois papers.[60]

Just as it named specific individuals and labeled their conduct, the *Écho de la fabrique* debated other newspapers and thus demonstrated to its own readers that their spokesmen could hold their own in the public arena. The *Écho de la fabrique* warned its readers against the *Courrier de Lyon*, which "presented itself to the public as a journal of *fusion*, of harmony, of concord, of preservation" but in fact, having been "founded, supported and edited by bankers, merchants, manufacturers, property-owners," represented views that were hostile to workers' interests. This was clear, the *Écho* asserted, from the terms in which the rival paper referred to the working-class publication.[61] When the *Précurseur* was reorganized under a new name after the April 1834 uprising, the *Indicateur*, the continuation of the *Écho de la fabrique,* explained the change to its readers and assured them that the new paper, the *Censeur,* would be even more supportive of workers than its predecessor.[62] By employing the same tactics of labeling and critiquing other press organs used in the "bourgeois" press, the working-class papers demonstrated that they were indeed full participants in the process of shaping public opinion.

The Lyon working-class newspapers tried to demonstrate the advantages of publicity most directly by surrounding the silk industry's arbitration council—the Conseil de Prud'hommes—with publicity, treating it as a miniature parliamentary body. The *Écho de la fabrique* endorsed candidates for election to the workers' seats on the council and condemned the deputies chosen by the silk merchants: "Where could one have found more hostile names?"[63] The paper urged the worker deputies, a minority of the council, to use press publicity to force the merchant deputies to accept their positions, on pain of blocking the working of the institution.[64] In mid-1835, the *Indicateur* concluded that publicity had fundamentally altered the functioning of the institution. "The master silk weaver has learned . . . to know his rights, the merchant to be a little less sure of his prerogatives." Press coverage had made arbitration council decisions more consistent: "Thanks to the press and publicity, we have enjoyed our full rights."[65] What the workers' press presented as a positive development outraged the conservative

Courrier de Lyon, which called the *Écho de la fabrique*'s reporting on the council "continual libeling of the council members, whom the master silk weavers consider their natural enemies. The worker's fury, stirred up by these papers, doesn't stop at the door of the council, it pursues the merchant council member even beyond."[66] The *Courrier*'s complaint was eloquent testimony to the efficacy of the articles published by the working-class press. Press publicity did serve the workers' interests, but by making a newspaper the centerpiece of their strategy, the master artisans who backed the *Écho de la fabrique* found themselves compelled to adopt many of the forms of bourgeois society and its ways of doing things. Like other newspapers, the *Écho de la fabrique* was constituted as a joint-stock partnership, with an *acte de société* duly drawn up and deposited with a notary.[67] It is true that the *Écho*'s founders adapted this structure to their needs: the shares were priced at only 100 livres, and each twenty-member *loge* of the Society of Mutual Duty was required to pay for one share, meaning that each *mutuelliste* wound up contributing five francs to the paper's working capital.[68] Nevertheless, when the Society of Mutual Duty ousted Chastaing from the editorship and took over direct control of the paper in August 1833, it did so by having its members exercise their legal rights as shareholders.[69]

An even more striking example of how the publication of workers' newspapers drove their backers to follow the habits of the society around them was furnished by an incident in early January 1834. An exchange of words between the editors of the two *Échos* having gotten out of hand, two staffers from the *Écho des travailleurs* had decided that "a reparation has become necessary, because there had been a personal insult," and had challenged their opposite numbers to a duel. The four of them had appeared on the field of honor at the break of dawn, where a reconciliation was agreed to in accordance with the rules that the nineteenth-century French bourgeoisie had inherited from the aristocratic past. This was a most uncharacteristic method of settling disputes among workers, but their status as journalists apparently convinced these staffers of the working-class press that they had to emulate the conduct of other newspapermen.[70]

The decision of the Society of Mutual Duty to oust the *Écho de la fabrique*'s editor, Chastaing, was a reaction to another consequence of the new dependence of the workers' movement on the press; it had given control over the public image of the working classes to the newspaper's editor, shifting power away from genuine workers to an outsider with special skills derived from his bourgeois education. Chastaing's firing demonstrated that the organized master silk weavers could defeat one particular editor, but they still had to replace him with another; and even if the new editor was a *chef d'atelier* himself, his success would depend on his having special skills like those of Chastaing, not on the authenticity of his working-class background. In 1835, when Chastaing's opponents were publishing the *Indicateur*, they actually asked one of Chastaing's supporters for help in finding an editor for their paper, explaining that the absence of a real journalist was "the reason the paper isn't always good."[71] Furthermore, Chastaing's ability to launch a new paper after his ouster demonstrated that the *mutuellistes* could control the *Écho de la fabrique*, but they could no longer control the working-class press. Protected by the legal framework of liberal society, a journalist could go over the heads of the working-class community's internal hierarchy and appeal directly to the mass of the working-class population. The rival working-class papers had become commodities competing in a small-scale version of the general press market—the *Écho des travailleurs* even employed the eminently commercial tactic of offering a reduced price to its first 300 subscribers[72]—and it was by no means certain that readers would favor the title edited by authentic workers.

Devoted to the working class and determined to defend its interests, the early Lyon working-class papers nevertheless reflected an ambivalent attitude toward their presumed audience. None of the working-class papers made any effort to translate the actual culture or vocabulary of the working classes into print; the humorous parodies of popular speech found in the republican *Glaneuse* had no equivalents in these publications. The articles in the working-class papers reflected a classic dual attitude toward their presumed working-class audience. On the one hand, they denounced the misery and oppression of the working class, but on the other hand they insisted on its need for education and moral improvement, and on their own right to guide its members. In answer to workers who claimed that they needed their children's income too much to let them attend school, for example, one editorialist, who claimed to have been "born proletarian," insisted on his duty to "speak firmly to the class that has entrusted us with its defense" and to counter its prejudices.⁷³ The second prospectus of the *Écho de la fabrique* included an appeal for contributions from "all men of letters, lawyers, doctors, artists. . . . The people is thirsty for knowledge . . . "⁷⁴ and the paper regularly printed excerpts from literary classics with the avowed aim of upgrading the intellects of its readers. "Our goal was to awaken the imagination, to exercise the judgment, and to give workers the taste for and the desire for reading," Marius Chastaing explained in an early number of the *Écho des travailleurs*, which continued the same policy.⁷⁵ Another editorialist urged imitation of the English movement to create lending libraries for workers. "Shouldn't we hope that the Lyonnais worker will finally come to understand that instruction overcomes common prejudices and the defects of a bad education?" he asked, while recommending "moral and instructive books" that would provide practical education.⁷⁶ The workers' papers called for better housing and more sanitary living conditions for workers, but simultaneously they urged workers to abandon much of their own customary behavior. "We think it is necessary that, in language, in instruction, in manners, in habits, in short, in every aspect of life, there should be nothing other than *hommes comme il faut*," one journalist wrote in 1835.⁷⁷ While representing the workers, the working-class press would also give them the cultivation necessary to participate in the broader culture from which they were excluded.

However problematic its relationship to the majority of the actual silk workers of Lyon, there is no doubt that Lyon's working-class press had the effect of reshaping the boundaries of public printed discourse in the city, and indeed in western Europe as a whole. It is not surprising that members of the Lyon bourgeoisie reacted angrily to the very existence of these enterprises. When the *Écho de la fabrique* first appeared, several silk merchants tore up copies of its prospectus in front of one of the public places where they normally gathered to read and discuss the press, the Café du Commerce.⁷⁸ Their gesture was a symbolic and unsuccessful effort to deny the working class the access to the public space of debate that the paper was claiming for it. Regardless of its editors' real social standing, the *Écho de la fabrique* established a precedent for the existence of a press claiming to speak for and to an audience defined as distinct from the rest of the population because of its working-class status. The mere existence of such a press was an assertion of the possibility of a proletarian form of publicity, which in turn suggested the impossibility of a unified public opinion, and therefore raised fundamental questions about the entire notion of print journalism as a means of recreating the classical civic assembly.

The Doctrines of the Workers' Press

The importance of the working-class press was not merely that it asserted workers' rights to employ printed publicity in their own interests, but that it articulated new ideas about the rights of workers and new strategies for defending them. Open to the doctrines of such theorists as Saint-Simon and Fourier, the Lyon papers nevertheless maintained a critical distance from them. Rather than becoming organs of any clearly defined system, the papers constituted a forum for discussion of many ideas put forward as possible remedies to the plight of the lower classes in early industrial society. And, instead of conditioning their audience to follow any of the period's multiplicity of self-appointed prophets, the working-class press trained Lyon's literate workers to recognize the existence of conflicting ideas about their situation and, ultimately, to realize that they were obliged to form their own ideas about how to better their lot.

Ironically, it was Marius Chastaing, the outsider, who most energetically promoted a class-conscious workers' journalism. The prospectus to the *Écho des travailleurs* denounced the older *Écho de la fabrique* for "returning to the special interest that we had thought ourselves obliged to set aside," by emphasizing the specific problems of the silk trade. "We wanted to make it a tribune open to all proletarians, and not just to this or that group, no matter how numerous it is, because in our eyes all proletarians share the same interests." The new paper would advocate "the physical and moral emancipation of the proletarians," a formula with Saint-Simonian overtones, but it would insist that this emancipation could be carried out only by the proletariat itself. "It is folly or treason to suggest that any kind of aristocracy will ever consent to destroy itself."[79]

The fundamental idea structuring the ideas expressed in all the workers' papers was the demand for equality. According to one of Chastaing's early articles, Jesus had brought the message of equality into the world, proclaiming the *"égalité morale des hommes."* The French Revolution of 1789 had marked the conquest of political equality. The mission of the 1830s was to extend this to "social equality," to ensure that "the poor man is the equal of the rich."[80] This demand for equality was rooted in a conception of natural rights. "Every man has the natural right to develop himself physically, morally, intellectually," the *Écho de la fabrique* announced. In practical terms, this meant the right to decent housing, a healthy diet, universal elementary education, free higher education for those who qualified for it, and old-age pensions—a set of social rights that French workers would not fully obtain until well into the twentieth century.[81] Almost a decade before Louis Blanc introduced the phrase into public debate, the Lyon workers' papers explicitly included the *droit au travail*, the "right to work," among their demands. "Man has an obligation to work. Society has an obligation to furnish man with work. Work should provide man with his subsistence," declared the *Écho des travailleurs*.[82] The capacity to work constituted the worker's "property" and deserved equal treatment with other forms of property: "*Work*, which is the proletarian's capital, equals *money*, which is the property-owner's capital, [and] merchandise, the merchant's capital. All these properties should be governed by the same law; otherwise, there is no equality."[83]

The existing social order, as depicted in the workers' papers, was obviously in contradiction to this ideal. Chastaing produced a long series of graphic articles, most of them based on incidents reported in the *Gazette des tribunaux*, illustrating "the miseries of the proletariat." Ill housed, ill fed, unable to support themselves on the pittances they received from the silk merchants, the weavers were also subjected to moral indignities. The papers repeatedly protested against the way the merchants confined the silk weavers delivering their cloth to a small space in their warehouses, separated by metal bars from the rest of the building. "If I were a silk merchant, I wouldn't have a *cage* where the humiliated worker is locked up like an orangutan," one editorialist wrote. Even master weavers delivering their work "often had to suffer humiliations at the hands of the merchants and the beardless youths who serve as their clerks."[84]

To give workers a chance to achieve equality with the possessing classes in living conditions and dignity, the workers' papers advocated a variety of strategies derived from the basic principle of collective effort or association. The selfish individualism of liberal society was not a reflection of basic human nature, the workers' papers argued. The *Écho des travailleurs* contended, "It is no more difficult to get men to join together in their common interest, than to persuade them to work against each other. The spirit of association is natural." Its editorialist noted that some form of cooperation was integral to all the reformist movements of the day—the socialisms of Saint-Simon and Fourier, the Catholic democracy of Lamennais, the republican movement, and even the free-trade doctrines of the English Utilitarians, who advocated "a holy alliance of all producers."[85]

As this article suggested, the papers interpreted the notion of association broadly. For some editorialists, it was compatible with an essentially individualistic social order. An article in the *Écho des travailleurs* in early 1834, on the eve of the silk weavers' general strike, firmly asserted that freedom of competition was a "conquest that ensures and guarantees the individual development, the personal freedom, . . . that form one of the principal characteristics of modern peoples." Nevertheless, association formed an indispensable "counterweight" to this competitive spirit. By "association," however, this author meant a general consensus that individuals had to be given equal chances to succeed in the race for social rewards. Specifically, he called for universal education and equal access to capital, so that individual success could clearly be seen to result from personal qualities and initiative.[86]

For the most part, however, association meant some form of cooperative action, usually on the part of workers alone. Not surprisingly, the journalists applauded worker-initiated mutual-aid societies. When the printers of Lyon set up a fund to help unemployed members, the *Écho de la fabrique* told them, "You have set an example for all proletarians."[87] By introducing a law in February 1834 to ban both working-class organizations and political groupings, such as the Society of the Rights of Man, the July Monarchy focused the press's attention on this form of association and provoked the two *Échos* into numerous dissertations on its importance. "By what right do you tell us . . . 'You don't have the right to come together, to associate yourselves, to discuss, in common and in public, your rights and your interests'?" the *Écho des travailleurs* demanded. Such groupings "make the proletarian class more moral and enlightened, they teach it the meaning of liberty."[88] The papers also argued for the right of workers to form unions and call strikes—the "right of coalition."⁸⁹ "We assert that the question of coalitions is nothing else than the question of *physical and moral emancipation of the proletarian class*," the *Écho des travailleurs* wrote, maintaining that this was the only way for workers to obtain their just demands: higher wages, shorter working hours, and the right to form producers' cooperatives to compete with the merchants.⁹⁰ The *Écho de la fabrique* publicized the numerous examples of labor militancy in the fall of 1833, France's first nationwide strike wave: "The whole working class rises up and marches to the conquest of a new world!"⁹¹

The workers' papers publicized and supported a number of other forms of associationist action, such as cooperatives. In late 1833, the *Écho des travailleurs* published the prospectus of a "central commercial firm of master silk weavers and workers for the fabrication and sale of silk cloth," an unsuccessful attempt to break the merchants' monopoly on silk sales.⁹² A more controversial initiative was the launching of France's first consumer cooperative, the *Commerce véridique*,⁹³ which was publicized in a series of lengthy articles in the *Indicateur* in late 1834. Michel Derrion, a silk merchant who had thrown himself into the movement to improve working-class conditions, argued that consumers, using their united buying power, had a better opportunity to change the social system than producers.⁹⁴ The rival workers' paper, the *Tribune prolétaire*, ridiculed Derrion for reducing the principal of association to something as trivial as a grocery store,⁹⁵ but the *Commerce véridique*, which operated for three years (1835–38) was the most tangible associationist project successfully promoted through the press during this period and laid the basis for France's entire consumer-cooperative movement.

The early 1830s was the period when socialist doctrines were beginning to circulate in France. Like the bourgeois press, the Lyon workers' papers took note of the doctrines spread by the followers of Saint-Simon and Fourier but never really identified themselves with either movement. The *Écho de la fabrique*, the only workers' paper publishing at the time, paid relatively little attention to the Saint-Simonian "missionaries," who attracted large crowds to their public lectures in the spring of 1832. In the fall, it printed a list of thirty-four Saint-Simonian titles available for purchase by mail order.⁹⁶ But that movement's elitist tendencies and its divorce from republican politics probably alienated the worker-journalists. The *Écho de la fabrique* reported that the toasts drunk at a Saint-Simonian banquet in December 1832 had included one "To the rich!" meant to emphasize that the movement "wants the happiness of all, without exclusion."[97] Such ecumenicism went counter to the insistence of the workers' papers that the existing social order was unjust. Marius Chastaing's determination to add the formula "Everything by the people" to the Saint-Simonian slogan, "Everything for the people," implied a certain criticism of the movement.[98] The Saint-Simonians' official publication, the *Globe,* had, for its part, immediately sensed the difference between its views and the class-based outlook represented by the *Écho.* Although it praised the Lyon paper's 1831 prospectus, the *Globe* remarked, "This piece does not reflect . . . a lively sympathy for the merchants; perhaps it is not without some injustice toward them," although it also asked, "Are the merchants always fair to the workers?"[99]

Fourier's doctrines attracted more comment in the two *Échos.* Announcing the creation of Fourier's journal, the *Phalanstère,* in 1832, Chastaing wrote that "even if this planned society is unworkable, it will be glorious to have tried it," and a self-proclaimed disciple of the master was given space to outline the virtues of Fourier's projected communes in a long series of articles in the *Écho de la fabrique.*[100] At the end of 1833, however, the *Écho des travailleurs* opened its columns to a local Fourierist 'revisionist,' Jules Dubroca, who suggested modifications in the master's ideas to make their implementation more practical.[101] In general, both papers argued that workers should borrow from the socialist doctrinaires without becoming slavish followers. "They should teach us . . . that the changes that we are called upon to carry out in the future are much broader and more radical than what had been imagined," a contributor to the *Écho des travailleurs* concluded, remarking that socialist arguments demonstrated the insufficiency of the paper's own earlier call for all citizens to be allowed to compete equally for society's rewards.[102] But the journalists had persistent doubts that either school's schemes could really be implemented. "All these systems have a basic defect, insufficient knowledge of human nature," the *Tribune prolétaire* wrote in early 1835.[103] The workers' press thus served to acquaint its audience with the ideas of the early socialists, but, by mixing critical comments with the expositions it provided, the press implicitly urged readers not to embrace any ideologue as a prophet but to subject all new ideas to a rational examination. The success of Fourier and Cabet among French workers in the years after 1835 suggests that this cautious rationalism may have suited the literate and fairly well educated journalists better than it did many of their readers.

Not all the press's discussions of burning social and economic issues reflected direct Saint-Simonian or Fourierist influence. The workers' papers were also open to discussions of the ideas of bourgeois economists and to debates growing directly out of workers' experiences. A number of articles addressed the impact of mechanization on the working class. In an early number of the *Écho de la fabrique*, Henri-Joseph Bouvery, a prominent *mutuelliste*, wrote a widely remarked critique of machinery. "How will [its advocates] support an immense population, all of whose means of existence depend on work, when they are thrown out of the workshops, which will no longer be employing anything except machinery?" he asked. Instead of leading a crusade on this issue, however, the paper followed up by printing a lengthy defense of the benefits of technological progress authored by Anselme Petetin, the editor of the republican *Précurseur*.[104] This initiated a series of exchanges between the two authors in which Petetin was finally allowed to have the last word. A later article in the *Tribune prolétaire* claimed that, as consumers, workers gained more from the lowering of prices and the expansion of markets brought about by machinery than they lost. "There is no case where a machine has been introduced in any industry without increasing consumption, and consequently without giving employment to more labor than those that it has replaced," the paper claimed, citing the examples of the printing press and the Jacquard loom.[105] At the end of 1834, the *Tribune prolétaire* and the *Indicateur* squared off over the relative merits of the economists J. B. Say and Léonard Simonde de Sismondi, with the former coming out for Say's advocacy of freedom and progress, while the latter sympathized with Sismondi's call for paternalist controls on employers.[106] On balance, Lyon's working-class press— itself a product of modern technology—urged workers to accept the inevitability and desirability of industrial progress.

Technically speaking, the Lyon working-class papers were not entitled to discuss explicitly political issues. By renouncing any aspiration to report on political debates, discuss government policies, or comment on legislative proposals, they avoided having to post caution money. In fact, there was little secret about the papers' democratic and republican sympathies. The *Écho de la fabrique* argued for universal suffrage, claiming that only then would the "fundamental principle of popular sovereignty" be realized. "The immediate result will be relief for the working classes, equal protection of all interests, recognition of all rights, in a word, the abolition of all privileges."[107] At the time when the *Précurseur* was publicly announcing its conversion to republicanism, the *Écho de la fabrique* forswore any intention of quarreling with its colleague: "We march under the same banner, its principles are ours, why should we be divided?"[108] The paper was equally sympathetic to the city's other republican journal, the *Glaneuse*, which it recommended to working-class readers, and to republican groups such as the Société des Amis du peuple.[109] The "educational" snippets Marius Chastaing selected for his readers often included texts with a clear political coloration, such as the excerpts from Robespierre published in the issue of 17 February 1833.

Although the workers' papers clearly sympathized with the republican movement, they also suggested that workers could advance their interests through various direct means. Although the violent insurrection of November 1831 had shown that workers could be a powerful force, the press insistently recommended peaceful forms of collective action. "The progressive emancipation of the working classes, as we understand it, without upheaval, without violence, without shocks, cannot be the product of a mass collective upsurge, but a series of changes, a chain of linked occurrences," editorialized the *Écho de la fabrique*.[110] But when *fabricants* threatened to reduce piece rates in the early months of 1834, the paper defended the justice of the strike the *mutuellistes* organized in terms that made it clear that it condoned what it saw as defensive violence. Accused of endorsing illegal actions, the paper responded: "What is legal about our situation, when, deprived of work, our most sacred right, and suffering daily under the law of necessity, we have to accept, for our pains, whatever salary it pleases our lords and masters to give us?" It went on to ridicule the preparations the authorities had made to contain potential disorder and to warn that "the workers now know how to make any injustice on the part of the masters impossible, and to inflict . . . a harsh lesson on anyone who still deserves it."[111] Only in the wake of the unsuccessful uprising of April 1834 did the workers' papers abandon this militant tone and insist more emphatically on the virtues of nonviolent action. Commemorating the third anniversary of the November 1831 insurrection, the *Indicateur* promised that the proletarian would no longer "demand his rights, with arms in hand . . . [that] from now on, it is with wisdom, in signaling and eliminating abuses, . . . that we will arrive at a more prosperous future."[112] The protestations of working-class journalists on this point were undoubtedly sincere; none of them had ever been shown to have participated personally in either of the Lyon uprisings.[113] But nothing they could do or say could eliminate suspicions about their real commitment to peaceful reform.

The most comprehensive statement of the ideas that the workers' press meant to promote was the series of toasts drunk at the public banquet celebrating the first anniversary of the *Écho de la fabrique* in October 1832. The toasts were carefully written out in advance and approved by the paper's board of directors, and subsequently published in a special issue of the paper. Falconnet, the paper's founder, led off by saluting *"l'industrie,"* using the term in such a way that it could be translated either as "industry" or as "industriousness." In either case, "the reign of idleness is about to pass away, that of *l'industrie* will replace it." As a result, productive workers were the class of the future: "Don't be ashamed, my colleagues, of being nothing but workers. On the contrary, we should be proud of it." Workers deserved equality because all men were by nature equal: "From now on, the difference between them will be that of virtue, of talent; let us not suffer any others," said another *chef d'atelier*. But attaining this result required what a third speaker called for: "Emancipation of the industrial classes!" Berger, the *gérant*, or director, of the *Écho de la fabrique*, reminded the audience that, because of existing social conditions, the workers' condition could be described as "work, misery, children without assistance, the poorhouse for the old, bodies not given proper burial."[114]

To improve conditions, workers needed to take action. Berger referred to the famous slogan of *Les révolutions de Paris* of 1789, telling the audience, "One walks badly and slowly as long as one walks on one's knees. Let us rise up! Face [the possessing classes] without fear, but without threats." Education was one path to emancipation, according to a toast proposed by a representative of the bourgeois republican paper, the *Précurseur*, but the worker banqueters also called for more explicitly political reforms. "What is the law? Is it not the expression of the general will?" one speaker asked. "There cannot be any true law worthy of that name, except that which we proletarians have helped to make. Let's not be afraid to say it: the workers need representation to keep their rights from being sacrificed." To attain this goal required "the union of workers," asserted another *chef d'atelier* associated with the paper: "Union makes us strong!" Eager to stay within the bounds of the law, the banqueters called for "moderation" and praised "concord," but they also found ways to remind one another that "it was the people who built the barricades in July [1830]. . . . The people achieved grandeur for three days. That's been forgotten!" Short of achieving their ultimate goals, the *Écho de la fabrique* banqueters called for a number of more modest but specific reforms, many of which were regularly promoted in the paper itself. These included "freedom of defense in the *conseil de prud'hommes*," a catchphrase that meant the right to be represented by legal counsel or by another worker in council hearings, "the destruction of abuses in the silk industry," and the formation of a broad-based mutual-aid society to provide sick pay and pensions for weavers.[115]

In the toasts at their 1832 banquet, and in the columns of their papers, the proponents of this first avatar of working-class journalism thus laid out a program with extensive social and political implications. They advocated genuine social equality and an extensive set of specific reforms to give workers the same chances in life as their betters. To attain this, they called for a democratic political system in which the laws would truly reflect the general will. As interim measures, they urged workers to demand practical reforms in working conditions and in the administration of labor law, and to use such strategies as union organization to defend their interests. Many details of this social-democratic program remained vague, including the extent of state involvement, but the general outline of a comprehensive strategy for inserting the working class into the larger stream of public life was clear. Alongside this practical program, the papers acquainted their audience with the more radical ideas of the Saint-Simonians and the Fourierists, but the working-class press itself refused to endorse any genuinely socialist program. The most articulate thinker among the Lyon working-class journalists, Marius Chastaing, remained confident that individualistic natural-rights doctrine could be extended to protect workers' property in their labor power.

The working-class press communicated its message as much by its tone as by its content. On the whole, the papers strove for a sober and serious mien, comparable to that of the bourgeois press. Their articles were written in standard, educated French, and, in their educational articles, the papers urged their readers to acquaint themselves with the classics of French high literature from the Renaissance onward. Heated polemics normally appeared only in signed letters to the editor, and the workers' papers made no effort to imitate the irony and the humor of the *Glaneuse*, or to integrate fictional techniques into their columns as that paper did. When the *Indicateur*, in an uncharacteristic outburst, wrote that "it would be hard to find a group more perverse, more selfish, more wasteful and more atheistic than our current wholesale silk merchants," the rival workers' paper, the *Tribune prolétaire*, responded: "There are enough prejudices against the popular press. Let's keep a clear line between the real defenders of the popular class and those who use only insults and violence for arguments. . . . Isn't the cause [of the people] grand enough, beautiful enough, just enough, to inspire those who defend it to something other than sordid insults . . . ?"[116] The goal of the working-class papers was to convey a convincing picture of the social reality of life in Lyon, with an emphasis on the sharp gap between the lives of the rich and the poor, and to underline the obvious moral message that emerged from that reality; emotionally charged denunciations undermined their fundamental message.

Emotion broke through most powerfully in the working-class papers when they commemorated the silk weavers killed in the November 1831 insurrection. The anniversary of the event became a journalistic ritual in which the moral pathos of the workers' cause could be given full expression, in language that strove for the sublime. For the occasion, the *Écho de la fabrique* and the *Écho des travailleurs* made a striking alteration to their usual layout: they printed heavy black borders around the memorials printed on their front pages, the only time either paper deviated from its normal format to make a special point (Fig. 6). In 1834, the *Tribune prolétaire* printed a blank column in commemoration, with a note at the bottom suggesting that words were insufficient to convey the emotions it wished to express: "Our silence will be understood. But on this sad anniversary, we ask that the victims who perished be remembered."[117]

In earlier years, the papers were more explicit. In 1832 and 1833, Marius Chastaing apostrophized the famous black flag carried by the silk

Fig. 6 L'Écho des travàilleurs. The workers' papers in Lyon, the *Écho de la fabrique* and the *Écho des travailleurs,* commemorated the anniversary of the November 1831 uprising every year by putting a funeral border around the text on their front page and using flowery language to call on readers to remember those killed in the fighting. In the absence of any physical reminder of the event, these special issues of the papers were the most important institutionalized memorial to the uprising, and a regularly recurring reminder of its significance. For a translation of this text, see Appendix 2.

weavers who had descended from the Croix-Rousse on 21 November 1831: "Solemn and sacred emblem, you were their only banner! A short inscription was your device: *Live working or die fighting!* ... Your blood has watered the soil where the tree of proletarian emancipation will grow." In a separate article in 1833, he underlined the moral lesson of the workers' short-lived triumph: "Misery stood side by side with wealth without saying 'Division of property,' and hatred was forgiven without any dream of revenge. Sublime example of morality in the popular Hercules!" The working-class press sought to make the November uprising the founding myth for a new era. "Proletarians of all classes, you were united, during those three days, by a fraternity of arms, by a community of dangers, don't ever forget it."[118] For this solemn occasion, the papers appropriated an exalted style, full of religious overtones, meant to appeal to the heart more than to the mind.

The relative linguistic restraint of the working-class papers was a necessary part of their broader strategy to prove that workers' interests could be defended without resort to violence. The collective violence of the November 1831 insurrection had forced the issue of proletarian misery into public consciousness, but Marius Chastaing and his colleagues were determined to show that it had been an aberration. The contrast between their relatively dispassionate language and the violent attacks directed at them in the conservative papers allowed the *Écho de la fabrique* and its successors to claim the moral high ground in a culture that claimed to value reasoned argument and dignified language over violence. The claim of the *Courrier de Lyon* that "only those who read these sheets know how many turpitudes and shameful excesses this ignoble literature, which we can stigmatize with the name of gutter press, permits itself"[119] hardly squared with reality, and allowed the editors of the working-class papers to assert that it was the defenders of the established order, rather than its opponents, who exhibited the most violent tendencies.

Their ability to provoke the conservative press to such paroxysms of verbal violence was one indication of the genuine impact of the working-class papers. If anything, conservative critics exaggerated their influence, repeating the maneuver by which Charles X's ministers had made the liberal press the scapegoat for their inability to silence criticism of that regime. Jean-Baptiste Monfalcon, editor of the *Courrier de Lyon* in the years between the two insurrections of the early 1830s, and the most influential nineteenth-century historian of those events, accorded the *Écho de la fabrique* a major role in them: "Its editors took it as their constant mission to incite the workers to hate the businessmen, to distort the *chefs d'atelier*'s already misconceived notions of their real interests, and to divide the silk industry into two hostile and irreconcilable camps. One must grant this paper a large part in our regrettable disorders. It addressed its slanders and its anarchic doctrines to excitable workers who read nothing else."[120]

Both the prefect, Gasparin, and his rival Prunelle, the mayor, the two officials principally responsible for maintaining the government's authority in the city, paid close attention to the papers. In November 1832, Gasparin told the interior minister that the *Écho de la fabrique*'s office was the center of all oppositional activities in Lyon, and in 1833 he kept his Paris superiors informed as he secretly tried to influence the choice of a new editor for the paper. "If the new man chosen isn't loyal to you, he is at least wholeheartedly devoted to moderation," the Comte d'Argout remarked in an overly optimistic letter on this occasion.[121] The prefect and the mayor clashed over the best tactics to follow against the papers, however, with Prunelle protesting Gasparin's weakness when the prefect refused to confiscate the prospectus of the *Écho des travailleurs*. The mayor did successfully prevent Chastaing's paper from putting up posters advertising itself for the first several months of its existence.[122] Because they did not print explicitly political articles, the two *Échos* were not routinely hauled into court before April 1834, as the republican papers were, but when they both reappeared under new titles after the insurrection, the prosecutors made them prime targets. Jules Favre, the Lyon republicans' workhorse lawyer, put up a losing struggle on behalf of both the papers, eloquently denouncing what he called "a systematic war declared on publications that carry instruction and the hope of a better future into the popular ranks," but both papers were hit with heavy fines for covering political issues. Both announced that they would try to raise enough money to post the bond required for political papers, but neither was able to do so, and both ceased publication even before the September laws of 1835 threw additional obstacles in the way of the dissident press.[123]

The attention the authorities and the conservative press paid to the working-class press was above all a reaction to these papers' symbolic invasion of the public arena. Their real effect on the attitudes of Lyon's working-class population may not have been as great as their enemies claimed or feared. Their press runs were certainly limited: the *Écho des travailleurs* claimed to have reached the level of 300 paid subscribers by February 1834, but this was well below even the modest figures reached by the "bourgeois" press. More significant is that the *Écho de la fabrique*'s much-advertised first anniversary banquet in 1832 drew only 240 to 300 participants, even though tickets were priced at just three francs, little more than a day's wage for a silk weaver.[124] In contrast, a banquet held around the same time on behalf of the republican *Glaneuse* drew more than 1,700 people.[125] Support for the *Écho*'s second anniversary celebration was so weak that the banquet was called off.[126]

Although workers would undoubtedly have sympathized with many of the views expressed in the working-class papers, their attitudes toward the world were shaped more by their immediate life experiences and by traditional forms of popular sociability than by the well-reasoned articles of Marius Chastaing and his consorts. German scholar Werner Giesselmann concludes, on the basis of his massive statistical study of popular protest during the July Monarchy, that most incidents of resistance were inspired only indirectly by the press, and that even the role of such organizations as mutual-aid societies and the secret republican groups that sought to recruit workers was limited. Faced with poverty and humiliation, French workers did not need journalists or organizers to spur them into action when circumstances—a cut in wages, for instance—provoked them.[127]

That the working-class press may have had more impact as a symbol in the minds of middle-class readers than as a guide to action in the minds of workers does not negate its historical importance, however. The *Écho de la fabrique* and its journalistic offspring may indeed have planted their banner in the proletarian camp, as they boasted, but their real impact came from their visible presence in the arena of public debate. The silk merchants could confine weavers to the "cage" in their warehouses, but, so long as they accepted press freedom as part of the legal structure of their society, they could not wall themselves off from the working-class papers. The newspapers could do peacefully and on a regular basis what the workers did physically when they rose in insurrection: invade the center of society, and use its own forms of discourse to challenge it. Worker insurrections and mass demonstrations were rare events, even in tumultuous Lyon, but the *Écho de la fabrique* appeared every week, giving the working class for which it substituted a permanent presence in the public sphere. The violence of the responses it inspired is a measure of the degree of disturbance this presence caused. The self-proclaimed workers' press inspired this response because it modeled a new form of social identity for the laboring classes. Symbolically, the *Écho* represented what the French bourgeoisie feared most: a self-conscious working class that was capable of articulating a coherent set of demands whose satisfaction would require a reshaping of society. Whether or not such newspapers had an immediate impact among the actual workers of the early 1830s, the possibility they represented became a major factor in the way all members of French society imagined their world.

5

Creating Events

Press Banquets
and Press Trials
in the
July Monarchy

Claiming Attention

Defining and exemplifying a repertoire of new social identities was one of the most important ways in which newspapers made an impact in the aftermath of the Revolution of 1830, but it was not the only way in which they exercised their influence. Another measure of their importance was their ability to make themselves centers of public attention and to mobilize both supporters and opponents into patterns of action that changed the balance of power within the public sphere. Public political banquets on behalf of newspapers, and public trials involving journalists, were two especially important occasions for such mobilizations, and in Lyon they were the most important forms of nonviolent political action in the early years of the July Monarchy. Both involved floods of public oratory that could be transmuted into newspaper texts, and opportunities for self-referential commentary that allowed newspapers the chance to assert their own importance. Opposition papers gave extensive coverage to the political banquets they organized and the public trials in which they were involved because these were two of the rare opportunities these publications had to tell stories about events they could control and present to their own advantage. Papers could anticipate these events well in advance and plan to exploit them as thoroughly as possible; banquets and trials were opportunities to show that the newspaper could do more than simply broadcast words; and both offered a chance to affect reality by rallying the faithful to defy the pressures of the government. These public events were also opportunities for the groups and movements for which the newspapers claimed to speak to demonstrate their existence and their ability to act effectively. On these symbolic occasions, opposition groups, gathered in the name of their newspaper, could temporarily make their principles appear to prevail over those of the regime in power. To exploit these occasions effectively, newspapers promoted the formation of support groups, which became another way for the press to shape public affairs.

On 31 August 1832, Lyon's leading opposition newspaper, the *Précurseur,* devoted most of its front page to a long account of a banquet held in honor of the prominent deputy H. C. Odilon-Barrot. The occasion for Odilon-Barrot's visit to Lyon was, not at all coincidentally, the trial of the *Précurseur*'s own editor, Anselme Petetin, scheduled for the following day. The turnout to honor this well-known critic of the government had "shown the force of public opinion in Lyon, where the *juste-milieu* [the Orleanist government] boasts of its undivided support." The 500 participants had included "representatives of every branch of industry, and one noted the presence of distinguished persons regarded as strong supporters of order and tranquillity," the paper claimed, stressing the respectability of its supporters and the fact that they were drawn from the groups that produced the city's wealth. By transcribing at length the toasts offered at the banquet, the *Précurseur* was able to offer itself the luxury of publishing extravagant encomiums to the opposition press. "Each blow struck against it strikes the representative system [of government], which cannot exist without the support of the press. To defend the press is to defend the constitution," one speaker thundered, before proposing "a double toast: to the freedom of the press and to the press as it exists today, to the writers . . . who have taken on a mission for the good of the nation, and are fulfilling it with conviction and devotion."

The Odilon-Barrot banquet was a classic example of what American historian and media critic Daniel Boorstin has defined as a "pseudo-event." According to Boorstin, a pseudo-event is a deliberately arranged happening planned "primarily (not always exclusively) for the immediate purpose of being reported or reproduced" and "intended to be a self-fulfilling prophecy."[1] Writing in the early 1960s, Boorstin intended his notion of the pseudo-event as a critique of American media practices that, in his view, had led to the substitution of manipulated occurrences for the straightforward reporting of reality. His acute analysis of the constructed character of most modern media content was coupled with what now seems to have been a naive faith in the existence of a realm of authentic events, the transcribing of which he saw as the proper mission of the press. In fact, the media are always deeply implicated in the identification of events *as* events. Demonstrating that the Odilon-Barrot banquet of 1832 fits Boorstin's definition of a pseudo-event is simple: it had certainly been arranged well in advance, and its main purpose was clearly to give the impression that there was broad support for the cause of Petetin and the *Précurseur*. The more interesting question is why this particular category of pseudo-event came to play a central role in the French press during the first half of the nineteenth century.

To appreciate the significance of the obviously contrived reports on the banquet, and on the press trial that provided the pretext for it, it is important to recognize how little control opposition journalists of the early 1830s had over the news events they covered. This was not because the journalistic canons of the period limited them to reporting the random flow of "authentic" events, but because the government they were contesting dominated most of the mechanisms for generating occurrences that fit the period's definition of news. In an era in which newspapers gave legislative debates more space than anything else, the king's ministers determined what issues would be discussed, and the regime's election laws barred declared republicans and members of the working classes from participation in those debates. National election campaigns involved only a tiny fraction of the population and offered only limited opportunities for media intervention. Politics in this "age of notables" was a matter for small coteries; parties in the modern sense, capable of generating news by holding congresses and press conferences, did not yet exist. The stringent legal restrictions on the formation of organized groups meant that the opposition movements had to operate semiclandestinely and avoid press coverage of their activities.

There were, of course, forms of news that escaped from these various control mechanisms to some degree. Louis-Philippe's government could not control events beyond its own borders, and part of the intense interest the left-wing press had in occurrences such as the Belgian and Polish uprisings of the early 1830s doubtless stemmed from the opportunities these disturbances provided for commentary on issues that the French authorities would have preferred to ignore. In the domestic sphere, outbreaks of collective violence, such as the Lyon insurrection of November 1831, gave the press the opportunity to demonstrate the existence of mass opposition to the regime. Although this demonstration was essential to the various radical critiques of the July Monarchy system, press exploitation of these occurrences was difficult. Not only was open identification with such movements illegal, exposing papers that encouraged them to instant retaliation, but popular uprisings, by their very nature, occurred at unpredictable moments, frequently catching journalists and newspapers off guard. They were part of a "repertoire of collective action," in Charles Tilly's terms, which was sometimes politically effective, but was not easily exploitable in journalistic terms.[2] Furthermore, violent insurrection was not the alternative to the existing order with which the opposition movements of the early 1830s wanted to identify themselves. The countersociety they wanted to portray for their readers was one in which genuine order and justice would prevail; in their scheme of things, collective violence was a consequence of the defects in the existing order, not a moment of genuine liberation to be counterposed to the structures of oppression.

The public banquet in honor of a newspaper or an editor was the purest example of the creation of an occasion for ritual celebration of the values the opposition press represented and a manifestation of the social reality for which the papers claimed to speak. Whereas the government forced public trials on newspapers, banquets resulted from the initiatives of opposition groups themselves. They could decide the timing of these rituals and define their content. One way the opposition could put the government on the defensive was to announce a banquet. As a result, the banquet became one of the political opposition's most effective forms of public action, and in 1848 a banquet campaign would become the occasion for the overthrow of the Orleanist regime.

The nineteenth-century French political banquet imitated English precedents dating from the eighteenth century. The adoption of the word "toast" to refer to the short speech and accompanying shared drink that formed an essential part of banquet ritual was clear evidence that the French banqueters were aware of the ceremony's origins. In nineteenth-century France, however, banqueting acquired a provocative symbolic significance that it lacked in the English context, where banquet rituals were common to supporters and opponents of the governments of the day and where nonpolitical banquets were also common.[3] In July Monarchy France, the public banquet came to be associated with political opposition for several reasons. In the first place, it offered a way of evading the legal obstacles to the public expression of political sentiments. Suspicion of political associations and public gatherings was deeply rooted in French law, going back to the old regime. The National Assembly had passed a law banning them in 1791. The Jacobin club network had flourished in spite of this legislation, but its example had led to renewed bans on open club meetings in the Directory period, and the Napoleonic regime had codified this prohibition in a law of 1811, which remained in force throughout most of the nineteenth century. Under the provisions of this law, prior authorization was required for any collective gathering of more than twenty people. Such authorization was never granted for anything resembling a political rally or protest.

Staging a political meeting in the form of a banquet provided a way of getting around the restrictions of the 1811 law without provoking a direct clash with the authorities. Dining in common was not on its face a political act. The fact that admission to a banquet required advance purchase of a ticket allowed the organizers to argue that their meeting was a private gathering, analogous to a wedding or a concert, since members of the public could not simply walk in off the street and attend. Because banquet participants paid for their tickets, they were obviously not penniless; cracking down on a banquet thus involved the authorities in a clash with respectable citizens. Before a strict new law on public associations was passed in March 1834—the law that sparked the protests leading to the April 1834 insurrections in Lyon and Paris—it was not even clear that prefects had the legal authority to ban a properly organized banquet.

The banquet's attractions were not merely pragmatic, however. Dining in common was an activity heavy with symbolism well adapted to the opposition movements of the period. The egalitarian experience of sharing the same meal created a sense of community among the participants and made the banquet a utopian anticipation of a republic of equal male citizens. (Women do not seem to have been admitted to these gatherings.) As Rebecca Spang has pointed out, the banquet, where food was the same for all participants, stood in deliberate opposition to the private restaurant meal, where each individual chose his own food and where one's participation in the pleasures of the table depended on how much money one had to spend.[4] By drinking to the series of elaborate political toasts that was always a part of the proceedings, the attendees of a banquet affirmed their common political faith. Sharing wine was a time-honored means of creating a symbolic community, with deep roots in the Christian tradition, in artisan rituals, and in daily life. As the editor of the Lyon workers' paper *L'Écho de la fabrique* wrote on the occasion of that publication's anniversary celebration, "Banquets are to the civilized man what Communion is to the devout Christian."[5] Banquets were part of the French republican tradition; memories of the "fraternal banquets" of 1794 were by no means dead in the 1830s. They were also an important aspect of the Masonic movement; the public banquet given in honor of Odilon-Barrot during his 1832 visit was followed by an unpublicized affair organized by the city's lodges.[6] In Lyon, the banquet organized in honor of Lafayette in September 1829 had made a lasting impression and demonstrated the effectiveness of such events. According to the *Précurseur,* 80,000 citizens had lined the general's route and 500 "heads of families . . . , the natural leaders of our city," had dined with him. The honors accorded the "hero of two worlds" had been "one of the most beautiful scenes of contemporary history."[7]

The banquet was also a means of achieving publicity for the cause in whose name the banqueters assembled. The convocation of a banquet to endorse republicanism or any other subversive political creed was, of course, out of the question, but it was possible to hold a banquet in honor of a newspaper whose sympathies were well known. When supporters of the *Écho de la fabrique* organized a public banquet in 1832, they presented their enterprise as a major symbolic breakthrough for the working class in general. "It was hailed as the program of a new era, and with reason, because it was the first to take place under the joint auspices of journalism and industry," its organizer later claimed. Traditionally, banquets had been the monopoly of the wealthy, but "why should the people not think of imitating them?"[8] Banquets also provided opportunities for journalists to raise their political profiles. Seated at the head table with the most prominent notables of their movement, journalists were saluted in the toasts of others, and given the limelight to propose toasts of their own. The promotion of a projected banquet offered newspapers the opportunity to fill their columns with ideologically tinged articles. Long before it took place, the upcoming banquet could be turned into the subject of discourse and commentary that allowed the newspaper to affirm its political profile. The inevitable critiques from rival newspapers, and the inevitable difficulties created by the authorities, generated occasions for supportive newspapers to pose as defenders of the good cause and as innocent victims of oppression. Coverage of the banquet could fill a newspaper's columns for several days after it occurred, and each such assembly was sure to leave in its wake a series of polemical exchanges with hostile newspapers bent on denigrating the affair.

Lyon's republican journalists promoted numerous banquets in the early 1830s. Some, like the festivity in honor of Odilon-Barrot held on 31 August 1832, were closely connected with press trials, but others were designed to demonstrate the opposition's ability to choose its own ritual occasions. In July 1831, the *Glaneuse*, the more radical of the city's two left-wing papers, responded to the call for an officially sponsored banquet commemorating the previous year's revolution by sponsoring an alternative celebration, the city's first collective expression of dissent from the Orleanist regime. At the banquet, the paper's editor proposed a provocative toast "to liberty, which we are supposed to have, which we don't have, but which we may have some day." Other toasts saluted the revolutionaries of Poland and Italy, "the ladies of Lyon and Paris," and "the hero of Arcole," evoking three emotionally charged symbols excluded from the discourse of the rival official celebration.[9] A banquet in September 1832, co-sponsored by the *Précurseur* and the *Glaneuse*, honored the visiting republican dignitary Étienne Garnier-Pagès. In

February 1833, supporters of the latter paper feted its editor on his release from prison. The self-proclaimed workers' newspaper, the *Écho de la fabrique*, celebrated the completion of its first year of publication with a "proletarian" banquet. The major press-related banquets of 1832 and 1833 drew large numbers of participants. The press and the prefecture agreed that the Odilon-Barrot affair drew some 500 diners, while estimates for attendance at the Garnier-Pagès banquet a month later ranged from 1,700 to 2,000.[10] The *Écho de la fabrique* attracted 340 people for its first-anniversary festivity.[11] The prefect Gasparin reported on 5 May 1833 that 6,700 tickets had been sold for a massive banquet planned for that day but banned on his orders.[12] These gatherings were certainly the largest public political assemblies organized in the city during this troubled period.

Political press banquets were elaborately structured rituals. Ticketholders to the Odilon-Barrot affair received a printed notice instructing them to present themselves at the gathering-place promptly at 1:00 P.M. with their signed tickets. They were to sit in assigned places, and there were to be no spontaneous demonstrations during the ceremony: "All the toasts to be drunk will have been chosen in advance by a commission.... The right to speak will not be granted under any pretext except to those invited to do so by the president."[13] Toasts at the *Écho de la fabrique* banquet also required prior approval, and they had to be nonpolitical. The organizers warned that "no songs will be sung," although they relaxed their precautions to the extent of announcing that "the reading of poems will be permitted."[14] These measures were meant both to ensure that the political sentiments expressed remained safely within the bounds of legality and to mask the tensions that inevitably developed behind the scenes during the organization of a banquet. At a closed Masonic banquet held following the public festivities in honor of Odilon-Barrot in September 1832, the Venerable of the sponsoring lodge admitted that "the public nature" of the earlier event "put limits on that political frankness that needs the veil of privacy"; only behind closed doors could the local republicans challenge Odilon-Barrot to explain why he had not explicitly endorsed their principles.[15] Reports in the opposition press following public banquets emphasized the solemnity and good order that had reigned during the proceedings. The *Glaneuse* described the Garnier-Pagès banquet of October 1832 as "grave and austere": "The toasts were heard with profound seriousness. There were no declamations. Nothing expressed an air of irritability or anger."[16] It was a triumph for the supporters of the *Précurseur* when the proministerial *Courrier de Lyon* had to admit that "order . . . reigned" during the Odilon-Barrot banquet.[17] In the face of enemies who denounced them as fomenters of disorder, Lyon's republican banqueters used their public gatherings to project an image of dignity and self-control.

The high point of a political banquet was the long series of toasts, often running to a dozen or more, which, if properly washed down, must have made the maintenance of gravity and solemnity a considerable challenge. Like nominating speeches at American political conventions, the toasts were important not only for what was said but for who said it. Journalists had the opportunity to take the spotlight and make ringing declarations of their devotion to principle. The participation of local dignitaries provided the opportunity to emphasize the respectability of the republican papers' audience. At the Odilon-Barrot banquet of 1832, those proposing toasts included one of the paper's editors, a merchant, an engineer, a chemist, a notary who was also a municipal official, and a politically active physician—a lineup that demonstrated the degree of bourgeois support for the paper's cause. At the *Écho de la fabrique*'s anniversary banquet, a number of *chefs d'atelier* delivered toasts that demonstrated both the reality of the paper's anchorage in the working classes and their ability to master the public rituals of bourgeois life.

Banquet toasts were an opportunity to reiterate the newspapers' political principles, to shower praise on the editors, especially if they were about to face the ordeal of a trial, and to sing the praises of press freedom in general. At the Odilon-Barrot banquet, Paul Villars, an editor of the *Précurseur*, used his toast to claim that the press was the true embodiment of the freedom for which the revolutionaries of 1830 had fought. Liberal journalists, "their good faith abused," had initially supported the Orleanist regime, but now open opposition was "the sole path that good sense and reason leave open to them, and, gentlemen, perhaps all of France will soon follow them."[18] Toasts at the Garnier-Pagès banquet in October 1832 were evenly divided between populist encomiums "to the people, . . . the thirty-two million French excluded from the rights of citizenship by a legal fiction," and salutes "to independent writers," with special mention not only of Parisian journalistic celebrities but also of "the brave editors of the patriot journals of our city."[19] The *Écho de la fabrique* devoted a special issue to the texts of the toasts drunk at its anniversary banquet, most of them ringing appeals "to the emancipation of the proletarians! . . . helots of industry, victims of the egoism that is devouring our old civilization." As at the other banquets, these programmatic toasts alternated with celebrations of the press. The *chef d'atelier* Blanc raised his glass "to the prosperity of the *Écho de la fabrique,*" asking participants to endorse his claim that the paper "is a good replacement for the *prud'hommes* whom an aristocratic law has allowed the merchants to use against us." On behalf of the *Précurseur,* Alexandre Bret hailed "the spread of education even to the poorest classes," and Marius Chastaing, the *Écho*'s editor, drank to the courage of journalists who had opposed oppression, including Jérôme Morin, the newspaperman-hero of Lyon's version of the July Revolution in 1830.[20] Banquets in honor of visiting dignitaries might end with a response from the guest of honor, which sometimes served to temper the ardor expressed in the toasts. Garnier-Pagès, for example, felt obliged to remind the enthusiasts gathered in honor of the *Glaneuse* of the need to respect the right of property, although he added that "it is men and not property that deserve political representation."[21] By agreeing to accept the banqueters' homage, however, such figures from the legal opposition as Odilon-Barrot and Garnier-Pagès lent a veneer of respectability to the more radical sentiments expressed in the toasts.

After the banquet came the journalistic struggle over its interpretation. Newspapers that had organized banquets on their own behalf naturally concluded that the ceremonies had been rousing successes. A "grand manifestation," the *Glaneuse* concluded after its banquet in honor of Garnier-Pagès,[22] and the *Précurseur* was equally satisfied with its celebration of Odilon-Barrot. But public banquets offered hostile newspapers an opportunity to challenge these expressions of self-satisfaction, and sometimes provoked the sponsoring papers to elaborate campaigns of self-justification. The *Courrier de Lyon,* the city's ministerial organ, tried to argue that the republicans' devotion to banquets was a way of hiding their actual powerlessness. "They are great accomplishments announced, detailed, and commented upon in certain papers the way the victories of the Grand Army used to be," it editorialized. "Defeated in its uprising, the republican movement is reduced to showing its power, its logical arguments, and its respect of liberty in such demonstrations."[23]

But the antirepublican paper nevertheless paid close attention to these events. It deconstructed every detail of the Odilon-Barrot banquet, asserting that the organizers had had to fill the hall with out-of-town guests because they could not attract enough residents of Lyon, that the bourgeois participants had not included "the outstanding members of the magistracy, the bar, and commerce," and that the toasts had been mealy-mouthed and equivocal, a reflection of the fact that the occasion had brought together "heterogeneous elements that balanced and neutralized each other" rather than a unified group. Odilon-Barrot, an avowed supporter of the principle of constitutional monarchy, had found himself out of place, according to the *Courrier,* and, in a final swipe at its opponents, it insinuated that the caterer had hesitated to lay out his best silverware for fear the guests might take it home with them.[24] This last allegation drove banquet participants to challenge the members of the *Courrier de Lyon*'s administrative council to choose between appearing on the field of honor or resigning from their own paper, as some apparently did.[25] But the *Précurseur* was forced to acknowledge that Odilon-Barrot had in fact refused to endorse the paper's republican sentiments.[26] In October 1833, the *Écho des travailleurs* cited the cancellation of a banquet planned to mark the second anniversary of the founding of its rival, the *Écho de la fabrique,* as evidence of declining support for that paper and for the version of working-class emancipation it represented.[27] The effort that newspapers exerted to build banquets into major public events thus created certain hazards: to the extent that a banquet became a subject of public debate, the sponsoring newspaper risked losing control of its interpretation.

The true impact of the banquets on public opinion is difficult to measure. The local authorities certainly took them seriously and reported the measures they had taken to control them to Paris, even though they tried to minimize their significance.[28] Faced with the imminent arrival of the deputy Odilon-Barrot in 1832, the prefect Gasparin had warned: "You know how much effect a famous name and the appeal of something new have in the cities of the provinces, and there is no doubt that the republicans will use this visit to create an important event, to honor the guest, and to surround him with ceremonies, transforming the visit into a triumph of opinion."[29] In the wake of the Garnier-Pagès banquet a month later, he claimed that the affair had weakened the opposition by exacerbating disagreements between moderate and radical republicans and by demonstrating their movement's dependence on support from "simple artisans." At the same time, however, the prefect wrote that "the dangers of such reunions are clear." "Surely the government will present to the Chambers a law to guarantee the social order," he continued. "Otherwise there is no doubt that every occasion will be seized . . . to open veritable clubs whose influence will be most dangerous."[30] On his own authority, the city's police commissioner surrounded the banquet site with agents, although no disorders took place.[31] After the *Écho de la fabrique*'s banquet a month later, Gasparin analyzed the toasts to show that they reflected a determination to justify the insurrection of the year before and a continuing effort to obtain higher salaries through "imposed coalitions" rather than hard work.[32]

The ultimate testimony to the importance of newspaper-related banquets in Lyon's public life in the early 1830s was the prefect Gasparin's successful decision to ban the largest of these events: the banquet planned to raise funds for the defense of the *Précurseur* and the *Glaneuse* in May 1833. When the *Glaneuse* announced the plan for a banquet on 21 April 1833, claiming that it would involve up to 6,000 participants, Gasparin promptly mailed a copy of the paper to his superior, the minister of the interior, asking rhetorically, "Can we allow such a federation to take place under our noses?" and three days later announced that he was prohibiting the meeting.[33] The republican papers questioned the legality of the ban and called on their supporters to defy it,[34] but Gasparin had correctly calculated that they would shy away from a violent confrontation. By 30 April, Gasparin could tell Paris that the organizers had announced a one-week postponement of the gathering. "If the republican assembly falls through, their party is through in Lyon, it will have gone under the yoke and won't dare to risk announcing another public demonstration, for fear of a new embarrassment," he crowed.[35]

As the prefect assembled his forces, Garnier-Pagès, invited as the guest of honor for a second time, pulled out.[36] Gasparin's informants told him that the banquet planners were becoming increasingly divided among themselves, with some insisting on an overt challenge to the prefect and others searching for a way of backing down without suffering total humiliation.³⁷ The *Courrier de Lyon*, the prefect's journalistic ally, demonstrated to the republicans that press publicity could be a two-edged sword. It challenged them to explain how a version of the banquet announcement bearing the forbidden motto "Liberté, égalité, fraternité, ou la mort!" happened to have been circulated, and asserted that the prefect's ban was more than justified. "No one can deny the grave difficulties that might result from such a large assembly of mostly unknown men. . . . How can one argue that there are no limits on the freedom of assembly, and who could claim that, in this case, they have not been largely exceeded?"³⁸

Two days before the fateful date, the republican papers finally gave up, asserting that police provocateurs were planning to incite violence if an attempt was made to hold the banquet.³⁹ Gasparin exulted, as though preventing a meeting in defense of the local press had been a military victory: "The great result has been obtained and without a blow being struck. . . . The entire city applauds the success we have achieved."⁴⁰ He was sure he had crippled the republican movement, not just in Lyon but throughout southern France as well: "If we had weakened here, the whole Midi would have imitated Lyon. Already a banquet was in the works at Grenoble."⁴¹ The legitimist press, which did not promote banquets itself, used the occasion to discredit both the republicans and the local authorities, the former because their "mountain has given birth to a mouse" and the latter for resorting to an "illegal edict."⁴² The insurrection of April 1834 was to demonstrate that the prefect's self-congratulation was premature, but he had effectively denied the city's opposition groups their most effective legal means of publicizing themselves.

Newspapers on Trial

The banquet in honor of or on behalf of a newspaper was the Lyon republicans' own way of creating a public ritual to emphasize the symbolic importance of the press. In theory, newspaper supporters controlled the timing of their banquets, and they certainly controlled the manner in which they proceeded. In reality, most of the Lyon press banquets of the early 1830s took place in close association with a very different public ritual, one in which journalists found themselves compelled to participate whether they wanted to or not: the public press trial. Staged by the Orleanist authorities at their leisure, and intended to intimidate or silence the journalists, press trials were not occasions that newsmen sought to provoke. Forced to fight on their enemies' terms, journalists were nevertheless often able to use these legal rituals to build up their own symbolic importance and that of their publications. Whereas the banquet allowed newspapers to offer readers a vision of an egalitarian republic, the trial allowed them to portray themselves as victims of injustice and martyrs to a cause, powerful images that could be counted on to evoke public sympathy and identification.

Press prosecutions under the July Monarchy were unquestionably a form of political justice, a calculated strategy of using the courts to intimidate editors, publishers, and printers. Although juries frequently acquitted editors brought to trial during the early years of the July Monarchy, defending such cases was costly and time-consuming. When the verdicts did go against the papers and their editors, the results were heavy fines and prison sentences, and royal prosecutors soon made it clear that they would drag hostile editors into court as often as necessary to discourage them and bankrupt their papers. Newspaper personnel thus participated in trials only under duress. Despite its forced character, however, the press trial was not a ritual completely under the control of the authorities. As Otto Kirchheimer pointed out in his classic study of political prosecutions, having defined the impartial administration of justice as one of the essential marks of their own legitimacy, nineteenth-century constitutional regimes like the July Monarchy could not manipulate court proceedings too blatantly without undermining their own claims. Public trials could not produce their intended effect of legitimizing the government's authority unless they had the character of a true contest.[43]

The July Monarchy, with its simultaneous commitments to rule by a narrow elite and the rule of law, faced special problems in this regard; even most of the government's own personnel failed to qualify for the right to vote and could not be counted on to show unfailing loyalty to it.[44] In Lyon, local juries regularly acquitted press defendants throughout the early 1830s, driving the magistrates to complain that "in general these matters are not well understood in Lyon."[45] Kirchheimer cites press prosecutions under the July Monarchy to support his claim that in many nineteenth-century political cases "it was the defendant's friends rather than the authorities who profited from the chance to create beneficial images and give their campaign a new élan at a time when it was most needed."[46] The public prosecutor in Lyon during the first years of the July Monarchy recognized this risk when he reminded the prefect that "it is important nowadays, when press crimes are submitted to juries and judged with greater solemnity, not to bring too many cases of this sort."[47]

One of the reasons these press trials were so risky for the government was that, in contrast to the defendants in many modern political cases, the journalists of the July Monarchy had the special advantage of being able to write their own accounts of their trials and to publish them in newspapers that dominated the media market.[48] Even though they could not prevent the government from disrupting their lives and damaging their papers financially, opposition journalists could subvert the communicative and persuasive functions that trials were supposed to perform for the regime.[49] Under the law, testimony and speeches made at trials could be published without fear of setting off a new round of legal proceedings; during the Restoration, François Guizot, one of the chief architects of the July Monarchy's restrictive policies, had written a memorable defense of the importance of such publicity in assuring citizens that laws were being applied fairly.[50] Not only were trials publicized in the columns of the newspapers themselves, but accounts of the proceedings also were regularly printed as separate pamphlets that could be distributed to sympathizers even beyond the paper's regular circulation area.[51] While the press trial was thus a hazard to the existence of a newspaper and the freedom of its editor, it was also an unparalleled opportunity to promote the causes for which the newspaper existed in the first place, and to exalt the editor's patriotism and dedication.

Like many institutions of France's nineteenth-century political culture, the press trial had time-honored precedents in the English-speaking world but represented an innovation on the continent.[52] Widely publicized trials of press offenses had taken place in England throughout the eighteenth century, and even in the American colonies, where the Zenger trial in 1735 had been essential in establishing a right of freedom from prior restraint on publication. In France under the old regime, however, those arrested for violating the laws on dissemination of printed matter had no guaranteed right to any trial at all. Most of the dozens of authors, publishers, and vendors imprisoned in the Bastille prior to 1789—a list that included most of the century's distinguished men of letters—were incarcerated by means of a *lettre de cachet,* an arrest warrant issued on the king's authority, and received no trial whatsoever. Cases that were referred to courts, such as the mass trial of suspects charged with distributing antiministerial propaganda during the Maupeou "coup" of 1771–74, were held in secret, although echoes of the controversies they generated among the judges eventually made their way into unauthorized news publications.[53]

Although freedom of the press and the right to public trial were both guaranteed in the Declaration of the Rights of Man and Citizen in 1789, the revolutionary era did not produce press trials in the nineteenth-century sense. Before the Terror, in the absence of any effective legislation concerning press offenses, prosecutors took some journalists, particularly the outspoken Jean-Paul Marat, to court for libel or inciting to riot. Rather than an orderly, ritualized judicial combat, however, the proceedings were usually arbitrary and tumultuous. Both the police and revolutionary activists frequently short-circuited the judicial process by invading and devastating printing shops and physically assaulting journalists. During the Terror, the prominent journalists put on trial were usually charged with some form of treason or criminal conspiracy, instead of with press offenses in the strict sense. Lumped together with nonjournalist defendants in mass trials, they were not allowed to cite freedom of the press as a defense, their papers were not allowed to print defense speeches, and hasty trials and executions deprived them of any chance to publicize their own positions.[54] The Directory's trial of Babeuf and his supporters was the first in which supporters of the defendants successfully used the press to contest the charges against them, although the death sentence imposed on Babeuf was hardly an encouraging precedent.[55] The Napoleonic period brought the return of prior censorship, which spared journalists the danger of public trial by preventing them from committing press offenses in the first place. In the French context, the public press trial was thus largely a new creation of the constitutional-monarchist period after 1814.

The details of the laws concerning the press were altered a number of times during the Restoration and the early years of the July Monarchy, but the basic principles that created the conditions for publicized trials remained intact. The 1814 *Charte* and all subsequent legislation provided for public trials for serious legal offenses, and guaranteed defendants and their counsel the opportunity to speak during the proceedings; the closed-door procedures of the old regime, and the one-sided prosecutions of the revolutionary period, were definitively abandoned. Newspapers were authorized to report these public proceedings and comment on them editorially, although the July Monarchy government tried to assert that if they wanted to cite what was said in court they had to publish a complete stenographic transcript, rather than picking out only those passages favorable to their cause.[56] Whether journalistic defendants had the right to be judged by members of the public as well as heard by them was another matter. The liberal press law of 1819 guaranteed jury trials in press cases, but the more conservative Villèle ministry suspended this right in 1822, a policy that remained in effect until the Revolution of 1830. The restoration of the right of jury trial in press cases on 8 October 1830 was one way the newly installed July Monarchy sought to distance itself from its discredited predecessor.[57] The combination of public proceedings and the intervention of the jury, which functioned as the public's representative, institutionalized the legal framework of the public sphere: newspapers and their personnel could be punished only after public proceedings and only with the accord of the public's representatives.

This new legal framework did not permit the press to operate without restrictions, however. Prosecutors still retained the right to bring charges for a wide variety of offenses, including critical comments about the king, allegedly false or calumnious statements concerning government ministers, and advocacy of a change in the form of government, a provision used impartially against legitimist supporters of the Bourbons and republicans. To ensure that prosecuted journalists actually appeared in court, newspapers that reported on politics, particularly the proceedings of the two legislative chambers, had to register with the authorities and designate a legally responsible director, or *gérant*. As we have seen, they also had to post a sizable bond, the famous caution money, to guarantee payment of any fines that might be imposed on the paper. The authorities were well aware of the difficulties in enforcing these rules effectively; various additional regulations attempted to prevent the naming of straw men, paid only for taking the risk of imprisonment, as *gérants*, and to ensure that the *gérant* was actually the person exercising the function of editor-in-chief. Because the risk of forfeiting the paper's caution money deterred journalists from taking flight to avoid prosecution, the system thus served to coerce defendants' participation in the legal ritual.

By loosening some of the restrictions on the press, while leaving the government a complex set of regulations that provided many opportunities for prosecutions, the July Monarchy's press legislation created conditions under which frequent press trials were almost inevitable. The satirical *Glaneuse* imagined local prosecutors exchanging ideas about how to drag out investigations and exhaust journalists by subjecting them to "endless harassments,"[58] but opposition journalists had no choice but to make the most of the opportunities for publicity that the system gave them. The most valuable opportunity press trials offered to prosecuted newspapers was a unique chance to dress themselves up as martyrs. A pending trial could be built up into a major news story. To make the trial even more visible, a celebrated barrister could be engaged. In the provinces, this frequently implied the arrival of a Parisian notable whose name would already be familiar from the newspaper's own columns and whose presence in the flesh would serve to underline the importance of the newspaper whose fate was at issue. Readers could be urged to identify publicly with their paper by attending the proceedings, and the newspaper could fill its columns with the text of its attorney's peroration and its editor's self-justification in court. In the case of a conviction, a campaign to collect money to pay the fine could become yet another means of building bonds between readers and their newspaper.

No French papers were more successful in using their trials as opportunities for this kind of defensive self-promotion than the journals of Lyon. Their tactics were manifold. Even the layout of the paper could be used to underline the iniquity of the trials. Stories about them were given more space than any other local news events covered in the Lyon papers; the legitimist *Cri du peuple* devoted its entire front page to one of its appearances in court.[59] Running titles such as "Latest Trial of the *Précurseur*," "Seventh Trial of the *Précurseur*," and "Eighth Trial of the *Précurseur*" reminded readers that individual prosecutions were simply part of a long-term strategy to wear the papers down. Although press trials were, strictly speaking, trials of individual editors and directors, the papers invariably titled their stories "Trial of the *Précurseur*" or "Trial of the *Glaneuse*," making the newspaper a fictive defendant.

Even before the government had brought the full weight of its judicial weaponry to bear on the press, the city's leading republican journalist, Anselme Petetin, had seized on the issue of freedom of the press as a unique tool for political organizing. On 11 June 1832, his paper, the *Précurseur*, announced the founding of the Association Lyonnaise pour la Liberté de la Presse, a defense fund for opposition papers. The Association's statutes contained the usual arguments about the supreme importance of freedom of the press: "In an advanced civilization, freedom of the press is the foundation of all social and political institutions. It is a right that subsumes all the others, . . . and guarantees them." In view of the measures that had just been taken against the Paris republican papers, as part of the repression following the disturbances of 5–6 June 1832, the members of the Association "have decided . . . to exercise the right of association guaranteed by the constitution to unite for the common purpose of ensuring the existence of the liberal press in Lyon." Members would contribute up to 11 francs a year to help pay fines that might be levied on local papers.[60] An initial meeting, held in the *Précurseur*'s own offices, drew 200 people.[61] The Lyon group served as a model for a national Association pour la Liberté de la Presse, founded in Paris in September 1832, which became the main organization of the moderate republicans in the early 1830s.[62] Reports on the formation of similar groups in other *départements* were one way the press demonstrated its importance to its readers. While the announcement of the founding of the Lyon Association in the *Précurseur* stressed its intention to work within the constitution, a separately circulated version of the statutes bore the date "26 prairial XL" in the republican calendar, an unmistakable signal of the group's true loyalties.[63]

The *Précurseur* did not have long to wait before it needed to call on the Association for support. In the tense climate following the Paris disturbances of 5–6 June 1832, the paper was seized five times in twenty-five days, and Petetin, its editor, was threatened with a lengthy prison term.[64] These indictments set up the first widely publicized press trial in Lyon since the July Revolution.[65] Urged by national leaders of the left-wing opposition, including General Lafayette, the reform-minded lawyer-deputy Odilon-Barrot agreed to come from Paris to handle the defense, providing, as we have seen, the occasion for the first major press-support banquet of the period.[66] The trial gave the *Précurseur* ample opportunity to publicize its principles, its editor's steadfastness, and its lawyer's eloquence. On the eve of the trial, a friendly deputy, Jean Couderc, published a lengthy "defense of the *Précurseur*" in the paper itself, insisting that if freedom of the press had any meaning, it had to extend to the discussion of political principles.[67] Accused of having managed to pack the jury with its own shareholders, the paper responded by publishing the list of their names. The lengthy list of lawyers, merchants, physicians, and notaries reminded jurors, chosen from the same milieu, of the paper's solid roots in Lyon's bourgeoisie.[68] Coverage of the banquet in honor of Odilon-Barrot, organized by the paper, merged seamlessly into reporting on the trial itself, which ended with acquittals on all charges.

In the *Précurseur*'s columns, the paper and its lawyers engaged in an orgy of mutual adulation. Odilon-Barrot fully lived up to his reputation, holding forth for four hours, and his colleague Alphonse Gilardin chimed in with a "magnificent peroration" of three hours.[69] Odilon-Barrot built the paper up into a symbol of liberty and of Lyon's cultural maturity. "The *Précurseur*, gentlemen, is a serious newspaper whose profundity, together with the energy and talent of its editors, has put it at the head of the provincial press and alongside the best papers in Paris. You can already see that, thanks to it, there is a growing recognition that there is thought and originality outside the capital. You see the Paris papers repeatedly copying its articles and thus rendering it a tribute that envy might have stifled." Petetin himself took the stand and set forth yet another version of his argument for the newspaper's central place in modern civilization. If his fellow citizens failed to rally to his defense, he warned, "they are forgetting that . . . it is their rights, their interests, that are under attack, and that they will have reason later to regret their blind cowardice."[70]

After the verdict, Petetin used the columns of his paper to magnify the significance of the victory and to denigrate the luckless public prosecutor. "Our triumph has been all the more complete . . . in that the prosecutor showed no restraint in his proceedings," he proclaimed, while admitting that he had made matters more difficult for himself because there was indeed "something bitter" in the incriminated articles, "which could have misled a less enlightened jury about their criminality."[71] Another contributor hailed the show of support for the paper during its ordeal, "a public demonstration so favorable to the press that we don't see how the government can keep it down."[72] In words, at least, the trial had been a complete victory for the *Précurseur*, which emerged consecrated as a symbol of liberty, civilization, civic courage, and the maturity of the provinces. Although the legitimist papers did not promote banquets the way the republican journals did, they used the same tactics of calling on famous defenders and showering them with praise when they performed in court. The *Réparateur* despaired of conveying to its readers the impact of Pierre-Antoine Berryer's oratory: "Whoever heard him will find only a pale image of this striking talent in a cold analysis. . . . Throughout the auditorium, on almost every face, in almost every expression, one could sense the effect of persuasion, follow the impact of the evidence, see the conviction."[73] Given the enthusiastic tone of the paper's reports, readers could be excused for not realizing that the prosecutor had won his case.

When their lawyers failed to carry the day, journalists were reduced to protesting the injustices inflicted on them. "Our director is on the way to prison, because today prison is the lot of all brave men," the *Glaneuse* announced in early January 1833; two months later it printed a "balance sheet" showing that it had accumulated five convictions, involving a total sentence of more than four years in prison and 9,500 francs in fines.[74] Convicted editors could still take parting shots at their prosecutors and judges, however. In response to a prosecutor who objected to his "incisive, brutal, and sometimes savage" style, the *Précurseur*'s Petetin riposted in print that the his enemy's "insipidity and trivial declamations" were "an excellent lesson for anyone who expresses himself too incisively."[75] Convicted journalists depicted the audience's shock at unfavorable verdicts, taking what satisfaction they could from the thought that "the stupefaction and astonishment" of the spectators at one trial "must have struck everyone, the jurors, the court, and even the prosecutor."[76] Hit with a massive fine in May 1833, the *Glaneuse* announced that it would not even comment on the jury's conduct: "The only sentiment the men on it could inspire in us is pity. They held the palladium of all our liberties, freedom of the press, in their hands, and, if they didn't understand the sanctity of their mission, we can hardly feel sorry for them, and we don't have the strength to hate them."[77]

As they departed for prison, convicted journalists took the opportunity to paint themselves in heroic poses, as Petetin did in March 1833:

> As far as we ourselves are concerned, surely no one thinks that [the outcome] has discouraged us. Bodily punishment for political reasons is more a useless but humiliating nuisance than a serious punishment. In reality, the true misfortune, at a time when so many devoted friends, who entered political life, like us, under the sun of July [1830] languish in dungeons all over France, would be the shame of being the only one not to suffer for a cause that we have seen, no doubt, with better prospects, but never more sacred and more worthy of devotion.[78]

Adolphe Granier, editor of the *Précurseur*'s republican rival, the *Glaneuse*, was not to be outdone when he came up for trial two months later. When he took up his pen, he announced,

> Calumny with its stinking breath, the police with their infamous machinations, the government with its indictments and its prisons, all that appeared to me with terrifying clarity. But behind this hideous tableau I saw Liberty, this daughter of heaven, showing me the goal to which our efforts must lead. From that moment, my role in the great drama that will end with the emancipation of the proletarian was settled. I accepted that role with all its consequences; to have refused it would have been cowardice.[79]

Such proclamations had little to do with reality—Petetin, in fact, was sick of appearing in court and spent much of the year 1833 maneuvering unsuccessfully to get someone else appointed to take the risk of legal responsibility for the paper[80]—but they reaffirmed a heroic image of the republican journalist and of the movement in general.

As in the case of banquets, press trials inspired hostile commentaries in the progovernment press, and anxious scrutiny on the part of the local authorities. Unconcerned about the principle of press freedom, the *Courrier de Lyon* celebrated each conviction with as much enthusiasm as its rivals showed for each acquittal. Even in the absence of specific cases, its editorials urged jurors to be prepared to take a hard line against journalists: "May the jury comprehend its mission to protect a society exposed to so many and such deplorable storms . . . the true public force is in its hands."[81] After a libel conviction against the *Écho de la fabrique*, the ministerial paper announced that the Lyon press had "pushed scandal and shamelessness further than anywhere else."[82] But even in reporting on this case, the ministerial paper was forced to reveal that the *Écho* had been successful in rallying supporters to its cause—so many supporters that the square in front of courthouse had been "covered with workers and *chefs d'atelier* in animated groups."[83] The *Courrier de Lyon* did not limit its involvement in press trials to hostile reporting; its editor later boasted that he had taken a leading role in lobbying the prefect for the appointment of a more effective prosecutor, after the first of the July Monarchy's appointees had failed to win convictions in several cases.[84]

In view of the widespread publicity they generated for the defendants, it is no surprise that the authorities worried as much about the reactions to press trials as they did about press banquets. Gasparin regularly reported the results of these trials to the Ministry of the Interior, with appropriate comments; after one conviction against the *Précurseur* in May 1833, he praised the "severity" of the jury, whose "courage has revived."[85] The presiding judges also sent reports to their superior, the minister of justice. "This affair was the occasion of the most intense arguments. All the means used in similar situations in Paris were employed: anonymous letters, threats, a packed audience to make noise, numerous incidents, a heated and animated plea, storms in the gallery that had to be cleared repeatedly. . . . It was a press trial with all the trimmings," one wrote after an 1832 trial.[86]

Both the trials and the banquets were public rituals that made it possible for a newspaper's views and statements about its importance to be asserted in a way that was usually impossible to do in its own columns. There was a close connection between these two rituals and the newspaper itself: both rituals achieved their effect only because they could be extensively publicized in the newspaper itself, and in the columns of its friends and rivals throughout the French press. Both could also be turned into major news stories, in which the newspaper printing the story and its editor usually played a starring role. In these instances, the

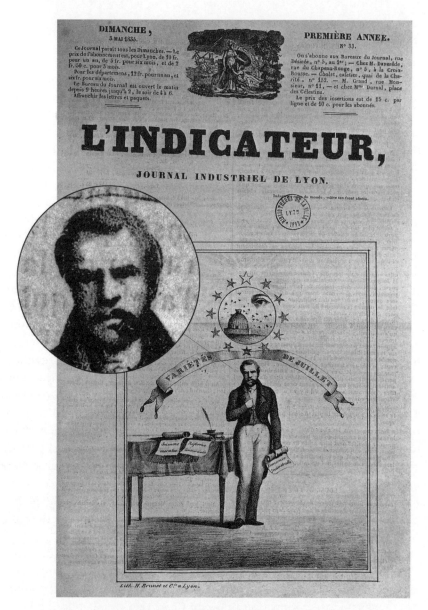

Fig. 7 L'Indicateur. This caricature, the only one published in any Lyon workers' paper during the early 1830s, comments on the repression that finally silenced the opposition press in 1835, symbolized by the padlock sealing the lips of the man holding a paper labeled "industrial economy."

newspaper's capacity for creating a social and political reality was employed for the purpose of promoting the paper itself and asserting its symbolic importance; words were used to glorify the importance of the newspaper's own words and to turn the paper's own claims about its importance into social reality. Through manipulation of banquets and trials, the press successfully challenged the authorities' control of public space, that symbolic arena in which newsworthy events took place. Promoting banquets allowed newspapers to create happenings in which their audiences, normally limited to the role of "invisible communities," could manifest themselves and act out, for the space of a few hours, the egalitarian republican principles for which the left-wing press stood. Trials allowed journalists to make ritualized reaffirmations of the connection between freedom of the press and general principles of liberty, reaffirmations that took on real meaning because the journalists' freedom was visibly at stake. By reporting on public reactions to the courtroom proceedings, newspapers used trial stories to undermine the regime's legitimacy. Journalists thus used their days in court to turn the regime's own mechanisms for generating newsworthy events against it.

The banquets and controversial trials of the early 1830s did not, of course, bring down the Orleanist regime. Faced with widespread evidence of contestation, the government reacted by extending its arsenal of repressive legislation. A more elaborate law on associations, passed in March 1834, threatened not only banquets but also workers' organizations, such as the *mutuellistes* who played such a significant role in Lyon. Protests against the law led to the republican insurrections of April 1834, which in turn permitted the government to arrest leading republican journalists and silence their papers. The September Laws of 1835, rushed through the chambers after a failed assassination attempt against Louis Philippe, further tightened the restrictions on the press, lengthening the list of offenses for which they could be tried and reducing the number of jurors needed for a conviction (see Fig. 7).[87] Papers as outspoken as those that appeared in the years from 1831 to 1834 would not be seen in France again until after the Revolution of 1848.

Nevertheless, the press banquets and trials of the early 1830s were steps in the evolution of a new kind of political culture, one in which oppositional viewpoints achieved public visibility not only in words, as had been the case during the Restoration, but also in symbolic gatherings. In the early 1830s, newspapers were uniquely capable of serving as points of crystallization for such occurrences. In later periods, political clubs, parties, trade unions, and other organizations took over this role, and newspapers, even as they gained in circulation, lost their special symbolic importance. The years in which newspapers served as the most important mechanisms for rallying currents of opinion were, however, an essential phase in the development of a modern political culture in France. They were also a critical period in the history of the press itself, leaving behind them a lingering nostalgia, both among journalists and the wider public, for a time when newspapers had been symbols of freedom rather than commercial enterprises.

6

Textualizing Insurrection

The Press
and the
Lyon Revolts
of
1831 and 1834

The Problem of Insurrection

Narrating symbolic public occurrences, such as banquets and political trials, gave the newspapers of the early 1830s opportunities to assert their own importance and to project the impression that the groups they represented could shape events. Furthermore, these were events that were largely made up of discourse and thus lent themselves easily to journalistic presentation. The significance of these formalized events remained limited, however. They allowed their participants to reaffirm their status and define their ideological positions, but they did not drastically change their sense of themselves or the way others saw them. The two Lyon insurrections of November 1831 and April 1834 were occurrences of a very different order. Both took a heavy toll in human lives—100 soldiers and 69 civilians killed in 1831, and 131 soldiers and 192 civilians in 1834—and the extensive fighting in the city center in 1834 resulted in widespread property damage.[1] These outbreaks of collective violence raised fundamental questions about the nature of social identities in European life. From the moment they occurred, journalists and readers were aware that the way the press presented these events would have momentous consequences, both for French society and for its journalistic system. Newspapers would determine how the participants in the insurrections were identified and how their actions were interpreted. A master weaver whose letter appeared in one of the local papers two weeks after the 1831 insurrection put the matter succinctly: "On the 21st and 22nd [of November], we were troublemakers who were not to be tolerated, on the 23rd and 24th, we were 'Lyonnais, upstanding workers.' What are we today? Insurgents, evildoers, rebels, we violated the laws, etc. etc."[2]

The insurrections returned the newspapers to their core function, the one they had inherited from their ancestors, the *canards* and broadsheets of the old regime: the recounting of dramatic and unexpected episodes, those that constituted true *nouvelles* or news. The philosopher Paul Ricoeur has outlined a three-stage process by which such events take on meaning: "First something happens, explodes, disrupts an existing order; then an imperious need to assign it a meaning makes itself heard, reflecting a need for things to be put back in order; finally the event is not simply integrated into the order of things, but . . . it is recognized, honored and exalted for creating meaning."[3] The role of the press in representing the Lyon insurrections of the 1830s both exemplifies and nuances Ricoeur's analysis. Newspapers did assign meaning to these events and integrate them into the ongoing flow of news. As we have seen, one particular newspaper article, Paris journalist Saint-Marc Girardin's commentary on the November 1831 insurrection in the *Journal des débats* of 8 December 1831, can fairly be said to have given that event its claim to world-historical significance by insisting that "the Lyon uprising has revealed a fundamental secret, that of the internal social conflict between the class that owns things and the one that does not." As another Paris journalist commented, the *Journal des débats* "has employed extremely clear terms that show the profound impression that the events in Lyon have made in the upper ranks of society. . . . There is a surprising appreciation of many facts that are elements of disorganization and upheaval in the heart of society."⁴ But the insurrections did not develop independently of the newspapers. The press was directly involved in creating the circumstances that led to the uprisings, which thus did not come as total surprises to the journalists or their readers; the contrast that Daniel Boorstin has drawn between "genuine" historical events and "synthetic happenings" deliberately planned for the sake of media coverage is not entirely justified.⁵

On the other hand, however, the newspapers did not find it simple to complete the process of assigning meaning to these events that Ricoeur's analysis assigns them. Faced with the kind of news that the media system supposedly existed to report, the press often proved incapable of providing readers a coherent representation of the social and political reality in which both it and its audience were caught up. Even as journalists perfected a rhetoric of realism that foreshadowed the work of the period's great novelists, they were plunged into a crisis of representation that forced them to renounce the goal of fully narrating history in the making.⁶ This was especially true for the papers in Lyon. They were the main constituents of the city's public sphere, but this direct connection to the city's public life compromised their ability to speak openly about events that threatened the premises of a politics based on rational discussion. The insurrectionary crises also underlined the degree to which Lyon's local public sphere was subordinate to the national one. Even as they narrated the dramas taking place around them, the Lyon papers had to contend with competing and ultimately more influential depictions in the Paris press. The role of the press in the "construction" of the insurrectionary events of the early 1830s in Lyon was thus a complex and often contradictory process that contrasted sharply with the depiction of the same events in other media.

Urban insurrections were not a new phenomenon in Lyon's history. The city had been shaken by many episodes of collective violence, going back to medieval times.⁷ In 1744 and 1786, conflicts within the silk trade had exploded into violence that was suppressed only by military intervention.⁸ The contrast between the way such events had been reported at the end of the eighteenth century and the media coverage they received in the 1830s reveals a great deal about the changes that had taken place between the two eras. Contemporaries had recognized the 1786 insurrection as a newsworthy event: Europe's "newspaper of record," the

Gazette de Leyde, published in the Dutch city of Leiden, printed five lengthy and detailed reports on the uprising and told readers that it was no blind outbreak of meaningless violence, but rather a reaction to the cyclical nature of the silk trade, whose downturns "inevitably throw many of the inhabitants into unemployment and distress."[9] The newspapers that covered the 1786 uprising, however, attached no long-range historical significance to it. Such urban "tumults" were seen as unfortunate but hardly unusual by-products of European civilization, not as harbingers of a change in the very nature of society. Even though the storming of the Bastille was only three years in the future, no observer in 1786 seems to have regarded the Lyon uprising as a sign of any wider threat to the social order.[10]

In Lyon itself, the 1786 revolt was no doubt the subject of much discussion, but none that found its way into print. The one local periodical, the *Journal de Lyon*, printed not a word about the disturbance, and there seem to have been no published pamphlets about it. Even in the last years before their collapse, the institutions of the old regime were still capable of limiting printed discussion of this kind of disturbance to newspapers and readers too far away from the scene to intervene in them. The city's public had not yet gained the right to debate in print issues that directly concerned them. The Revolution of 1789 completely changed this situation. The abolition of censorship allowed the creation of local newspapers that spoke directly about local affairs, including outbreaks of violence. Whereas the old-regime press had affected a stance of distant objectivity, the revolutionary press was unabashedly partisan, making no effort to distinguish between journalistic and political truth. As a result, the revolutionary press of Lyon was also a press whose fate was completely tied to the events it reported. Each local political crisis in the long and troubled decade of the 1790s resulted in the disappearance of newspapers that had backed the losing side, and the creation of new publications explicitly identified with the new dominant faction. None of these papers survived long enough to become a significant local institution in its own right.[11]

The press of the early 1830s differed from both the papers of the old regime and those of the Revolution. Unlike the papers before 1789, those of the nineteenth century were no longer restrained from publishing the details of controversial local events. The events of 1831 and 1834 were reported immediately, both in the several competing newspapers published in Lyon itself and in the national press in Paris, as well as in regional newspapers all over the country. In place of the detached tone of the *Gazette de Leyde*'s newsletters, nineteenth-century readers were offered coverage that captured the tension and drama of the violence while it was actually happening. On the other hand, however, the very existence of the nineteenth-century press was no longer directly at the mercy of sudden events, as the newspapers of the revolutionary era had been. The legal framework that both protected and regulated newspapers was sufficiently established that it could not easily be overthrown, and the newspapers themselves were no longer the fragile enterprises of forty years earlier. Not only were individual newspapers now usually more solid and difficult to uproot, but French society had come to accept the permanence of a journalistic field made up of publications with conflicting points of view.

In 1786, the press had described Lyon's insurrection from a distance and treated it as a phenomenon that could not affect the basic nature of society. The Revolution of 1789 had shown that a total transformation of society and institutions was indeed possible. The situation in the early 1830s fell between these two extremes: the Lyon insurrections raised fundamental questions about the social and political order and showed that revolution was a real possibility, but they did not burst the framework of nineteenth-century society asunder. In representing them, the press had to operate simultaneously on two different levels, that of the existing order in which the newspapers themselves were rooted and that of the possible revolutionary situation that threatened to replace it and in which the function of the press would be completely different. Furthermore, journalists were acutely aware that the words they chose to publish were themselves among the actions that would determine the outcome of these crises, as well as their personal fates and those of their journals. Particularly for the provincial journalists of Lyon, the events of November 1831 and April 1834 were moments of existential crisis in which they seemed to hold the destiny of European society in their hands, but in which the survival of their newspapers was also at risk.

Nowhere was the intensity of response to these events more manifest than in the single issue, dated 25 November 1831, that Lyon's radical paper, the *Glaneuse*, managed to put out during the first insurrection.

Because this single number of the paper was composed at intervals during the week of the crisis, but without any revision of the earlier sections to take into account the changing situation, it became a text that displayed, in both its form and its content, the movement from shock to revolutionary euphoria and then to hurried retreat as the old order reasserted itself. The first pages of the paper's special edition were dramatically altered from the *Glaneuse*'s normal format to convey a sense of the magnitude of the events it was reporting: instead of its usual two columns of small type on the front page, it bore one column in unusually large letters. Readers, accustomed up to that moment to a satirical paper aimed at theater-goers, found that its editorial tone had also been completely transformed. The front-page story, datelined to the first day of the insurrection, the 21st, began: "At the moment when we write, killing is taking place in the streets. Blood runs over the pavement!" In melodramatic prose, the paper declared its allegiance to the cause of "the most numerous and the poorest class" and called down maledictions on those who had "refused to hear the heartrending cries of poverty and hunger! They stolidly cleaned their rifles, and then they waited until the anguish of despair had finally driven these unfortunates to take up arms so that they could kill them like wild beasts." In a few paragraphs, the paper managed both to convey the intensity of the crisis and to assign a meaning to it: the insurrection was the justified revolt of an oppressed class against its depraved and indifferent exploiters.

The first installment of the paper's narrative left the outcome of events in doubt; the continuation, datelined 22–23 November, reflected the situation just after the army's evacuation of the city, early in the morning hours of the 23rd. The silk workers now controlled the city, and the paper identified itself completely with them, pointing out to the property-owning classes that they had not committed any of the atrocities their opponents had predicted. "Don't slander the people; yesterday, they held you in their hands, they could easily have crushed you, instead they generously let you go." In an address "To the Workers of Lyon," the paper sought to make the workers conscious of what they had accomplished, inspire them with a new sense of themselves, and put itself forward as their guide. "Victory is yours; the noble usage that you have made of it proves your strength and your moderation," the paper editorialized, before warning that "intrigue will try to steal the fruit of your courage from you." To protect themselves, the workers needed leadership: "You know our rights to your esteem and your confidence."

Rather than following this proclamation of triumph with the outline of a revolutionary program, however, this peculiarly constructed issue of the *Glaneuse* then narrated an episode that marked the definitive failure of the one attempt made in Lyon to give the November uprising a political coloration. The editor of the *Glaneuse*, Adolphe Granier, knew the details only too well, because he had been involved in it and subsequently had the responsibility for the attempt thrust upon him. On 23 November, a proclamation, written by Granier but known as the Lacombe *affiche*, or poster, because it bore the signature of one Lacombe, identified as the "syndic" of the workers' commission that was maintaining order in the city, had been printed. It announced that the insurgents intended to name new administrators for the city and the *département*. Such a step would have been a genuine bid to take over political power. A local artist's depiction of this moment of exaltation among the victorious workers shows them posed like the figures in Jacques-Louis David's "Oath of the Tennis Court" as one of the *Glaneuse*'s editors exhorts them (Fig. 8). The *Glaneuse* printed the text of the inflammatory manifesto, but also explained the circumstances under which it had been written (without mentioning Granier's role) and the fact that its supposed promulgator had quickly withdrawn his endorsement. The paper thus managed to report the one truly revolutionary moment of the November insurrection and simultaneously to efface it. Granier made no effort to fan the revolutionary flames in the style of Marat or the *Père Duchesne*. Instead, his paper served to document the swift passage from exaltation to the abandonment of revolutionary aspirations. Granier subsequently claimed that the silk weavers occupying the city hall on 23 November had solicited his involvement in preparing this manifesto, and that he had only belatedly realized that the invitation was a trap meant to discredit the republican movement.[12]

By the time the *Glaneuse* put out its next issue, two days later, its main concern was to defend itself against the inevitable accusations of having incited either the insurrection itself or the brief effort to turn it in a revolutionary direction. "Only humanitarian feelings guided our pen when we expressed concern, when we sought aid for those dying of hunger," the paper claimed, "yet they say that we set the fire! That we should have stifled our pain!" Anticipating the arrest warrant he was sure would

Fig. 8 "Serment de l'Hôtel-de-Ville (de Lyon)." Michel-Ange Pèrier, one of the editors of the *Glaneuse,* leading the insurgents who had seized Lyon's Hôtel-de-Ville in a revolutionary oath on 23 November 1831. Although the members of this "provisional command" had not kept their promise to "allow themselves to be buried under the ruins of the city" rather than allow the troops to reenter, the caption for this engraving celebrated their victory over the army and the fact that they had taken 500 soldiers prisoner during the fighting.

come his way, Granier cast himself as a hero who had responded to the challenge posed by the extraordinary events the city had just experienced. "At a moment when great questions were being settled by cannon fire, to write literature would have been to neglect our duty as men and citizens. One had to take sides, when our brothers were killing each other in the streets; one had to say out loud what one thought.... We are not phrase-making machines, but men with convictions, and for those convictions, we know how to fight and die." Even as he indulged in this bombast, however, the *Glaneuse*'s editor announced the paper's retreat to regulated language. Although he promised to remain faithful to his convictions, he explained that the paper would "return to the lighthearted vein that the gravity of circumstances made it throw off for a moment." In other words, the *Glaneuse* would strive to stay in business by returning to its ostensibly nonpolitical form and claiming that only the overwhelming force of circumstances had made it behave otherwise. This retreat from advocacy was undoubtedly the better part of wisdom for the paper and its editor, both of whom succeeded in living to fight another day. But it was also an admission that the November 1831 insurrection would not bring about a revolution; the paper's tone told readers that the event's outcome was already determined, even though the workers still controlled the city at the moment when these pages were published.

In the course of a few pages, the *Glaneuse* thus managed to demonstrate the range of roles that the insurrectionary situation imposed on the press. The opening pages of its special issue immediately elevated the events of 21 November to the level of a great event and assigned the participants roles in a melodrama in which workers and employers confronted each other as innocence and evil. As the week moved along, the paper briefly transformed itself from impassioned witness to revolutionary oracle, identifying itself with the workers' movement and seeking to direct it. As the potential revolutionary spark guttered out, however, the paper abruptly shifted to a defensive posture, seeking to safeguard its future existence within the legal order that was clearly going to reassert itself.

In the Midst of the Action

The tortuous maneuvering exemplified in this special issue of the *Glaneuse* was an extreme demonstration of the difficulties the press faced in dealing with the two Lyon insurrections. Part of this difficulty stemmed from the dual nature of these events. In one sense, the two insurrections were unanticipated occurrences, whose outbreak and outcome could not be predicted in advance. Newspapers had to improvise their reactions to these events as they developed, as the *Glaneuse*'s text demonstrates. In another sense, however, everyone understood that violent mass insurrections were always a possible outcome of the period's social and political tensions; just such an outbreak in Paris in July 1830 had led to the establishment of the Orleanist monarchy. Journalists in particular were well aware that everything they published in periods of such tension would influence the course of events. In both 1831 and 1834, the Lyon papers faced those difficulties even before the outbreak of actual violence. On both occasions, Lyon's press—but not the papers published in Paris—gave ample indication of an approaching crisis. Press coverage in 1831 was extraordinarily confused because so many of the city's papers were in the midst of internal problems that hampered their ability to react effectively to the unfolding crisis; the incoherence of the bourgeois press's reaction to the workers' campaign may well have encouraged the latter to take a hard line. The city's most radical political paper, the republican *Sentinelle*, which had never paid much attention to local affairs in any case, ceased publication on 19 October 1831, just as the campaign for a *tarif*, or fixed piece rates, for silk products began to develop. The city's less-important liberal paper, the *Journal du commerce*, had advertised the fact that it was up for sale just as the crisis started.[13] And, as we have seen earlier, the *Précurseur*, the city's most important newspaper, was in the middle of an internal crisis that made its coverage of the affair completely schizophrenic. Most of the paper's leader articles about the silk weavers' campaign in the month before the outbreak of fighting insisted that the low pay the weavers were protesting was an inevitable result of the economic condition of the industry and that imposing a *tarif* would be a futile gesture. The paper conceded that the condition of the workers was "indeed miserable" but insisted that the only solution to their problems was to adapt to the harsh reality of the international market.[14] Although the paper dismissed the idea of a *tarif* as unworkable, it considered the threat of working-class violence all too likely. The mass meetings held in the place de la Croix-Rousse were making the suburb into "another Spitalfields," the paper warned, conjuring up images of recent disturbances among weavers in England.[15] A few days after this warning, the paper printed another article that was even more specific, alleging that the silk workers had formed a structured organization with a definite chain of command, that they had formed links with workers in other trades and even in other manufacturing cities, and that their organized force was intimidating the local authorities into granting their demands.[16] The brewing crisis thus led the bourgeois press to attribute a new identity to workers. Normally treated as passive and incapable of meaningful action, they were now suddenly seen as militant and dangerously able to influence events.

These explicit and obviously well-informed articles were only part of the *Précurseur*'s discourse about the looming threat of a confrontation, however. Their author—probably the paper's assistant editor, Jean-Baptiste Monfalcon, soon to become the editor of the progovernment *Courrier de Lyon*—had to share its columns with other writers who took a more sympathetic view of the workers' goals and tactics. Meanwhile, the paper's representatives were negotiating with the Parisian journalist Anselme Petetin, who agreed to become the paper's editor at the end of October but took up his position only on 18 November 1831, three days before the insurrection. Under Monfalcon's direction, the paper had condemned the very idea of a *tarif* as "neither legal nor reasonable"; less than a month later, Petetin was telling its readers that "it is time . . . to do something about the material interests of the masses, so long neglected because of the vanity of the castes."[17] But Petetin arrived on the scene too late to have had any real impact on the direction of the crisis. The *Journal du commerce* also gave a confusing notion of the workers' movement, augmented with commentary that underlined the incoherence of the more influential *Précurseur*'s editorial conduct. At the height of the demonstrations in favor of the *tarif*, the smaller paper praised the workers' moderation and good conduct and printed a letter accusing the *Précurseur* of trying to goad them into "misbehavior that would lead to the rejection of their just demands." A week later, however, the same paper warned workers against outside agitators who were supposedly inciting them to unjustified violence.[18] Although it would claim the role of a revolutionary paper during the insurrection, the four-month-old *Glaneuse* could hardly be accused of having advocated social violence before 21 November. The paper sympathized editorially with the plight of the workers, who were too poorly paid to afford basic necessities, but at the same time, it echoed standard liberal criticism of government-imposed economic regulations. Its only positive suggestion was a charity campaign on behalf of workers to which it proposed to donate 100 francs in its own name.[19]

Although none of these papers actively promoted an insurrection, the possibility of collective violence haunted their columns for weeks before the actual uprising. Opening its charity campaign, the *Glaneuse* stated: "If riots ferment in our cities, if a useful and numerous class suffers . . . , it is neither the employers nor the workers who should be blamed; the evil is deeper and goes back further."[20] In the columns of the *Précurseur*, the question of whether the workers' demonstrations were verging on violence was a major subject of debate between rival editorialists. "A tumultuous group of 500 or 600 individuals emitted the most intolerable vociferations against some of the best known silk merchants," a story on 22 October 1831 complained, but another writer replied three days later by asserting: "The demonstrating workers control themselves strictly. . . . They have not yet disturbed the public peace and the law; you have no right to claim that they will in the future."[21] The paper's divided assessment echoed the deeper split in the city's bourgeoisie—soon to be reflected by the creation of a second bourgeois paper—over whether a social order based on public opinion could survive if the lower classes successfully forced their way into the public sphere. Meanwhile, reports on the riot in the English city of Bristol at the end of October, in the course of which a crowd demanding parliamentary reform had attacked opponents and set fire to public buildings, made the threat of urban violence all too real.[22]

The discordant coverage in the bourgeois press reflected a very different kind of prerevolutionary situation from that which had occurred in July 1830, when the *Précurseur*, speaking for a relatively united public, had been able to anticipate the crisis by laying out a revolutionary script in which paper and readers had clearly defined and complementary roles. Editorialists in the bourgeois press invoked the prospect of working-class violence for several contradictory purposes. At times, it served as a pretext for exhorting the authorities about their duty to maintain order and to avoid concessions that might lead the workers to believe that menacing behavior would serve their cause. Equally often, however, journalists reminded the *fabricants*—members of the bourgeoisie, whom this press supposedly represented—that their intransigence risked inciting an explosion. Finally, their expressions of sympathy for the sufferings of the poor, and their calls for charity, were designed to pacify the workers and to demonstrate that the bourgeois society the press represented was concerned about their fate. In making these arguments, the papers posed as disinterested observers responsible for reminding the public of a danger that the decision-makers in the prefecture and in the silk merchants' circles seemed oblivious to.

The confusion reigning in the columns of the bourgeois press may well have encouraged the leaders of the workers' protest movement. The workers' own paper, the *Écho de la fabrique*, had not been a factor in initiating the *tarif* campaign—which was well under way before its first number appeared—nor did it advocate insurrection. The paper was willing to evoke that possibility much more directly than the bourgeois press, however. The hypothetical scenario for an outbreak of fighting that it outlined in its second issue, two weeks before the uprising, proved so accurate that it probably reflects plans formulated among militant silk weavers who feared that the *tarif* would be subverted. Although the paper blamed the rumors it reported on *agents provocateurs*, it said that there was talk "that the masses were ready to act; that the Croix-Rousse was marching under a black flag to attack the Hôtel-de-Ville, and then the houses of certain merchants whose names were on a proscription list." In response to these rumors, bourgeois National Guards had been stationed around the Hôtel-de-Ville. A crowd had gathered to witness this unusual deployment of force, and the paper claimed that provocateurs were trying to turn them against the guards.[23] The scenario outlined in the paper closely resembled the events of 21 November, the first day of the insurrection. On that day, workers carrying a black flag did indeed try to march from the Croix-Rousse to the Hôtel-de-Ville, and violence did indeed erupt between them and the National Guard. But the edition of the paper that actually appeared on the eve of the fighting emphatically denied any intention of inciting trouble. It was only the "men of *fear*, who always see the *masses* ready to ravage everything, and society on the point of dissolution," who had spread alarming rumors, the *Écho* claimed. For its part, the paper disavowed any sense of alarm. The Paris papers had been filled with panicky reports, "while we enjoyed the most complete security," it concluded.[24] The *Écho*'s strategy was to try to substitute the evocation of possible violence for the reality of insurrection; the threat of an uprising, advertised in the paper's columns, would discredit the bourgeoisie and obtain the concessions the workers sought. In this respect, the legitimist *Gazette du Lyonnais* made common cause with the *Écho* against the bourgeois press: "We hear that certain people are very irritated with the workers and that they would welcome an armed confrontation. It's remarkable! These are the same people who, in July 1830, pushed the workers into public action."[25]

The journalistic buildup to the insurrection of April 1834 differed from what had happened two-and-a-half years earlier, both because the journalistic system had become more complex and because the experience of the earlier insurrection had changed the public's notion of how likely such an outbreak was. In the second crisis, several papers openly suggested that confrontation might serve the purposes of the groups they claimed to represent. The most outspoken of these was the *Courrier de Lyon*, mouthpiece of the city's more intransigent bourgeoisie, which reacted to the weavers' mounting activism, especially the strike of February 1834, with statements that were widely read as incitements to violence. "An actual insurrection would be a thousand times better than this continual fear of one, we would know where we stand, and at least we would rapidly settle matters," one article on the February 1834 strike exclaimed, and another called on the authorities to give demonstrators who appeared in the streets "a vigorous lesson," a phrase the legitimist *Gazette du Lyonnais* singled out as a "criminal provocation."[26] On the other side, the *Écho de la fabrique*, which had come under the direct control of the *mutuelliste* organization in August 1833, used equally emphatic language. In February, it explicitly justified the general strike as a warning to the fabricants to make concessions before they found themselves facing a repetition of November 1831. Turning the *Courrier de Lyon*'s inflammatory language against the enemy paper, it warned that, even if troops were called in, "from now on, the workers will know how to render any injustice on the masters' part impossible, and how to inflict on anyone who chooses to ask for it, if not a *vigorous*, at least a severe lesson." A month later, the paper threatened that workers would resist the new law against associations. "Those in power sow the storm. . . . Let them reap the tempest."[27] Whereas the *Courrier de Lyon* and the

Écho de la fabrique both resorted to a language of class conflict, the *Glaneuse* used a language of republican liberty. Its call for resistance to the law on associations—actually an editorial reprinted from the Paris republican paper, the *Tribune*—argued that both "pacific" and "insurrectionary" means were appropriate to resist such an infringement of popular sovereignty. "Those in power have declared civil war themselves. . . . Who will stand in the way of a mass uprising by the people? To take over the public space (*la place publique*) is the most valuable security for the right of association."[28]

As in 1831, the city's best-known paper, the *Précurseur*, attempted to position itself between the warring factions and ward off catastrophe. "What matters is that the city be protected from scenes of carnage . . . , that the workers and the merchants resume the relations whose interruption can only damage both," the paper wrote during the February 1834 strike. The paper did denounce the Orleanist government for violating the rights of citizens. "It cannot complain about riots and seditions; it has resorted to violence and taken the offensive," an editorial against the associations law asserted. But the paper held back from endorsing any actual mobilization, and it denounced rumors that workers might hold a mass demonstration to support the *mutuellistes* on trial.[29] Privately, Petetin had consistently sought to head off the mounting conflict. At the time of the February strike, he had set up a meeting between *mutuelliste* representatives and some of the silk merchants, and on the night before the uprising he had approached the prefect personally to warn him about a possible outbreak of violence the next day.[30] Petetin's worst journalistic enemy, Monfalcon, offered a public testimonial to his rival's sincerity, writing that "the men of the *Précurseur* were pained by this insurrection, which compromised them. . . . They represent the republic of principles, the intelligent democracy, that which has always been so easily overwhelmed by anarchy."[31] The city's two legitimist papers took some pleasure in denouncing the Orleanist authorities for having let a crisis develop, but their tone was not as provocative as it had been in 1831. Having gained power themselves through a popular insurrection, the *Gazette du Lyonnais* remarked, the "men of July" should have known better than to risk provoking similar resistance against their own regime. Its rival, the *Réparateur*, did not know whom to support in such a crisis; a victory for the workers would certainly not restore "respect for religion, for morality and the laws," but a victory for the government would "strike a mortal blow against our most precious liberties."[32] The alternative press that had developed since the 1831 uprising was even more cautious in its comments. The *Conseiller des femmes* and the *Papillon* both called for reconciliation between classes during the February strike, but, as the latter paper remarked, "this question is off-limits to us."[33]

In both 1831 and 1834, the Lyon press thus clearly raised the specter of insurrectionary violence for weeks or months before it actually occurred. Although none of the papers took as direct a role in promoting collective violence as the revolutionary journals of the 1790s had,[34] they certainly made insurrection appear as something well within the realm of possibility. In their columns, insurrection was not an abstract, theoretical matter but a concrete eventuality, likely to break out under specific, easily anticipated circumstances; indeed, they were freer in their comments about the subject before the uprisings started than they were once they had actual news to report. At the same time, however, the papers worked to distance themselves from the violence they imagined. In their columns, that violence was categorized as something almost beyond human control, or as a possible weapon in the hands of groups over which the newspapers claimed to have no influence. The more they evoked the possibility of violence, the more the city's papers insisted that they were doing so only in order to render it less likely. The role of the press in the "preconstruction" of the events that put Lyon in the spotlight in the early 1830s was thus a contorted one. The language in which the press anticipated catastrophe was hardly ever a language of advocacy; instead, the papers asserted that they raised grim possibilities only because of their duty to face facts, and to steer the community away from looming perils. This approach was completely different from the one the papers used in building up anticipation of political banquets and trials, events they could treat as fulfillments of their ideological projects.

For the Lyon papers, the violence that did erupt near the place des Terreaux on the morning of 21 November 1831 and on the place Saint-Jean on 9 April 1834 was thus simultaneously expected and unexpected, the fulfillment of predictions and a contradiction of the papers' expressed hopes. In 1831, a column of workers descended the Croix-Rousse slope bearing a black banner inscribed with the memorable phrase "Vivre en travaillant, ou mourir en combattant," to protest the merchants' continued resistance to the *tarif*. At the bottom of the hill, they encountered a National Guard unit whose members were primarily drawn from the silk merchants' district. Exactly what happened to start the fighting remains unclear, but some of the heavily outnumbered National Guards fired on the workers. Word of the shooting spread through the Croix-Rousse, drawing additional workers to the scene. As the fighting was spreading, the prefect, Bouvier-Dumolard, and the commanding general of the National Guard tried to enter the Croix-Rousse and calm the situation. Workers angry at what they saw as an unprovoked attack seized them as hostages. In 1834, the trial of the leaders of the silk weavers' *mutuelliste* organization provided the spark for the insurrection. Crowds gathered in the streets on 9 April 1834, the day the verdict was expected, confronting troops surrounding the courthouse near the city's cathedral of St. Jean. As usual in such situations, the exact circumstances in which the first shots were fired were never determined, but once fighting broke out, militants from the republican groups, many of them also silk weavers and artisans, built barricades, and a full-fledged insurrection was under way.

Because of the different patterns of the fighting in 1831 and 1834, the role of the Lyon press during the two insurrections differed significantly. In 1831, the three days of fighting spared the center of the city, where the papers had their offices and printing shops; the papers then had to decide what to say during a week when regular government in the city was suspended. Newspapers and journalists had to take real decisions that had a definite possibility of affecting the direction of events; whatever they did would make them and their papers participants in the crisis and risked exposing them to all-too-easily imagined perils in its aftermath. In 1834, however, the insurgents found themselves blockaded in the central district, where the newspapers' offices and printing shops were located. The fighting, together with the state of siege imposed by the army, prevented the local newspapers from appearing at all. When his paper was finally able to reappear, the *Précurseur*'s Anselme Petetin explained to readers: "We were . . . cut off from our printing shop. Most of our workers, caught by surprise a long way from this neighborhood when the crisis broke out, were trapped away from home like everyone else, because every street was under a state of siege."[35] In both cases, the Lyon press was also immediately reminded of its subordinate status vis-à-vis the national papers, which not only were more influential but also had a freedom to comment on the insurrections that the local publications lacked.

The local press coverage of the 1831 insurrection amply demonstrated how the contradictory pressures acting on the papers prevented them from either giving a full chronicle of the spectacular events occurring in the city or providing coherent leadership to any of the contending forces. At the point when Lyon's main daily paper, the *Précurseur*, first tried to give the story some shape, in the afternoon of 21 November, its editor, Anselme Petetin, faced the challenge of simultaneously conveying the gravity of the crisis while trying to assure readers that it was not completely out of hand. He had to think not only of his local readership but also, and perhaps even more important, of those throughout France, especially journalists and politicians in Paris who might make decisions that would determine Lyon's fate on the basis of newspaper reports. What Petetin wrote mingled observation and prediction, factual narrative and efforts to shape interpretation. He unequivocally acknowledged the scope of the event: the workers "had consolidated the organization of the uprising and were absolute masters of the Croix-Rousse," which, thinking of his out-of-town readers, he parenthetically identified as "the home of the largest part of the working population." Even as he communicated to readers the fact that Lyon was in the midst of a full-scale insurrection, however, Petetin tried to avoid further inflaming the passions and fears of his readers and of the actual participants in events. He reported the seizure of the prefect and the National Guard commander, but expressed the certainty that workers would not harm them. He stated that there were conflicting reports about who had fired first, and refused to endorse any of them. He reported deaths on both sides, including women and children—a detail that implied a moral appeal to stop a conflict that was claiming the lives of innocent victims.[36]

Petetin concluded his report with a carefully balanced statement that was designed to position the paper in such a way that it was safe from criticism from either side and to give it a potential role in ending the conflict. "No matter how convinced we are of the legitimacy of the workers' demands, and no matter how strongly we sympathize with their misery, all we can and should say today is that it is impossible in a civilized country to tolerate armed complaints, petitions expressed through revolt." But Petetin had not been content to function only as an observer: during the day, he had a special version of the paper printed up in large type, designed to be posted on walls in the city. In it, he reiterated in emotional language his sympathy for "these hardworking families . . . always tormented by the uncertainty of the future," but restated his appeal for order. "It is a soldier of the barricades who signs this paper," Petetin's poster concluded, referring to his role in the Paris events of July 1830. "LYONNAIS! you have shown great courage; now you need to show that you, like your Parisian brethren, have a love of order, moderation and generosity after a victory."[37]

The *Précurseur*'s relatively restrained account of the uprising, clearly calculated to avoid inflaming the situation, was only one journalistic strategy. Other Lyon papers reacted more emotionally. Whereas the *Précurseur* carefully avoided placing blame anywhere, the *Journal du commerce* criticized both the authorities, for their lack of preparation, and the violence of the workers: "Words of vengeance and devastation have afflicted the ears of peaceful citizens." Rather than suggesting that the worst was over, as the *Précurseur* sought to do from the outset, the legitimist *Gazette du Lyonnais* used highly emotional language: "Civil war continues its horrible ravages among us. . . . They stab each other, they tear each other apart, blood gushes all over. Terror is everywhere."[38] Whereas it obviously served the interests of the legitimist press to exaggerate the extent of the catastrophe unleashed by the policies of the Orleanist government, the city's liberal papers were more concerned to minimize the crisis. Publishing in a city whose military garrison had fled, and in which order in the streets was being ensured by patrols organized by the silk weavers themselves, the *Précurseur* hastened to announce to the world that "seeing the order that reigns in our city, a stranger would not believe that it is the day after a civil conflict."[39] The press of all persuasions formed a unanimous chorus to emphasize one vital point: the workers had not identified themselves with a political doctrine of any sort. "Any political move in the interest of any party whatever, Carlist, republican, or Napoleonist, would find no sympathy among the men whom we have seen fighting with such great valor," the *Précurseur* announced. The *Écho de la fabrique* was equally firm on this point: "They tried to give a political point to what was nothing but misery revolting against its rags," and the *Journal du commerce* adopted the same line.[40] In the midst of the event that was about to be reinterpreted as a warning of a social danger more fundamental than any political conflict, these journalists clung to an older dogma, in which challenges to the political constitution were still seen as the most dangerous of evils. A political insurrection would necessarily provoke the government to drastic measures, as the republican uprising of April 1834 would show.

Although the workers actually gained control of the city when the army garrison withdrew, only the republican *Glaneuse* flirted with the idea of making itself a herald of revolution, and the issue in which it did so was not circulated until the possibility of success was already past. Lyon's liberal press organs omitted any mention of this possibility, moving quickly from deploring the outbreak of violence to insisting that it had ended and minimizing its consequences. The *Précurseur* had missed only one edition because of the uprising. Its issue dated 25 November 1831 emphasized the nonpolitical character of the silk weavers' movement, dismissed the Lacombe poster as "an affair of a forged signature," and announced that the city council had voted a relief fund for the wounded and their families. The following day's issue devoted itself to defending both the prefect, Bouvier-Dumolard, and the weavers themselves from responsibility for the outbreak, while the number of 27 November claimed that the workers and the silk merchants now both accepted the necessity of a "peaceable agreement." Anyone who expected the city's newspaper of record to make any further effort to elucidate what had occurred during the insurrection must have been disabused of that notion by Petetin's article on 29 November, in which he explained why he did not intend to print the many conflicting contributions being sent to the paper. In his view, the newspaper "is forbidden to take any particular role in an affair that concerns all the citizens of Lyon; it has no right to open its columns to any new quarrels, now that the war is over." Above all, he swore that he would make no effort to determine who had started the fighting. "No matter how this sinister question is presented, we will always reject it, and, glad of our own ignorance, we don't want to see in any of our fellow citizens men who plotted murder." In effect, the city's leading newspaper declared that the spectacular events of the previous days were unrepresentable: any serious attempt to reduce them to narrative form would necessarily set off further conflicts. The *Journal du commerce* was equally determined to bury the episode. Its number of 27

November 1831 marveled at the "speed with which the reign of order and peace has succeeded awful anarchy," and the subsequent number rejoiced that "the ending, after such cruel catastrophes, was much happier than one might have expected in the presence of such excesses and misfortunes."

Because it appeared only on Sundays, the *Écho de la fabrique* enjoyed the accidental good fortune, from the point of view of its journalistic survival, of not publishing an issue until the end of the week of fighting; unlike the *Précurseur*, it does not seem to have put out any kind of special edition during the insurrection. Despite its close connections with the *mutuellistes* who had taken control of the city on the 23rd, the paper was thus able to anticipate the return of the authorities and keep a safe distance from the now completed events. Its emphasis was on the tragic aspect of the movement, and its editorial stance one of dignified sorrow rather than commitment: "It is with a broken heart and the head covered in mourning for our brothers and friends that we are going to recount the events that occurred in our city; we swear before God that our tears are not exclusionary, they will be for all." The narrative of the fighting that followed, not surprisingly, put all the blame for the conflict on the bourgeois troops of the National Guard and praised the heroism and restraint of the workers.[41] Subsequent issues made a point of urging workers to obey the laws, and the subject of the uprising quickly disappeared from its columns. By February 1832, the *Écho* could virtuously denounce the parliamentary deputy Fulchiron for raising the subject in the Chamber: "Out of love for peace and unity, the workers haven't wanted to disturb the bloody ground of these events."[42] The only Lyon publication that was in no hurry to put the insurrection behind it was the legitimist *Gazette du Lyonnais*, which faithfully pursued its strategy of *politique du pire*. In the midst of the week of the insurrection, it thoughtfully reminded local readers of a law dating from the Year IV making local taxpayers responsible for damages resulting from such troubles, and rather than joining the other papers in insisting that everything was returning to normal by the weekend after the uprising, it asked "whether this painful provisional situation is ever going to end."[43]

The cautious conduct of the Lyon papers during the week following the November insurrection demonstrated that they were not anxious to play the role of genuinely revolutionary press organs, like the radical papers of the 1790s. Creations of the postrevolutionary liberal order and organs of a provincial public sphere isolated from the seat of power in the capital, the papers assumed that the workers' victory would prove temporary. Although they succeeded in communicating the fact that an extraordinary event was taking place around them, they deliberately hesitated to assign it any clear meaning. Rather than constructing an event by assigning a meaning to what was happening, they served to "deconstruct" the insurrection by emphasizing its contradictions, refusing to shape it into a coherent narrative, and insisting that the real news was the absence of any significance in the details they reported. The incoherence of the Lyon papers during the 1831 insurrection was underlined in the best-informed Paris paper, *Le Temps*. In a letter dated 28 November, its Lyon correspondent, Jean-Baptiste Monfalcon, told his readers; "Our city is still controlled by the workers; they hold power, and continue to be masters of the city hall and the prefecture. . . . One understands that in these circumstances the Lyon papers are not in a situation that allows them to express fully their ideas about the causes that led to the unhappy catastrophe of 21 November, or about ways to prevent such grave disorders in the future."[44] Monfalcon's insinuation that the papers were restrained by fear of the victorious workers was only half the story, however; even during the brief revolutionary interregnum, the journalists were all too conscious of the danger they would face when the government reasserted its authority over the city. In a situation where any words they published were only too visibly bound to be considered as actions, the city's journalists sought refuge in noncommittal formulas that concealed as much as they revealed.

In 1834, the absence of the Lyon papers in the midst of insurrection was even more literal; none of them was able to publish until the fighting had ended. In the detailed retrospective accounts that several papers published after the republicans' defeat, the city's journalists emphasized their helplessness and passivity during the crisis. "Our offices, like all the city's houses . . . , were surrounded by sentinels and soldiers. Anyone who tried to set foot in the street found a gun pointed at him," Anselme Petetin wrote. "For the same reason, we were without any information."[45] Petetin and his colleagues, anticipating the technique Stendhal would use so effectively in his fictional reconstruction of the Battle of Waterloo a few years later, reported the Battle of Lyon as experienced from the off-center viewpoint of helpless individuals forced to interpret whatever clues they could gather from their apartment windows. "The tocsin sounds, the cannon roars, and balls are whistling through the air as I write you these lines," wrote Jean-Baptiste Monfalcon in a chronicle subsequently published in the Paris daily *Le Temps*. "A violent north wind adds to the horror of the scene. . . . I don't know what is happening elsewhere, sometimes the bells are silent, sometimes the tocsin rings out. Should one assume that the republicans and the troops have kept gaining and losing control of the bell-towers? . . . Shut up in our apartment for 36 hours, we have no news of our friends and family, no one knows what has happened beyond his immediate surroundings, and all fear for the fate of those they care about. What anxiety!"[46] The anonymous editor of the legitimist *Réparateur* struck the same note, complaining that he had "no more idea of what is going on five hundred paces away . . . than if we were on the other side of the ocean."[47] Even when the fighting had ended, the papers hesitated to try to reconstruct the details. "Information is abundant; a prudence that will be easily understood in the midst of the passions that are still aroused all around us does not allow us to take advantage of it," the *Réparateur* remarked, and the *Précurseur* editorialized: "One should not exacerbate passions that are already too inflamed on both sides."[48]

Such restraint did not affect the national papers published in Paris during either insurrection. Unlike the Lyon papers, Paris journals such as the *Journal des débats* and the *National* gave only limited coverage to the events that preceded the two outbreaks of violence, which therefore burst upon their readers—but not their journalists, who regularly read the provincial press—as unanticipated shocks. In another contrast to the Lyon papers, the Parisian ones, separated by several hundred kilometers from the scene, had to content themselves for several days with fragmentary and often conflicting reports from disparate sources. The first details of the 1831 insurrection, which appeared in papers dated 25 November, came via the government, and their credibility was seriously undermined when initial claims that order had been promptly restored were followed by admissions that in fact all communication with the local authorities had been cut off. "Now that the ministry can't tell us what has happened to its prefect, its mayor, its military commanders, and who actually controls Lyon, any conjectures are possible," the

National wrote on 26 November 1831. Like the other Paris dailies, it presented readers with a confusing mosaic of reports gleaned from official sources, private letters, rumors circulating at the Bourse, articles from the Lyon newspapers themselves, and bulletins from journalists on the scene. Although the heavily politicized Paris papers normally relied on like-minded publications in the provinces when they had to cover events there, the urgency of the situation in Lyon forced Paris editors to take information from whatever sources they could get. The republican *National* included long excerpts from the Lyon legitimist paper, the *Gazette du Lyonnais*, while the Orléanist *Journal des débats* found itself relying on the oppositional *Précurseur*.

Their distance from the scene and their inability to piece together a coherent narrative of the events in Lyon did not discourage the Paris papers from imposing their interpretations on the insurrection. Indeed, uninhibited by any concern that the violence would surge into their own offices, they were quicker to accord symbolic meaning to the story than the Lyon journalists. The Saint-Simonian paper, the *Globe*, was the first to assign world-historical significance to the uprising. Its editorial of 26 November 1831 proclaimed: "Providence has brought about one of those terrible events that strikes men and makes them tremble to the marrow of their bones." Subsequent articles reiterated the paper's claim that the Lyon uprising proved the urgency of implementing the Saint-Simonian program of reconciliation between classes. The conservative *Journal des débats* immediately categorized the uprising as an "attack against industriousness," saying: "Property, the alpha and omega of social rights, has been denounced to the working classes as an odious privilege." The paper was certain that the outbreak of fighting had been caused by the workers, not by the bourgeois National Guards.[49] The *National*, the main organ of the opposition, was equally certain that the National Guard had fired first, and equally quick to categorize the movement's causes: "The state of desperation into which the working class has been plunged by a system that completely neglects the needs of the poorer [part of] the population is a more than sufficient explanation."[50] A paradoxical result of the greater freedom the Paris press had to report on the crisis was that the most outspoken writing about the insurrection penned in Lyon itself appeared in the Paris press, in the form of anonymous letters. The *National* of 27 November 1831, for example, included a vivid eyewitness account from Lyon dated 23 November, whose author, generally sympathetic to the workers, was far less inhibited in describing events than any of the Lyon journalists. According to him, the fighting on 21 November had begun when a silk merchant serving in the National Guard had disrupted attempted negotiations with the workers by crying out, "No more concessions to this *canaille!*" None of the Lyon papers printed such "fighting words" while the fate of the city still hung in the balance.

Although Saint-Marc Girardin's celebrated article in the *Journal des débats* was destined to be the most influential commentary by the Paris press on the 1831 uprising, the most detailed and comprehensive narrative actually appeared in another Paris liberal paper, *Le Temps*. The paper's anonymous Lyon correspondent was in fact one of Lyon's most important journalists, Jean-Baptiste Monfalcon, who, having failed in his bid to become editor-in-chief of the city's main daily, the *Précurseur*, was temporarily without an outlet for his writing in his home city. Monfalcon, who had been the main journalistic proponent of a hard-line policy against concessions to the silk weavers in the weeks before the uprising, used the *Temps* to continue his campaign. In his anonymous letters to *Le Temps*, Monfalcon asserted the special authority of his account by underlining his personal presence on the scene and his extensive knowledge about the city. His very first bulletin gave an almost minute-by-minute chronicle of the fighting on 21 November, specifying, for example, "6:30 P.M. I leave the army headquarters. . . . A deputation of insurrectionary workers has just been admitted." A long "backgrounder" article published several days later shifted to an anonymous and authoritative tone, purporting to explain the economic developments that had led to the workers' movement and the impossibility of countering their effects by the imposition of a *tarif*. Monfalcon admitted that the workers "had not exactly planned an armed attack" and that their conduct after their victory had generally been civilized, although he claimed that the chorus of praise for their restraint had been exaggerated. He mentioned particularly the experience of "a doctor from the Hôtel-Dieu [Lyon's charity hospital]" who had been surrounded and threatened by workers when he attempted to deliver a message to their headquarters in the Croix-Rousse and who "had been miraculously saved by a worker who recognized him."[51] The protagonist of this story was in fact Monfalcon himself, and the episode, which he subsequently recounted at least five times in published works on the history of Lyon and on his own life, crystallized for him the lesson of the workers' insurrection.[52] In his own person, or so he claimed, he had truly experienced what it meant for a bourgeois to be at the mercy of a working class in revolt.

Monfalcon's detailed contributions to *Le Temps*, though certainly hostile to the workers' movement, formed the most thorough and best-informed account of the November 1831 uprising published at the time. Because he was so focused on the specific peculiarities of the Lyon situation, however, Monfalcon's articles failed to make as much impact as Saint-Marc Girardin's article in the *Journal des débats*. Saint-Marc Girardin, a Parisian journalist who happened to pass through Lyon immediately after the uprising, succeeded in giving the event a universal significance, in good part because he was not weighed down with any real comprehension of the city's situation. Monfalcon, himself the son of a silk merchant, knew that the master silk weavers, the *chefs d'atelier*, whose organization had taken control of the insurrection, were not penniless proletarians but rather petty entrepreneurs in their own right, and this inhibited him from comparing them to slaves or barbarians, as Saint-Marc Girardin did. Monfalcon also knew too much about the internal divisions among Lyon's bourgeoisie to represent them as simply as the Parisian journalist did.

Saint-Marc Girardin chose his melodramatic terminology consciously and deliberately, with the intent of using the Lyon insurrection to change the French bourgeoisie's own conception of itself. He wrote: "The middle class needs to know the state of things; it needs to understand its own position clearly. It has beneath it a population of proletarians that acts and boils with anger, without knowing what it wants, without knowing where it is going." In comparing the workers in modern society to the barbarians at the time of the Romans, his intention was to warn the bourgeoisie against dangerous concessions to them, such as granting the right to vote or allowing them to serve in the National Guard. The Romans had "taken [the barbarians] in, saying . . . that they would defend the Empire once they had been incorporated into it. We know what happened as a result."[53] His assertion that "each manufacturer lives . . . like the planters of the colonies in the midst of their slaves, one against a hundred, and the Lyon revolt is a sort of uprising of Saint-Domingue," was even more inflammatory, conjuring up fears of rape and massacre. More than twenty years later, he still defended his rhetorical strategy. There had been, he claimed, "neither hate nor injury in this comparison"; it had simply been the most effective way of getting his message across.[54]

If Saint-Marc Girardin felt compelled to use such extreme imagery, it was in part because he had to counter a prevailing discourse that minimized the extent of the social danger. Indeed, six months before the Lyon uprising, he himself had written that "in France there is no longer either people or bourgeoisie. There are parties, sects, coteries; there are no more classes, tribes, castes."[55] At that time, his strategy had been to insist that, since the abolition of legal privilege in 1789 and 1830, the entire French population had become a single community. Whether the Lyon insurrection genuinely changed his thinking, or whether it merely convinced him that his aim of preserving the liberal, property-based society created in 1789 had to be pursued by other means, he now adopted a very different language. His vehement imagery sought to convince bourgeois readers not only that they were a group quite distinct from the lower classes, but also that their interests were those of civilization itself. By way of contrast, he simultaneously proposed a new image of this bourgeoisie's Other. Out of a specific group of silk weavers in a single city, Saint-Marc Girardin's article made a proletariat, a group to be found throughout European society. In Saint-Marc Girardin's text, this group took on threatening characteristics and world-historical significance.

Saint-Marc Girardin's article was a uniquely effective example of the press's power to mold the collective understanding of events and to propose new definitions of social identity. Historians have continued to cite it regularly down to the present day, and no other single piece of writing about France's nineteenth-century civil conflicts has had as much impact. The articles in the Paris press on the 1834 Lyon uprising were much less memorable. As in 1831, the Paris papers made Lyon their main story for several days, fitting the available news into a pattern that fit their preexisting ideological assumptions. The republican *National de 1834*, for example, assigned blame for the crisis even before it knew what had actually happened. At the moment when its issue of 12 April 1834 went to press, the paper knew only that fighting had broken out and that communication with the city had been cut off. It was nevertheless certain that the disaster was the result of despair on the side of the workers, who were infuriated at seeing their organization directly targeted by the new law against associations, and of progovernment circles bent on "a striking revenge for the days of November." Without the Lyon papers as source material, however, the Paris papers had much less to print than they had in 1831, and the Lyon story soon had to share space with coverage of the short-lived republican insurrection in Paris itself.

As in 1831, the most detailed coverage of the 1834 insurrection appeared in *Le Temps*, which was once again able to rely on Jean-Baptiste Monfalcon, the conservative bourgeois journalist with the knack for losing his journalistic outlet in his own city just when major events were about to break out. Although his personal viewpoint colored his reports, Monfalcon's bulletins, pieced together from notes jotted down in the midst of events, were remarkably effective in conveying the atmosphere of the fighting. The vividness of Monfalcon's detailed reports gave weight to his general assertions. His overall goal was to impress upon Parisian readers the seriousness of the situation: "Paris cannot imagine the audacity of the republican party and the insolence of most of the workers. . . . Either the constitutional monarchy must go under or the political associations formed to overthrow it must be destroyed." Monfalcon clung to this position even though he also admitted that "in general the working classes aren't taking much part in what is happening."[56]

Although the Paris press thus intervened powerfully in the imposition of meanings on the Lyon insurrections, they also quickly lost interest in them when the fighting ended. The *National*'s last editorial comment on the November 1831 affair was on 7 December 1831; Saint-Marc Girardin's famous article, published in the *Journal des débats* on 8 December 1831, concluded that paper's involvement with the subject. The situation after 1834 was different: the aftermath of the Lyon insurrection merged with that of the fighting in Paris, and memories of both events were kept alive throughout 1835 by the "monster trial" held to punish the alleged republican instigators of the two movements.[57] The trial, however, was a Paris event; the Lyon republican movement was reduced to an epiphenomenon of the national movement, and press coverage was concentrated in the Paris papers.

In Lyon itself, the end of the two insurrections was by no means the end of press coverage. On both occasions, the city's papers marked the end of the fighting by publishing detailed retrospective accounts of what had just happened. In 1834, in particular, they had to compensate for having been unable to publish during the struggle. Even as they made up for lost time, however, their overriding concern was to emphasize that the story was well and truly over with. The memorable events had to be both narrated and consigned to oblivion as quickly as possible. This tendency was clearest in the case of the *Précurseur*, the paper whose cause had the most to lose as a result of the fighting in both 1831 and 1834. After the latter uprising, Petetin walked a narrow line, sympathizing with the silk weavers' grievances but insisting that violent resistance to the law was not justified. Both sides had fought heroically: "Frenchmen against Frenchmen! Tragic heroism which the combatants on both sides cannot refuse to recognize." The only possible line of conduct for the future was for "all good citizens . . . to mourn together and to unite to ward off the deplorable consequences, both for the silk industry and for liberty, which the government will try to blame for our misfortunes." In any event, as Petetin stressed, his newspaper bore no responsibility: "We did everything humanly possible to ward it off."[58] The legitimist *Réparateur*, which had been able to print its papers during the 1834 fighting but not to distribute them, eventually gave a long and detailed account of the fighting, but was careful to insist that it did not know who had started the fighting—the same claim the *Précurseur* had made in 1831.[59] Eugénie Niboyet's feminist paper, the *Conseiller des femmes*, felt compelled to break its normal silence on political affairs in 1834, but its editor's heartfelt appeal was also for immediate forgetting: "May the tears of those who survived wash away the blood, and may the fact that it flowed be remembered only as a way of saving it!"[60] The *Journal du commerce*, converted into a progovernment paper by 1834, denounced that year's insurrection, but, like the opposition papers, its main message was the need to move on: "In the midst of partisan fury, let us try to spread words of union and forgetting, of sympathy and pacification."[61]

Although the Lyon newspapers shared the common goal of putting the two insurrections behind them as rapidly as possible, the magnitude of these events was great enough to inspire the appearance of other genres of journalism that challenged the monopoly on representing local affairs that the papers normally took for granted. In both 1831 and 1834, a few broadsheets combining illustrations and texts appeared soon after the events. Broadsheets differed from newspapers because they were a discontinuous medium, appearing only when something sensational took place. The pictorial elements in these representations made some information about these events accessible even to those on the borderline of literacy. Although the broadsheets about the Lyon insurrections were all the products of artists with at least some formal training, and of publishers with specialized equipment, at least some of the lithographs and engravings generated by the uprisings were intended for a popular audience, and some, such as the "Combat du Pont Morand" (Fig. 9), explicitly celebrated the workers' conduct. The caption for this engraving reads: "The workers of La Guillotière and the Brotteaux [two working-class suburbs of Lyon] took up arms to support those from the Croix-Rousse. After a bloody battle, they emerged as victors. By the moderation they showed after their victory, they astonished Europe and gave it indisputable proofs of their honesty and their respect for the laws."[62]

In the newspapers, the workers who participated in the two insurrections were almost always anonymous, and even sympathetic journalists tended to describe events from the bourgeois side of the barricades; even the authors of the *Écho de la fabrique* had to maintain the impression that they had not been involved in the fighting. In contrast, several of the broadsheet illustrations, including the one reproduced here, are drawn as if seen from the workers' side of the barricades or explicitly show uniformed soldiers fleeing in defeat.[63] In depicting the Lyon uprising from the workers' point of view, artists were following a set of pictorial conventions adopted in popular prints of the July 1830 revolution, but the effect of this was to put the silk weavers in the position of the victorious "people" and thereby to legitimize their actions.[64] The broadsheet illustrations showed individual workers in characteristic dress, and "Combat du Pont Morand" showed an identifiable individual, "Stanislas the Negro," a black man whose role in the November 1831 events is described in several period sources.[65] This difference in perspective reflected the differing nature of the two media. Whereas newspapers, as ongoing enterprises, had to act as though they expected the legal order to endure and therefore had to represent events as framed by that order, each broadsheet was an isolated publication and therefore enjoyed greater freedom. As a result, these visual images could show insurrectionary events in ways the print press could not. On the other hand, however, broadsheets did not serve as ongoing stand-ins for social groups in the way periodicals did. Their impact is difficult to measure because we know virtually nothing about their circulation; the prime minister

Fig. 9 "Événements de Lyon, Combat du Pont Morand." Following conventions used in illustrations of the fighting in Paris during the July Days of 1830, this anonymous artist shows the November 1831 insurrection from the viewpoint of workers from Lyon's suburb of Brotteaux firing across the Rhône at the warehouses of the silk merchants near the place des Terreaux. The isolated figure in the central foreground is probably "Stanislas the Negro," whose role in the fighting at the bridge is mentioned in several sources. Like most of the surviving broadsheet illustrations, this one celebrated the workers' victory, stating, "They astonished Europe and gave it unequivocal proofs of their honesty and their respect for the laws." It was offered for sale both in central Lyon and in the Croix-Rousse, the workers' stronghold.

Casimir-Périer and Gasparin did exchange letters about the importance of repressing the images circulating after the November 1831 uprising.[66]

Pamphlets, although they were, like newspapers, printed texts, enjoyed some of the freedom of broadsheets. They also had the advantage over the newspapers of being able to narrate the story all the way to its end. Like the newspapers, pamphlets had distinct and usually quite explicit ideological points of view.[67] Particularly in the aftermath of the 1834 uprising, when the authorities arrested all the leading republican and legitimist journalists in Lyon, pamphlets became the refuge for polemics that could no longer appear in the newspapers. "Neither the truth nor the harsh reflections that it brought with it were welcome" after the insurrection, one anonymous pamphleteer charged. Even though the besieged newspapers had done their best to defend their points of view, "the press was not able to say everything."[68] The pamphlet genre allowed the city's leading woman journalist, Eugénie Niboyet, to discuss a subject she could only touch on in her *Conseiller des femmes;* in an anonymous pamphlet, she was able to express her horror at the violence the insurrection had unleashed. "It is a spectacle that one has to have seen in order to believe; that's how much this armed struggle has changed men. For one side, it was enough to see a worker's clothing, for the other, a soldier's uniform, to aim and shoot, and yet these are Frenchmen fighting on both sides, men who supported the revolution of July." She noted that it had truly been a men's struggle: under the state of siege maintained even after the fighting ended, "only women circulate in the city, the men are still prisoners."[69] Pamphlets thus served to express reactions to the insurrections that the press excluded; they also kept discussions of them alive for some months after the newspapers had turned to other subjects. As in the case of broadsheets, their impact is difficult to assess, because they too created no lasting bond among their readers.

In 1834, in contrast to 1831, there was one local paper that felt free to comment extensively on the significance of the events for some time afterward: the *Courrier de Lyon,* which had helped incite the confrontation, called the bloody suppression of the uprising "a great measure of public security" and urged new legislation to prevent any future subversive movements.[70] The *Courrier,* the most combative and class-conscious bourgeois paper, thus integrated the uprising into its overall ideological strategy. This was a logical consequence of the paper's own history, and its special relationship to the previous insurrection. The paper's original prospectus, circulated in December 1831, had justified its creation by condemning all the city's existing publications because they had "helped spread the flames lighted by ignorance and popular passions." Plans to create the *Courrier* had been under way even before the uprising; it was a project of the conservative faction among the *Précurseur*'s stockholders and contributors, those who had wanted Monfalcon to become the paper's editor rather than Petetin. But references to the November events provided a clear justification for the new title, and the *Courrier* evoked them whenever it referred to its own history. Over the years, the fact that "the ground still trembled under our feet" when it had first appeared allowed the paper, despite its identification with the government, the chance to present itself as a heroic venture launched in a time of peril.[71] The *Courrier de Lyon* was also the local paper that was most insistent on invoking the memory of the November uprising in subsequent moments of tension. At the time of the city-wide silk workers' strike of February 1834, the paper called on strike supporters to remember that "M. Dumolart [the conciliatory prefect of 1831] is no longer around, and the lesson of the past has not been forgotten."[72]

Whereas the *Courrier de Lyon*, alone among the Lyon papers, regularly summoned up the memory of the insurrections to justify its very existence and to imbue its readers with the appropriate mixture of anxiety and militance, the workers' press developed their own formula for prolonging memory of the November 1831 uprising and shaping their readers' attitudes. As we have seen in an earlier chapter, the *Écho de la fabrique*'s issue of 25 November 1832 established a tradition of annual commemoration that was repeated in the local workers' press until its suppression in 1835, and one that anticipated the construction of memory in the national workers' movement for decades after the Paris Commune uprising. Until the 1835 press laws silenced them, the workers' papers thus tried to institutionalize a certain memory of the November events, one in which the workers appeared as innocent victims whose sacrifice was a continuing reproach to society, but one that would some day bear positive fruit. Lyon's most "class-conscious" newspapers thus strove to keep the memory of insurrection alive; for both the *Courrier de Lyon* and the workers' papers, these events had revealed the hidden nature of modern society and defined the historic roles these papers wanted their audiences to take on. In contrast, those publications that were less eager to see French society defined in terms of mutually antagonistic classes tried to avoid evoking the insurrectionary events that had done so much to legitimize such a conception.

Before, during, and after the fighting, the press as a whole thus played a vital role in the construction of the revolutionary insurrections of the early 1830s and the definition of their significance. The insurrections in turn changed the nature of the press. The 1831 uprising accelerated Lyon's post-1830 "media revolution," giving the working-class press an essential position within the city's "journalistic field" and generating a niche for an equally class-conscious bourgeois publication. The defeat of the 1834 uprising had as its sequel a government-imposed restructuring of the journalistic system, leading to the disappearance of the workers' papers, the republican organs, and of the city's lively alternative nonpolitical press. The authorities clearly understood that a multivocal press was inherently an element of disorder, an obstacle to imposing any stable set of social identities. Their repressive efforts were only temporarily successful; even in Lyon, despite a limited market and tight surveillance, the press system never returned to the simplicity of the days before 1830. Throughout the decade and a half leading up to the Revolution of 1848, the city continued to have a proregime paper, a crypto-republican paper, and at least one legitimist title. As we have seen, workers' papers in the tradition of the *Écho de la fabrique* resurfaced after 1840, and Léon Boitel's *Revue du Lyonnais* represented the alternative cultural press.

Although the experience of the two insurrections was thus fundamental to the development of the press after 1830, the nature of the newspaper medium militated against a continual focus on the subject in the calmer atmosphere after 1834. Political banquets and political trials were wars of words, perfectly suited to fill the columns of the newspapers and reinforce journalists' sense of their own importance. Violent insurrections were products of a different order of reality. Newspapers, apparently so well suited to textualize them, in fact proved ambivalent about doing so. The Revolution of 1789 had created its own press, one that spoke from the midst of the revolutionary whirlwind; the papers of the 1830s, anchored in the legal order and the order of discourse produced by the "bourgeois" Revolution of 1830, rejected such a role. Their highest priority was to maintain an order of things in which they would retain their central role—an order of things in which debate carried on in the public sphere determined the direction of public policy. The newspaper, with its predictable format and regular periodicity, conveyed a message of stability even in the midst of the unpredictable violence of insurrection. This was particularly evident in the responses of the Paris press to the two Lyon insurrections. Safely removed from the scene of the action, the national papers had little difficulty fitting these outbreaks into the preestablished category of sensational and violent news events. They were equally quick to pin appropriate ideological interpretations on what had happened, even before they knew how the story was going to end.

The difference in the responses of the Lyon and Paris papers to the experience of insurrection reflected the subordinate character of Lyon's public sphere, which could not impose meanings on events with the same power as its national analogue in Paris. Paradoxically, it was only when events of truly national significance occurred in their own city that the journalists and newspapers of Lyon lost their ability to determine how they would be represented. The difference in the response of the two sets of newspapers was also due to the fact that Lyon's press system had truly found itself in a revolutionary situation, in which the nature of the press itself was necessarily at stake, whereas the country as a whole had remained stable. The inability of the Lyon press to reduce the city's two most spectacular collective experiences of the early 1830s to orderly journalistic narratives reflected the collapse of the distinction between words and actions characteristic of a revolutionary situation. At the time of the revolutionary rupture in Paris in 1789, a similar situation had led to the rise of a new press that became one of the most powerful agents in the creation of a new public order. In Lyon in the 1830s, all parties had realized that this could not happen so long as the national government remained capable, sooner or later, of reimposing its rules. As a result, the press had been a powerful indicator of the fact that, even as violence imposed its mark on the city's people and buildings, there would be no genuine revolutionary upheaval.

Although they did not become insurrectionary newspapers, the periodicals of the early 1830s did serve to propagate a tradition that linked revolution and the pattern of social identities that the press itself had done much to define. Most other journalists eschewed Saint-Marc

Girardin's explosive imagery, but all of them accepted the basic argument underlying them—namely, that European civilization was endangered by a conflict between two potentially hostile social classes. Depending on their ideological outlook, journalists could interpret the two Lyon insurrections as urgent warnings of the need for social reconciliation, or as proof of the need for a more vigorous defense of class interests; the insurrections could be used to argue for political democracy, or for a return to Catholic monarchy. Even in periods when the papers' content did not explicitly evoke the danger of violence, the polarized configuration of Lyon's journalistic field was a constant reminder of that danger. Journalists had only to make the briefest allusions to past events to evoke highly charged memories. Inherently engaged in conflictual relationships with one another, rival newspapers with such distinctive social identities were constantly reenacting the confrontations that had led to the battles of 1831 and 1834. It was left to other kinds of publications to take up the task of mastering and exorcising the cleavages the press had constructed out of stories of insurrection.

7

From Newspapers to Books

The Recasting of Revolutionary Narrative

Turning News into History

Inherently wedded to the ongoing flow of events, newspapers necessarily shifted their attention to new stories after the tumultuous years of the early 1830s. The traces of that revolutionary episode were incorporated in the new structure of the press system, but specific references to the events of the period soon faded. The representation of even such spectacular occurrences as the two Lyon insurrections and the consolidation of the new forms of social identity generated in the early 1830s became a task for other media. The way in which these stories were recast in other, less ephemeral genres reveals much about the process by which social memory was institutionalized once the revolutionary moment of the early 1830s had clearly ended. The contrast between the journalistic narratives of the period and the other forms of writing that subsequently took their place throws light both on the press and on the functions of such genres as history, autobiography, and the novel.

Of the book-length publications generated by the Lyon events, two merit special attention. The first was Jean-Baptiste Monfalcon's *Histoire des insurrections de Lyon,* published just two months after the April 1834 uprising, which immediately established itself as the most authoritative account of the city's dramatic history since the July Revolution. It retained that status until the appearance of the archivally based historical monographs of Fernand Rude, Maurice Moissonnier, and Robert Bezucha in the mid-twentieth century. Using the apparently neutral apparatus of the period's historical and social science, Monfalcon told the story of Lyon's troubles in a way that made the triumph of the bourgeois order appear both inevitable and morally justified. Although the *Histoire des insurrections de Lyon* offered what appeared to be a stable and closed representation of the revolutionary episode of the early 1830s, its author went on to produce several alternative narratives of the same events, which shows that he himself was aware that he had not succeeded in containing its full meaning in his initial account. Monfalcon's continued wrestling with this story showed that it remained problematic, even when it was put into the framework of "bourgeois" literary genres. That lesson was also demonstrated by a novel published anonymously in 1835, *La révolte de Lyon, ou La fille du prolétaire,* which claimed to tell the story from the point of view of the workers; it may well be the earliest French "social novel." Whereas Monfalcon's history found a modest but enduring niche in libraries, *La révolte de Lyon* disappeared virtually without a trace. It prefigured important literary strategies for the representation of social reality that would subsequently be developed by more talented writers, however, and it raises important questions about what happened when authors tried to employ the literary genre of the novel to give a voice to the lower classes.

After the 1834 uprising, Jean-Baptiste Monfalcon, who had been assistant editor of the *Précurseur* in 1831 and then editor of the *Courrier de Lyon* from 1832 to early 1834, as well as Paris correspondent for the Paris daily *Le Temps* during both the insurrections, accepted a local publisher's invitation to write a history of the sensational events that had rocked Lyon since 1830. Monfalcon separated himself from the other journalist-pamphleteers in the city by the scope of his project and by his deliberate reconceptualization of the story. Even before the 1834 uprising had ended, Monfalcon had framed the two insurrections as connected episodes in a larger narrative. In the first of his bulletins about the 1834 revolt for the Paris paper *Le Temps*, he lamented the effect that the previous uprising had had on the weavers. "The victory in November [1831] gave Lyon's workers a pride and an audacity that made it impossible to deal with them. It will be for centuries the favorite story in the workshops, tradition will transmit this great event from generation to generation, and the last weaver will enthusiastically tell the last apprentice: 'Once upon a time we drove out the silk merchants and the garrison, and took the city.' "[1] His intention from the start was to recast the November uprising as part of a larger story whose lesson would be the reverse of the one he feared the workers had formulated for themselves.

Although Monfalcon certainly understood the importance of narrating the Lyon insurrections in a way that would justify the Orleanist regime and the liberal bourgeois social order, his account was the initiative of a private citizen, a situation he took full advantage of. His friend the prefect Gasparin had begun to compile an official report after the 1834 uprisings, but had been discouraged from publishing anything by his superiors in Paris.[2] Monfalcon profited from this situation to enlist Gasparin's active cooperation in his own project; in return, he let the prefect read over the proofs of the book and suggest corrections. At the same time, however, Monfalcon did not hesitate to remind Gasparin that he had other sources of information, even telling him that he had been given copies of secret reports addressed to the prefect without the latter's knowledge. When Gasparin questioned Monfalcon's conclusion that the number of republican combatants had been minimal, Monfalcon reminded him that his status as a physician had allowed him to visit their headquarters during the fighting, an opportunity the prefect had not enjoyed. He defended his right to include details that were embarrassing to some of the public officials involved in the insurrections, such as his claim that the commanders of the Lyon garrison had nearly decided to abandon the city in April 1834, as their predecessor had in November 1831. "I have some curious details about the war council of the 10th, where there was serious discussion about the evacuation of Lyon. . . . If I could tell everything, I would write some interesting pages on this strange discussion, . . . but one must not give weapons to the republic and hopes to a new insurrection. I'll take refuge in oratorical precautions," he told Gasparin.[3] Despite his close connections to officialdom, Monfalcon thus claimed the historian's prerogatives of omniscience and independence.

In shifting from journalism to history, Monfalcon moved from one of the least esteemed forms of nineteenth-century literature to one of the most prestigious.[4] Journalists wrote for the moment; historians wrote for the ages. "I wanted to write a work that would endure as a historical production," Monfalcon told Gasparin.[5] Newspapers could not escape from their involvement in a competitive field of partisan rivalries; by writing as a historian, Monfalcon could hope to create a narrative that would achieve hegemonic status. In the introduction to his book, Monfalcon justified his venture by saying: "Lyon's upheavals, and the numerous industrial and political questions connected to them, have not yet been the subject of any complete and impartial history." His claims to have achieved completeness and impartiality were meant to emphasize the distinction between his book and the newspaper and pamphlet accounts that he intended to supplant. The other accounts that had appeared, he asserted, were both biased and inaccurate, and, above all, they failed to address the question of "the causes that have led, step by step, to the complete overturning of the country's most important commercial city" (ix–xi). He could achieve factual completeness, Monfalcon claimed, because "the special circumstances in which I have found myself since the July Revolution have allowed me to follow closely the movement of men and events." He had kept daily notes on events as they unfolded and had consulted relevant documents, some of which he inserted into his narrative. If readers still had any doubts about his qualifications to tell the story fairly and completely, he reminded them: "This essay is written in Lyon, in the presence of all the witnesses to the events, in the face of rival parties that would be interested in publicizing any errors. It would therefore have been difficult for me to err, even if I had been willing to sacrifice my conscience to promote my opinions" (ix–xii).

In his preface, Monfalcon explicitly claimed the status of a historian: unbiased, scrupulous in searching out facts, willing to sacrifice his own presuppositions to the lessons of his research. In short, he presented himself as a participant in the broader nineteenth-century effort to create a culture based on objective, scientific observation.[6] When he dealt with the Lyon press in the course of his history, Monfalcon discreetly identified the editor of the *Courrier de Lyon* only as a "Monsieur M.," a ruse that would not have misled any reader in Lyon, but one that probably sufficed to obscure for most Parisian readers at the time, and for most subsequent readers wherever they have been, the fact that the author of the *Histoire des insurrections* was describing a series of events that he had been intimately involved in shaping.[7] Informed readers in Lyon found this self-portrait hard to swallow. They knew Monfalcon as the outspoken editor of the *Courrier de Lyon*, the newspaper that, in February 1834, two months before the violent April uprising, had editorialized that "an actual insurrection would be a thousand times better than this continual fear of one, we would know where we stand, and at least we would rapidly settle matters."[8] Monfalcon's fellow journalists found his pretensions infuriating. "It would take a volume twice as long as the one you have written to point out all the false assertions, the misrepresentations, omissions, and perversions of the facts, the odious calumnies, the underhandedly dishonest insinuations, the ridiculous reasoning of which your book is full," the republican editor Anselme Petetin wrote, and the legitimist *Réparateur* complained of Monfalcon's "revolting partiality."[9] But these critiques, discredited by the obvious passion of their authors, remained buried in the quickly forgotten pages of their newspapers, unable to stop the success of Monfalcon's book. That success was not entirely undeserved. As Monfalcon's correspondence shows, he had made a serious effort to document the roles of the major actors on the government's side. He consulted, among others, the prefect, the public prosecutor, the military commanders, and the head of the local hospital for statistics on casualties caused by the fighting.[10] He had read the opposition newspapers, and he had circulated among the republican fighters during the April 1834 insurrection. In the preface to a subsequent historical work, Monfalcon would castigate an eighteenth-century predecessor for "treating history like a novel" by "putting imaginary speeches in his characters' mouths, . . . making them say what he would have said in their place."[11] By contrast, his own work relied on primary sources, some of them incorporated directly into the text, and followed a strict scientific protocol.

In addition to his consultation of witnesses and documents, Monfalcon was able to give a convincing description of the Lyon silk weavers' milieu. As a doctor at Lyon's charity hospital and at the Perrache prison, he often had firsthand contact with the city's poor. Indeed, Monfalcon had to some extent established his career on the basis of his claim to special knowledge about the Lyon working class. As a young physician, he had sought to distinguish himself by his scholarly publications, one of the first of which had been an account of "the illnesses of silk workers," reprinted several times during the Restoration; for the remainder of his career, he recycled descriptive passages from this account in his journalistic and historical writings. Monfalcon's article exhibited the split vision of the working class he would retain throughout his life. He was fully aware of the physical effects of the conditions the silk workers lived under, which produced "pallor, emaciated limbs, or limbs swollen by lymphatic fluids" and *"niaiserie"* or simplemindedness. But he also attributed their degraded condition to their own behavior: "The corruption of morals is very great among them." His conclusion mixed both registers: the solution to the workers' problems would be "to get them away from the powerful influence of living in an unhealthy situation, to organize their pattern of life, and to give them morals."[12]

The one thing that most clearly separated Monfalcon's early descriptions of the silk workers from his writings after the insurrections was his assessment of their ability to organize collectively. In 1820, he had written, "While the workers of Manchester give themselves over with great violence to extremely reprehensible excesses whenever manufacturing falls off, the eighty thousand Lyonnais silk workers that the inactivity of the looms reduces to misery don't commit any disorder, and oppose poverty only with the force of inertia."[13] The events of the early 1830s had decisively disproved this claim, but they had, to Monfalcon's mind, merely underlined his earlier assertions that no good could come, either for the workers themselves or for society in general, from having a large, poverty-stricken population of dubious moral character concentrated in an urban area. He had drawn on his familiarity with the city's population repeatedly in his journalism, published in the *Précurseur* before his departure from its staff in the fall of 1831, and then in the militantly bourgeois *Courrier de Lyon*, and he had no hesitation about doing so again as he shifted to the mode of history.

To make his work convincing as history, Monfalcon realized that he had to recast his materials, even though he drew heavily on his earlier publications. In his autobiography, written almost twenty years later, Monfalcon recalled that at first the task of compiling a history seemed simple: "Everything was at hand; it was simply a matter of coordinating my articles from the *Courrier de Lyon* and my letters to *Le Temps*." He quickly discovered, however, that "what might have been fine as an ephemeral article written under the impressions of the moment in a newspaper could be worthless as a chapter of a book."[14] The restricted view from his apartment window, which had added so much intensity and authenticity to his newsletters on the 1834 uprising, was insufficient for a historical account. The credibility of his history thus came from its claim to be a total narrative based on a wide variety of sources, rather than from its being based on his own personal observations. As he transformed his story into its new form, he felt compelled to fill in background, describe major characters such as the prefect Gasparin and the mayor Prunelle, provide information about the various political movements in the city, and do additional research to answer questions he had not dealt with the first time. The most important difference between the book and the journalism that had preceded it was that Monfalcon could now present the whole sequence of events from 1831 to 1834 as a closed narrative with a clear lesson. The two insurrections "are two parts of the same event: the action and the reaction," and their significance was already fixed. The sense of the different units of narration woven together in the book came from their place in this larger scheme, and Monfalcon could allow himself to state that the narration of "occurrences isolated from their causes" (a good description of what he had provided in his best newspaper articles) "teach[es] little" (2).

In Monfalcon's newspaper articles, time had been structured by the spectacular events that constituted news. Monfalcon's occasional "backgrounders," in which he introduced sociological and economic descriptions of the *fabrique* and the workers, appeared only after the description of current events that justified his presumption of readers' interest in the subject. In his history, he reversed the sequence. The first chapters constituted a social and economic history of Lyon, starting with the Middle Ages and fleshed out with a wealth of detail that he had never been able to introduce into his journalistic writing. Monfalcon's *History* was one of the first works to approach a historical problem in this way, suggesting that the concrete details of a popular group's living situation explained its collective behavior; despite its polemical hostility to the workers, it foreshadowed later Marxist and *Annales*-school historiography. Like the physician that he was, Monfalcon in his history emphasized the causes of the social malady revealed by the uprisings, rather than the spectacular symptoms that had triggered his newspaper articles. Monfalcon concluded: "The grave and complicated issue of Lyon's industry . . . is a problem without a solution, in the sense that the worker's income is sometimes insufficient for his needs and that the state of the *fabrique* makes it impossible to increase what is paid for labor" (35). Nevertheless, it was worse than pointless to stir up agitation against the wealthy among the workers: "All our efforts should tend toward cementing the union between these two classes" (41). He then turned to a detailed narrative of the November 1831 insurrection. His story had none of the idiosyncratic firsthand observations that had created an *effet du réel* in his journalistic writing. Like a classic historian, he wrote as though he had been everywhere at once. He was now able to describe the simultaneous actions of the prefect, the military commander, and the silk workers' commission that had taken charge of the city for several days.

The middle section of the *Histoire des insurrections*, covering the period between the two uprisings, corresponded to the period when Monfalcon was editor of the conservative *Courrier de Lyon*. Drawing extensively on the articles he wrote for the paper, he described the gradual fusion of republicanism and the working-class protest movement. Monfalcon paid close attention to the role of the local press, attributing to it a leading role in transforming the city's political culture between the two insurrections. His pose as an objective historian allowed him both to acknowledge that the local press had influenced events and to discredit the way the papers had represented those events, by calling their interpretations partisan and misguided. His judgment of the moderate republican *Précurseur*, for which he had once written, was relatively evenhanded, but his verdicts on the more radical *Glaneuse*, the workers' papers, and the legitimist press were unequivocally hostile (132–33, 143, 145, 150–54). The *Histoire des insurrections* culminated with the story of the insurrection of April 1834, but its account of this dramatic event differed from Monfalcon's hour-by-hour bulletins in *Le Temps*. He did present himself as an eyewitness to the first outbreak of fighting on the morning of 9 April, but after that his historian's narrative took a very different direction from the journalistic account he had penned only a few weeks earlier. As a journalist, he had not known how much of the city the insurgents had managed to seize; as a historian, he knew that, at the end of the first day of fighting, "the insurgents had been cut off, concentrated in several streets of the central city, and prevented either from receiving aid from outside or from communicating among themselves." He now knew also that the military commanders, in spite of their success in containing the uprising, had considered evacuating the city, as they had in 1831, and he paused to meditate on the catastrophic consequences such a withdrawal might have had (229n, 239, 248, 250). After describing the defeat of the insurrection, Monfalcon turned once again to his analytic mode. He weighed the responsibility of the various groups involved in the fighting, considered the morality of the army's tactics, compared the April uprising to its predecessor in 1831, and commented on the role of leading personalities and the city's newspapers. He ended by drawing what he saw as the lessons of the disaster: the city needed a better-organized police force, and workers needed to be taught that they could use only peaceful means to improve their condition. "Effort and foresight, that is the slow but sure way that leads the worker to well-being" was Monfalcon's sententious conclusion (334).

Monfalcon's account of the insurrections was clearly a "calculated and distinctly partisan version of the events," as the twentieth-century American historian Robert Bezucha has written.[15] Despite occasional reminders that "it would be imprudent to deny to oneself the deep malaise of the poorer classes, and unjust to not take rapid action to improve their condition," he insisted that "the distress of the Lyon workers has been ridiculously exaggerated." Monfalcon vigorously challenged the claim for a homogenous working-class identity put forward by the *Écho de la fabrique*, pointing out the divisions between master weavers and journeymen, male and female workers, young and mature ones, and even between the two rival working-class papers in 1833 and early 1834. He claimed that the leaders of the organized protest movement were far better off than the mass of journeymen and women workers. "The Lyon silk worker is not dependent on the merchant, like someone hired for wages or by the day. Owner of his looms and tenant of his apartment, he is a free man and complete master in his own home" (33). If these workers felt underpaid, it was because they had acquired exaggerated expectations: "The master worker in Lyon does not live as his ancestors did; he has developed habits that they never dreamed of, a taste for the theater and for costly pleasures. He hangs out in cafés, and is usually elegantly dressed on Sundays" (40–41, 36–37). Whatever their condition, workers needed to understand that it was the result of inexorable economic laws and foreign competition, not exploitation by the wholesale silk merchants. They could not blame the existing order, because they had been "emancipated for a long time," and they had only to practice the virtues of "thrift, work, effort . . . to move into the class of the well-off." And, of course, violence would get them nowhere: "Every time violence has promised a redistribution of goods, this manner of establishing equality of fortune has ruined those who employ it" (39, 40).

Although he clearly opposed any assertion that the working class was oppressed, Monfalcon made his claim to impartiality credible by sprinkling the *Histoire des insurrections* with critical comments about members of the bourgeoisie. As entrepreneurs, members of the bourgeoisie should realize that they were the only truly productive class, the one on which all others depended: "The real producer is not the worker; he carries out the entrepreneur's idea, as a mason carries out the architect's idea." But this class was not aware of its true interests: "There is no real cohesion or unity among these entrepreneurs, and their class hardly understands its real interests any better than the workers do theirs" (45, 46). The rest of the middle classes were no better: "In general, . . . the men who support the *juste-milieu* in Lyon are egoists, concerned above all about their tranquillity, unwilling to put themselves forward when the occasion demands it, ungrateful to those who take risks on their behalf, and not overly endowed with courage for action" (101). The only manifestation of the local bourgeoisie that Monfalcon praised was the *Courrier de Lyon*, which he had edited, although he conceded, "This newspaper could have been better" (160). Monfalcon's comments on the local authorities, with the sole exception of his friend the prefect Gasparin, were equally critical: the police force was divided and incompetent, the mayor Prunelle had provided no leadership during the two insurrections, the bourgeois members of the local jury had failed to do their duty, and the army commander in 1831 had abandoned the city and his successor in 1834 had nearly repeated his error (172, 161, 121, 248–50). Monfalcon could thus insist, with some justification, that he had been as hard on the bourgeoisie and the authorities as he had been on the workers.

Although there was therefore something in Monfalcon's account to offend everyone, there were also elements in his narrative that laid the foundation for its success. Even as he denied the existence of a distinct working-class identity in Lyon, Monfalcon emphasized what the city's residents shared. The two insurrections had highlighted the uniqueness and importance of Lyon. "From now on," Monfalcon concluded, "they will fill some of the most interesting pages of our city's annals, and of the history, so worthy of interest, of the French in the nineteenth century" (334). His book was intended above all to correct Parisian ignorance about what had taken place: "In Paris itself, public opinion knows little about how the silk trade is organized; it doesn't really understand the terms *entrepreneur* and *workers*" (ix–x). Whatever the issues dividing the inhabitants of Lyon, they could all take a certain pride in the notion that their city's insurrections had made Lyon the center of national attention. Paradoxically, Lyon's civil conflicts thus became building blocks for a distinctive local consciousness that united the population against the capital. As a recent historian of the city has written, "If the canuts failed to gain anything in 1831, Lyon, on the other hand, made considerable progress in constructing its political identity."[16]

In spite of the hostile reception it met with from the Lyon press, Monfalcon's *Histoire des insurrections* had the necessary qualities to establish itself as bourgeois society's accepted account of these disturbing episodes. Monfalcon's detailed description of Lyon's working-class milieu refuted simplistic claims about the weavers' misery; it foreshadowed the sociological investigations of Villermé and others, who broadened the focus to France's other cities but followed Monfalcon's lead in mixing careful observation with moralistic judgments. Villermé, in fact, quoted extensively from Monfalcon in his survey of the condition of the French working class, published in 1840.[17] Monfalcon's narrative of events was so much more complete and convincing than any other source that even writers who vigorously disagreed with his views, such as Louis Blanc, relied on and sometimes simply plagiarized from his work for details about Lyon.[18] Because of its continuing interest as social history, the book was reprinted as recently as 1979.

Although he made some deprecating comments about the work in his

Fig. 10 Jean-Baptiste Monfalcon (1792–1874). The political writings of Jean-Baptiste Monfalcon made him appear to be the personification of Lyon's Orléanist bourgeoisie. His autobiography reveals his personal links with the city's silk-weaving population, as well as the intensity of the personal experiences that drove him to write obsessively about the conflicts of the early 1830s. This undated portrait suggests another aspect of his personality, his desire to be taken seriously as a cultural figure; during the early 1830s, he spent time preparing elaborate multilingual editions of the works of several classical authors.

autobiography, Monfalcon recognized the importance of what he had accomplished in providing a representation of the insurrections that taught the bourgeoisie the necessity of taking proper measures to defend itself, taught the workers the necessity of accepting their fate, and taught the community of Lyon to recognize the critical role it had played in the history of modern society. Throughout the rest of his long career as a local notable, Monfalcon returned obsessively to those themes. It is not surprising that he inserted long excerpts from the *Histoire des insurrections* into several subsequent works on the city's history.[19] Two other versions of the story he produced raise more interesting questions, because they show that he himself realized that his ostensibly coherent narrative had neither exhausted the subject nor provided a definitive representation of it. One of these accounts was supposedly addressed to working-class readers; the other transformed the story into the form of autobiographical recollection.

Two years after the publication of the *Histoire des insurrections*, Monfalcon published a far larger volume entitled *Code moral des ouvriers, ou Traité des devoirs et des droits des classes laborieuses*.[20] The title suggested a work intended to be circulated among the working classes, with the purpose of inoculating them against dangerous ideas, although the form of the book—"a veritable deluxe edition," as one reviewer commented[21]— hardly lent itself to such a project. In reality, Monfalcon had put the work together in order to build his own reputation. He had a limited edition of only thirty-five copies printed, which he circulated privately to a list of dignitaries, starting with King Louis-Philippe; in return, he received both the Prix Montyon of the Académie française and an award from the Royal Academy of Gard. The first half of the *Code moral* was devoted to an elaborate recital of the usual liberal clichés about the need for workers to better themselves through application and effort, and the impossibility of granting them political rights. Much of the second half consisted of a curious rewriting of Monfalcon's earlier book on the two insurrections, in a form that Monfalcon imagined would make it palatable to a working-class audience. Monfalcon's inspiration was to tell the story in the form of a dialogue between two fictitious workers, one of whom supposedly participated in the two uprisings.

Although his fictitious workers spoke a distinctly educated French—he managed to incorporate long paragraphs from his *Histoire des insurrec-*

tions directly into the dialogue—Monfalcon allowed his characters to make an articulate case against the bourgeois order. His invented insurgent spoke warmly of the revolutionary doctrine that the workers had imbibed by 1834, which promised them that "labor, glorified, would be the fundamental form of wealth; the capitalist landowners and moneyholders would be on their knees before labor; it would be their turn to beg, . . . and the worker would dictate his conditions," language that showed Monfalcon's accurate understanding of the appeal radical ideas had for them. He also understood the emotional attraction of revolt: "There is, I know, something about a struggle with power that makes a man feel bigger and makes his heart beat with more energy." It is true that the dialogue form also allowed Monfalcon to put into the mouths of his working-class characters speeches blaming republican agitators for misleading them and concluding sadly that they would never have been able to put the silk industry back on its feet if they had won, and he was able to attribute to his invented workers the realization that "long suffering will be the price of these illusions of a day!"[22] Despite the reassuring nature of his conclusion, Monfalcon had demonstrated that the story of the Lyon insurrections could in fact be told in a very different manner from the way he had constructed it in his *Histoire des insurrections*. In a generally favorable review of the work, a committee of members of the Royal Academy of Lyon praised his intentions but wondered whether it would actually attract a popular audience: "It probably is not in a sufficiently appropriate form for this hope to be realized."[23] Whether the Academy's committee reached this conclusion with disappointment or relief is unknown.

Having told the story of the insurrections from the point of view of the established order, and having tried to imagine how it might have looked from the other side of the barricades, Monfalcon was still not ready to abandon it. Some twenty years after the events, he created a third and much more personal version of the story in his *Souvenirs d'un bibliothécaire, ou Une vie d'homme de lettres en province,* an unpublished autobiography that he had duplicated in a small number of copies.[24] Monfalcon did not renounce the doctrinaire liberal views he had put forward in his earlier writings, but he shed a new light on his relationship to the nineteenth century's class conflicts by revealing the complexities of his own identity and his unique experiences during the troubles of the 1830s. The

Souvenirs revealed that Monfalcon was himself the son of a silk weaver, albeit one who had managed to achieve economic independence by selling his own products rather than working exclusively for wholesale merchants. Monfalcon's father had been the calculating *homo economicus* his son urged the silk weavers of the 1830s to emulate: "In his whole life, he had only a single idea, making money, and his entire intellectual and moral existence had been tied up in the details of his business." His family experience testified to the fact that hard work could lead to prosperity, but Monfalcon's comment on his father shows that he was also aware of the price that had to be paid for such success. Pushed by his mother, Monfalcon had risen above his father's station, but he was aware that he too had paid a price for his success. It had evidently estranged him from his parents, and he counseled workers to "make your children workers: they will be better off, and you as well."[25]

Monfalcon's narration of his own rise to middle-class status both confirmed and subverted the bourgeois pieties with which he had larded his historical writings. He had worked hard to get through medical school and to make a name for himself, but his great breakthrough, as he humorously recounted, came through a bit of fraud. Hoping to be hired to contribute to the prestigious *Dictionnaire des sciences médicales* launched by the Panckoucke publishing house in 1812, but realizing that, as an unknown twenty-year-old medical student, he lacked the qualifications, he made up a story about an uncle, retired and living in the provinces, who was willing to submit some sample articles for free. The editors liked the "uncle's" work and before long had offered him a regular contract. Monfalcon was able to leave his student garret for a respectable apartment and happily "gave up the dinners priced at one franc, the hallmark of students who didn't have the gumption to make up an uncle."[26] After completing his degree, he returned to Lyon, married, and set up his medical practice. He took little part in the liberal agitation that preceded the July Revolution, but he was friends with several of the medical men who did play leading roles in the movement. After 1830, one of these friends became the deputy mayor and named Monfalcon to several positions. He also pushed Monfalcon to join the staff of the *Précurseur*. The 1,500 francs Monfalcon was offered as deputy editor considerably exceeded the salaries from his various medical jobs; in addition, he discovered that he liked the work. Since the previous editor,

Jérôme Morin, wanted to leave the paper, Monfalcon had hopes of becoming editor-in-chief. His autobiography thus revealed how deeply and emotionally involved he had been in the events of the early 1830s, and how quickly he had realized the power of being in a position to determine how they would be represented.[27]

As we have seen, Monfalcon's ambition to take over the city's leading newspaper was thwarted, but the November 1831 insurrection propelled him into an even larger role in public affairs. According to his depiction of the first hours of the fighting, he had seen firsthand how shaky the bases of the bourgeois social order really were. The army garrison was thoroughly demoralized, and "the ranks of the bourgeois militia [the National Guard] included many rifles ready to go over to the revolt that was about to start." The commanding general was furious at the silk merchants, whom he blamed for provoking the fighting, and "most of the businessmen and property-owners, overwhelmed, fled to the countryside; there was an almost universal panicky terror." The middle-class National Guards who did try to fight the insurgents felt abandoned by the troops, who seemed reluctant to confront the workers.[28] Monfalcon himself treated the wounded, and on the second day of the fighting he volunteered to carry a prefectoral proclamation to the Croix-Rousse. Monfalcon had described the incident that followed on multiple occasions, starting with a short paragraph in one of the dispatches he had sent to *Le Temps* at the time,[29] but his autobiography gave him the opportunity to expand the narrative into a scene worthy of Victor Hugo:

> Everything is silent, and nothing moves all up and down the Grand'Côte: no sound of a loom, no human sound can be heard in this street, ordinarily so crowded and so noisy. Men, women, and children stand at the windows watching us without saying a word. There is nothing really hostile about this, but somehow the unaccustomed silence at that hour and in such critical circumstances fills me with a vague unease about the possible outcome of my mission. . . . seeing me pass by her shop, an old woman whom I recognize makes a movement of astonishment that she accompanies with a gesture of pity.

Before he could reach the top of the long street, Monfalcon found himself surrounded:

> Forty men armed with a few bad rifles encircle me. They swear at me, they attack me from all sides; my rifle, my sabre, my officer's epaulettes, my uniform are torn from me. Menaces give way to blows. My proclamation is stamped underfoot, and from all sides I hear cries of vengeance: "He's a merchant; let him pay for the others. . . ." Strong hands seize me by the neck and drag me to the gutter, and I realize how this violent scene is likely to end, when, over the shouts, I hear these words: "Don't kill him, he's my doctor, let him go." It is the voice of a lame silk worker who is not my patient but whom I know quite well.

Monfalcon's savior persuaded the angry silk weavers to inspect their intended victim's rifle. When they saw that it had not been fired recently, they let him go.[30]

Monfalcon's obsession with the story of how he had been "miraculously saved," as he wrote in his first version of the incident in *Le Temps*, shows how deeply personal the ostensibly objective narrative of the Lyon insurrections in his history was. Monfalcon had faced the possibility that the bourgeois order would be annihilated, not as a theoretical issue but as a life-or-death question in the most immediate sense: he had been in the hands of a working-class mob, moments away from being lynched. He was not merely an apologist for a system that rewarded him; he was a survivor who had nearly paid with his life for his involvement in the uprising and had been spared to tell the tale. He had an urgent warning to convey to his middle-class readers, but at the same time he had a real, personal connection with the workers. Unlike Saint-Marc Girardin, he knew that they were more than a population of hostile barbarians. Not only was he the son of a Lyon silk weaver, but his narration of his escape emphasized that he owed his life to their ability to act both humanely and rationally, even in the midst of a revolutionary situation. The woman who made a pitying gesture toward him, and the man who quick-wittedly invented a story to save his life, showed sympathies that extended beyond class boundaries; the weavers who controlled their anger long enough to examine his rifle showed a genuine sense of justice to which Monfalcon did credit.

Monfalcon's autobiographical account thus revealed complexities of the events in Lyon that were concealed in his historical narrative. He portrayed himself as neither a bourgeois completely incapable of understanding the behavior of the workers nor as an objective observer, but as a man who had grown up in the silk weavers' milieu and who had been shaken to the core of his being by the violent confrontation of 1831. His account is thus in some ways less ideological than Joseph Benoît's *Confessions d'un prolétaire*, the autobiography of a silk worker who participated in the 1831 and 1834 uprisings and in Lyon's revolution of 1848. For Benoît, the writing of his life story, undertaken between 1860 and 1871, was an opportunity to repeat the miserabilist explanation of the uprisings and the heroic rhetoric of the workers' press from the 1830s: "The workers of Lyon determined the character of all future struggles . . . in characterizing their own with the immortal slogan that summed up in three words the needs and aspirations of modern societies. They inscribed on their banner: 'Live working or die fighting!' "[31] Monfalcon's narrative has more in common with the curious first-person text published by Jean-Claude Romand, who claimed credit for inventing that slogan and whose *Confession d'un malheureux* was published in collaboration with the prosecutor responsible for sending him to prison. Romand's book was presented as an act of repentance by a man who had come to understand the error of his ways, but, like Monfalcon's story, it revealed more than the author, or at least his collaborator, may have intended. Whereas Benoît insisted on the sterling morality of the workers and prided himself on having given up his first lover, a penniless woman who engaged in prostitution, Romand admitted that he had had multiple affairs in his youth and that hunger had driven him to steal. Although his text was supposedly a lesson about repentance, he sounded proud of his role in the 1831 insurrection, claiming that he had inspired the workers to seize the prefect and the National Guard commander as hostages and that "it was also I who proposed to put on the workers' flag these words that have since become so famous: 'Live working or die fighting!' " He expressed disillusionment with the republican agitators who tried to profit from the movement, not because of their revolutionary ideas but because they treated the workers with contempt: "To them, I was nothing but a simple tailor, a proletarian, an instrument they could use and then discard."[32] Like Monfalcon's narrative, but from an opposing perspective, Romand's shows how autobiography could challenge the pieties of standard bourgeois and working-class self-presentation.

Although he may have expressed himself clumsily in his *Histoire des insurrections* and *Code moral des ouvriers,* Monfalcon's message of class reconciliation was undoubtedly a heartfelt one. His *Souvenirs* is the story of a man torn between the world of his parents and the bourgeois world he had entered. In this personalized narrative, Monfalcon highlighted the virtues and shortcomings of both classes and underlined the urgent need to come to compromise between them. Implicitly, he presented the story of his own life—dedicated both to the pursuit of his personal ambitions and to genuine efforts to improve the lot of the poor through his work in the hospitals and his commitment to bettering urban living conditions—as a model of how this might be done. Read by itself, Monfalcon's *Histoire des insurrections de Lyon* appears to demonstrate how the scientific protocols of nineteenth-century history and sociology served bourgeois ideological purposes. Read in the context of the other versions of the story Monfalcon composed, his history appears less as a manifesto of class warfare and more as a genuine attempt to find common ground between them and to make sense of a painful personal dilemma.

Monfalcon's viewpoint was limited, as were the outlooks of most of the bourgeois social reformers of the first half of the nineteenth century, by his conviction that inexorable economic laws made any real improvement in the condition of the workers impossible.[33] His father's economic success served him as proof that individual workers could, by dint of hard work, improve their conditions, but there was a contradiction Monfalcon could not reconcile between this isolated example and his recognition that the mass of Lyon's working population was doomed to remain poor. He differed from the majority of these writers, many of whom shared his medical background, because his experience in Lyon had forced him to confront the possibility that the working class would not simply accept its fate, and that indeed it was capable of organizing itself into a genuine political threat. Other writers on the subject worried about the signs of social breakdown among impoverished workers: the spread of disease, crime, and undisciplined sexual activity. Many of these reformers of the late 1830s and early 1840s drew on Monfalcon's descriptions of the Lyon silk workers' living conditions, but few were willing to accept his warning that such people were capable of effective collective action. Despite its clear bias in favor of middle-class values, and its explicit message that insurrections were doomed to failure, Monfalcon's history did not succeed in exorcising the social fears that the journalism of the early 1830s had raised.

The First French Social Novel

For more than a century, Jean-Baptiste Monfalcon's history was the most influential book about the Lyon insurrections, and the one that most effectively replaced and occluded the multiple narrations of those uprisings provided by the period's press. The alternative versions of the story that Monfalcon himself produced in his *Code moral des ouvriers* and his *Souvenirs*, circulated in a handful of copies, were not widely enough read to challenge his *Histoire des insurrections*. Monfalcon was not the only author to tell the story of the Lyon uprisings, however. In 1835, an anonymous writer announced that the two movements were only the prelude to a complete restructuring of society or to "the great final revolt of the proletariat." This warning appeared in the preface to a work that paralleled Monfalcon's in transmuting what had been a news story into book form, but in a very different genre. *La révolte de Lyon en 1834, ou La fille du prolétaire*, published seven years before Eugène Sue's *Les mystères de Paris*, has a claim to be considered the earliest French "social novel"; it was almost certainly the first French novel to include the word *prolétaire* in its title, and one of the first to cast a figure representing the bourgeoisie as its villain.[34] One of its rare reviewers claimed that the author "has recounted the misery and the pain of the people like a man who has seen them close up."[35] In its pages, workers from the Croix-Rousse are exhorted to recognize that "all of you belong to this class of men tempered in the waters of suffering. . . . By turns slaves, vassals and workers, those who have been under masters, lords and industrialists have had to suffer the harshest treatments, the most humiliating contempt, and the most flagrant injustice. But the slaves had a Spartacus, the serfs a Marat."[36] The novel, whose plot revolves around the rape and murder of an innocent working-class girl by an odious bourgeois, had much less impact on opinion than Monfalcon's history; only a handful of copies are known to have survived. The very fact of its appearance was significant, however. As Roddey Reid has written, the publication of "novels about urban workers inadmissibly conferred upon the latter something of an aura of cultural legitimacy; the new public sphere produced by the commercial print media granted the laboring classes a *droit de cité* they had not enjoyed before."[37] The novel's appropriation of the Lyon insurrections for construction of a counter-memory directed against bourgeois society as a whole was prophetic.

La révolte de Lyon appeared anonymously, and its authorship remains a mystery. The most likely candidate to have written the novel is a minor romantic author of the period, Louis Couailhac (1810–85). Couailhac, born in Lille and educated in Paris, had given lectures at Lyon's Lycée in the early 1830s, and in 1832 he had published a short story, "Le Canut, histoire contemporaine," that was similar in theme and tone to *La révolte de Lyon* and had in fact been serialized in the *Écho de la fabrique*. Like the novel, "Le Canut" was set in Lyon and integrated one of the city's insurrections—in this case, the uprising of November 1831—into its story line. As in *La révolte de Lyon*, the story strongly emphasized the sexual vulnerability of working-class women. The other stories Couailhac published in the same volume as "Le Canut" show his sympathy for radical causes at this phase in his career, although they indicate that he did not confine himself to "proletarian" themes; they also demonstrate his predilection for condemning innocent female characters to unpleasant fates. Subsequently he published several other novels and a number of stage works; he also had an extensive career as a Parisian journalist and eventually as a supporter of the Second Empire.[38]

As its dual title suggests, *La révolte de Lyon, ou La fille du prolétaire* attempts to tell the story of a collective social experience by focusing on a particular character, the "fille du prolétaire" Angélina, the lovely young daughter of the destitute silk weaver Jean. Angélina's tragic story is interwoven with those of numerous other characters: the villainous banker Durand; his daughter Emilie; Durand's accomplice Christine Dubois; the dashing young army officer Edmond; the militant weaver Francdidier, and an anonymous person who serves as the author's mouthpiece and turns up at various points in the narrative as a socially conscious preacher, a charismatic workers' leader, and a voice of warning among the city's bourgeois elite. Like the plot of Eugène Sue's *Les mystères de Paris*, that of *La révolte de Lyon* is full of improbable coincidences, and the characters are linked by unexpected relationships. At the outset, the story seems to be structured symmetrically around the two motherless households of Jean and Angélina, on the one hand, and Durand and his daughter Emilie, on the other. The two young men, Edmond and Francdidier, appear initially to be outside these two family complexes, but it will turn out that the first is Angélina's brother and the second is the illegitimate son of Christine Dubois and the banker Durand. Edmond will, of course, fall in love with Emilie; and Francdidier, no less inevitably, with Angélina.

La révolte de Lyon, like so much early French social fiction, is a simplistic melodrama, a straightforward confrontation between good and evil. Good is represented by characters who are either from the working class, or young and in love. The villains are the middle-aged bourgeois Durand and the socially ambiguous Christine Dubois, the only figure of any psychological complexity in the novel. Although the title, with its references to social violence and the proletariat, suggests an attempt to dramatize the theme of class conflict, the dominant element is in fact sexual victimization. Durand, a notorious seducer of working-class girls, is determined to use his economic hold over Angélina's father to get his daughter into his clutches. Frustrated in his designs by the heroine's honesty and virtue, he turns to outright crime. At his behest, Dubois kidnaps Angélina and drugs her. After Durand brutally rapes her, Dubois murders her to cover up the crime. The April 1834 insurrection, which has been building up in the background of the story, serves as a convenient device by which the lives of the other characters can be brought to a suitably tragic conclusion: Durand is killed by an unidentified avenger; Jean, Edmond, and Christine Dubois are mortally wounded in the fighting; and Emilie pines away from grief. The final character, Francdidier, simply disappears without explanation.

Because there is little real emphasis on the proletarian experience in La révolte de Lyon, an attempt to label it a "proletarian novel" in any of the three senses proposed by such critics as Michel Ragon (a book written by a proletarian author, a book giving a detailed description of proletarian life, or a book meant for a proletarian audience) would be misleading.[39] Although Angélina and her father are referred to as silk weavers, their working lives do not figure in the story. Angélina's father's face reveals him to be "one of those creatures that the fates amuse themselves in precipitating from a respectable status to the lowest rungs of society"; later in the story, readers learn that he came from a well-to-do family but lost his inheritance because of political persecution during the Reign of Terror (1:22, 2:46). Francdidier, Angélina's lover, is also something other than an ordinary worker. His primary identity is as an ex-soldier and political militant; the fact that he is also foreman in a silk-weaving workshop is almost irrelevant to his role in the story (1:154–55, 159). There is no mention in the book either of the highly structured organization the Lyon silk weavers had maintained from 1828 to 1834 or of the workers' press that had appeared there, in which Louis Couailhac's story "Le Canut" had been reprinted.

Although the details of the plot indicate that the novel's author had difficulty translating the actual experience of proletarian life into a story, there is no doubt that the book was intended to draw attention to the plight of the working class. The book begins with a short essay titled "Aperçu sur la question du prolétariat" and signed "L.S.," which is paginated separately from the novel's text and may or may not be by the same author as the novel.[40] This is followed by an "Avant-Propos," paginated as part of the novel, consisting of several citations from the eighteenth-century works of the Baron d'Holbach denouncing the injustice of societies that fail to provide for the basic needs of all their members, followed by excerpts from the widely circulated pamphlet of Lyon republican Jules Favre, "De la coalition des chefs d'atelier à Lyon," a defense of the Lyon silk workers arrested for striking in the spring of 1834. The "Aperçu sur la question du prolétariat" is remarkable enough to deserve attention in its own right. This manifesto proclaimed the Lyon insurrections to be a vital revelation about the nature and direction of nineteenth-century history. "The Lyon revolt is just a small symptom of a general crisis: the proletariat is the social question of the century. . . . The proletarian has intervened in public at the time of July [1830], . . . but above all in the two Lyon insurrections" (i), its author declared on the opening page. The argument that followed anticipated many of the ideas Karl Marx would develop more than a decade later. Like Marx, "L.S." insisted that the growth of the proletariat was inseparable from the development of "our industrial society." He continued: "The proletariat is the foundation of our society. It is everywhere; it lives with us; it never leaves us; he is our shadow" (1:ii, iv).

At a moment when the workers' papers in Lyon were trying to disassociate that class from any involvement with social upheaval, "L.S."

argued, as Marx would also do, that the laws of development inherent in its very nature would inevitably lead the proletariat to become a revolutionary force: "The march of this revolution of the masses will never stop. The proletarian does no more than follow his innate and natural law of development, just like every other organized being that tends to an indefinite perfection. The phenomenon will be with us until, having surrounded us without our having noticed, the question, heavy with the makings of a revolution, explodes with the sovereign power of a universal fact" (1:vi). As in Marx's doctrine, the outcome of this revolution would be a society with no class divisions and no distinction between rulers and ruled: "This will be the organization of the proletariat, based on itself. At once ruler and ruled, it will no longer divide a uniform whole into two isolated nations" (1:vi). Marx and Engels would write in 1848 that "a specter is haunting Europe"; "L.S." wrote in 1835: "The proletariat has risen up and dominates everything. But the idea is so unbelievably radical that it has scared the whole of European society, in revealing the very basis on which it rests" (1:viii).

Although the rhetoric of the "Aperçu" anticipates some of the most striking passages of the *Communist Manifesto*, it lacks the theoretical consistency of the work of Marx and Engels. The 1835 text offers no real definition of the proletariat, and its evocation of social conflict is not wedded to any analysis of its economic origins, even though the elements of such an analysis were already largely available. (In the same volume in which it reviewed *La révolte de Lyon*, for example, the left-wing *Revue républicaine* of 1835 offered an account of the expropriation of surplus value in terms very similar to those Marx would later employ, showing that "every man's work must . . . produce a surplus; this truth is demonstrated by the gradual accumulation of wealth. . . . The capitalist takes not only the part that is due to him, but that which should belong to the worker.")[41] Despite its striking rhetoric, the "Aperçu" also offers a surprisingly anodyne prescription for social change. Rather than preaching a proletarian revolution, "L.S." called for a peaceful transformation along the lines of what would later be called social democracy. "It is necessary to recast everything in a new order with the universal proletarian element. The issue this time is the great organization of equality and liberty for the masses, who should be initiated, at least gradually, into the great community. Their interests should be the soul and goal of government, and their perfection its function" (1:xii–xiii).

For all the vagueness of its conclusion, the "Aperçu" unmistakably anchored the novel it preceded to public discussions of the plight of urban workers. So did the citations from d'Holbach and Favre included in the novel's own "Avant-Propos." The passages from d'Holbach's eighteenth-century writings had been composed before the beginnings of industrialization, but such radical sentiments as the claim that "a society whose leaders and laws don't result in any good for its members obviously loses its rights over them" (1:5) had clear application to the conditions of the nineteenth century. The citations from Favre's defense of the Lyon silk workers' leaders, whose trial for illegal strike activities had set off the April 1834 insurrection, stressed particularly the gendered effects of economic exploitation, and especially the way in which it drove single women to prostitute themselves (1:10). The various "para-texts" that preceded the actual novel thus made it clear to readers that they should see *La révolte de Lyon* as a protest about the condition of the contemporary working class.

That the novel was meant to evoke sympathy for the plight of the suffering poor is obvious even without the items prefixed to it. The message is brought out most strongly in the interventions of a mysterious stranger, who delivers impassioned speeches at a church service, at a gathering of militant silk weavers, and at a banquet of the city's wealthy elite. Preaching to a crowd in Lyon's cathedral on All Saints' Day, he delivers a ringing denunciation of social injustice. "Struck by the workers' abjection and misery, and carried away by the natural irritation of his blood, he was no longer a priest summoning the bad man of wealth before God's tribunal. . . . A fiery tribune, he appealed to the ferocious but equitable justice of the people" (1:67). Having shocked his audience with his warning that "the day of reckoning is not far off" (1:68), he abruptly disappears, leaving the crowd to speculate that he was no real priest but some kind of Saint-Simonian (1:84). The same man appears later at a gathering of workers in Lyon's working-class suburb of La Croix-Rousse, where the November 1831 silk weavers' insurrection had started. There, in a scene clearly modeled after the political banquets so extensively covered in the Lyon press of the epoch, he gives an incendiary speech before leading the workers in a round of toasts. After reminding the workers of their misery, comparing them to slaves and serfs, and summoning memories of Spartacus and Marat, the speaker tells them:

> The workers have reached the pains of childbirth. The struggle will be a long one, comrades, because you are attacking the basis of social injustice, whereas until now it was a matter of healing its wounds. So what do we see? Human exploitation as in earlier times, not by patricians, not by feudal lords, it is true, but by manufacturers, by the rich who have resuscitated slavery and feudalism under other names. Look around you and you'll see the proof. Do you exercise the political rights of citizens? Do you vote on the laws that crush you? Your murmurs tell me that you don't. Is your physical survival safeguarded? Isn't the caprice or stubbornness of your masters enough to condemn you to death? Remember November 1831, don't forget those bloody days; our age dates from them. Don't forget those self-sacrificing fighters whose sublime courage wrote such a glorious opening page for your annals. Know, when the time comes, how to follow their example and die for the cause of the oppressed. (1:170–71)

Like the "Aperçu" prefaced to the novel, this text mixes anticipations of subsequent socialist rhetoric with other elements of the French revolutionary tradition. Workers are put in the same category with the slaves and serfs of earlier eras; the fact that these earlier oppressed groups had eventually achieved their emancipation is cited as evidence that workers too will achieve liberation. The struggle will be a difficult one, because it means attacking the root cause of oppression, not just its symptoms. It may well require violent means, as the references to Spartacus, to Marat, the revolutionary journalist known for his extreme rhetoric, and to the insurrection of November 1831 indicate. Despite its calls to end economic exploitation, however, this speech contains no suggestion of a socialist or communist challenge to the notion of private property; its content is fully compatible, as most French rhetoric about the plight of the poor from this period was, with the vision of a society of independent producers. The economic injustices that it specifically identifies are the absence of a safety net for the worker's subsistence and the fact that an employer's arbitrary decision can throw a worker into destitution. The specific remedies suggested are purely political, in the Rousseauist and Jacobin tradition: the "political rights of citizens" and the right to participate in the making of laws.

Despite his heated rhetoric, the mysterious speaker is not actually an advocate of violence. In a subsequent scene, where he speaks to a crowd of workers on the eve of the April insurrection, he urges moderation: "The progress of enlightenment and reason will bring us to our goal. Don't believe that armed struggle would be more direct and easier." This moderation leads one member of the group to call him an "enemy of the people" and to say, "You just try to put them [the people] to sleep with flattering words," but this critic is promptly exposed as an *agent provocateur*, thereby reinforcing the speaker's gradualist message (1:281, 284–85). The book's vague and incoherent depiction of the April 1834 uprising is equally equivocal. The author of *La révolte de Lyon* made no effort to narrate the actual events of April 1834 in any comprehensible fashion; he dwelt on "the horrors of civil war, a war more atrocious than that waged in enemy territory," and on the mutual sufferings of workers and soldiers, underlined by the fact that the silk weaver Jean and his soldier son Edmond recognize each other as they both lie dying in an improvised hospital (2:186, 276–77). In *La révolte de Lyon*, violence was presented not as a possible strategy for the conquest of power but as a spontaneous and uncontrollable outburst, morally justifiable but incapable of bringing about significant change: "It is the cry from on high that simultaneously communicates itself to all strong souls, as a spark sets off gunpowder" (2:300). The story actually warned potential proletarian readers against republican propagandists, "dandies with mustaches, gloved, corseted, spurred, who talk continually about the republic and spend their lives at dance halls or bars"(2:282), who might try to recruit workers for a planned movement with a political purpose. For all his talk of workers' rights, the novel's author was as incapable as Jean-Baptiste Monfalcon of seeing insurrection as a meaningful form of collective action, or of imagining how workers might actually change their situation.

The content of *La révolte de Lyon* thus reflects the atmosphere of the early 1830s, before socialist ideas had begun to spread among France's workers. The novel's workers are abstractions, and the program it offers shows considerably less real knowledge of their problems than the articles in the workers' press of Lyon during these years. It would be misleading to put *La révolte de Lyon* in the same category as Émile Zola's *Germinal*, a work written by a bourgeois outsider who nevertheless makes a serious effort to depict the actual lives of manual workers. Its distance from the lives of genuine proletarians was characteristic of the better-known "social novels" of the period, such as Eugène Sue's *Les mystères de Paris*. In fact, Roddey Reid, in his careful analysis of Sue's novel, has pointed out that it lacks both working-class and bourgeois characters. Reid has also noted that Sue's novel uses the themes of family and domesticity more than those of economic class to distinguish those characters who have embraced the values of bourgeois society: its lumpenproletarian and aristocratic characters alike are saved by their conversion to the bourgeois model of the household.[42] *La révolte de Lyon* propagates a similar model of domesticity, but one that is explicitly turned against the novel's one unmistakably bourgeois character, the villainous banker Durand. As we have seen, the novel is structured around the two symmetrical households of Durand and his daughter, on the one hand, and the *prolétaire* Jean and his daughter, on the other. The plot turns on the question of whether and how these motherless families can be reconstituted, and, indeed, if it were not for the interventions of the bourgeois villain and the events of April 1834, the marriages of Angélina and Francdidier, and of Emilie and Edmond, would not only have resolved this problem but also would have linked the two households into a larger family complex, as each father's son would have married the other's daughter.

This happy solution is not to be, however, because of the refusal of the bourgeois character to play the role of family patriarch. Durand—who, typically, is not even a "real" bourgeois, but "the son of a lousy tailor and a fruit vendor" who aspires to and finally does obtain a noble title (1:132, 2:85)—is an economic exploiter, but he does his real damage in the story because of his repeated violation of the norms of bourgeois domesticity. As the novel unfolds, we learn that Durand has rejected marriage in favor of sexual predation. He maintains a boudoir whose walls have been "decorated at great expense" with obscene paintings and in which he regularly seduces innocent young women procured for him by his accomplice, Christine Dubois (1:113–15). Mme. Dubois's intimacy with

Angélina and her father is thus exposed as a plot to add the young woman to the list of Durand's victims. Mme. Dubois plays this role because of the pair's earlier involvement, which involved the deliberate destruction of both of their marriages. Under the Directory—that regime of bourgeois corruption par excellence in the French collective memory—she had been the young wife of a wealthy *commissaire des guerres* but had fallen in love with Durand, a poor clerk. To smooth the way for their affair, they conspired to frame her husband for embezzlements Durand had actually committed. When she became pregnant by Durand, he did not marry her, and instead provided her with only a modest pension, while finding himself a wife who could help him in his aspiration to gain social status (1:123–28).

Many years later, Durand seeks Mme. Dubois out again. This time he wants her help in disposing of his wife, whose background is now too modest for his ever-increasing social aspirations (1:132–33). Having thus originally destroyed Dubois's family, the two accomplices now destroy Durand's as well. Rather than finding a new wife, Durand sinks into his life of depravity. Angélina's purity inspires him momentarily with thoughts of a genuine marriage based on love—as in Sue's *Mystères de Paris*, the formation of a companionate household is presented as a possible form of redemption—but her refusal to be won over by his wealth returns him to his original designs. The incestuous and thus antifamilial nature of his criminal passion is unmistakable, both because of the strongly underlined identification between Angélina and Durand's own daughter Emilie, who is the same age, and because, unbeknowst to him, Durand is competing with his own son, Francdidier, for Angélina's affections. Angélina's brutal rape and murder deprives her of both her life and her proper destiny as the partner of an honest man of her own class.

Durand is thus presented as a villain, not only because he exploits the poor economically, but also because his unregulated sexual and social ambitions lead to the destruction of honest families—his own and those of his victims. This had also been the message of Louis Couailhac's 1832 short story "Le Canut," in which an unemployed silk weaver's wife goes to ask a merchant for work for her husband. The merchant propositions her, the outraged husband attacks him and is sent to prison, and their family is plunged into complete destitution. The husband is released from prison just in time to take up arms in the November 1831 insurrection, in which he is shot and killed by the merchant. As in *La révolte de Lyon*, the aspect of proletarian vulnerability stressed in the story is the inability of male workers to protect their wives and daughters against this kind of predation. Sexual exploitation of vulnerable working-class women was certainly part of the reality of French society in the 1830s—even Jean-Baptiste Monfalcon had noted that women were more likely than men to be unable to earn enough to live on[43]—but the decision to present this as the essence of the conflict between bourgeois and workers showed how deeply these early social novels were embedded in the framework of bourgeois society. The "innocent" women in *La révolte de Lyon*, Angélina and Emilie, enjoy that status because they are sexually chaste, submissive to their elders, and aspire only to become lawfully wedded wives. The villainess, Mme. Dubois, is evil because she violates the bonds of marriage and family, killing first her honest husband, then Durand's first wife, and finally Angélina, her own son's true love. It is true that she is granted the benefit of some extenuating circumstances that do not apply to Durand: the story reveals that her downfall began when she was seduced by a priest, and she experiences a moment of hesitation and remorse before she smothers Angélina (2:151–56). In the end, however, even the revelation of her own victim status reinforces the message that those who engage in sex outside of marriage are inevitably demoralized as a result. The novel thus reinforced nineteenth-century stereotypes about women's roles; it was far less challenging to conventional gender hierarchies than Eugénie Niboyet's *Conseiller des femmes*.

The author of *La révolte de Lyon* accepted the bourgeois novel's valorization of companionate marriage as the normative model for human relations, but tried to turn this stereotype against the middle class by presenting his bourgeois characters as violators of the rules of domesticity, and his working-class characters as their upholders. He was sufficiently self-aware to recognize the less savory aspects of the uses his own text made of the story's sexual element. After Durand's accomplice Mme. Dubois has anaesthetized poor Angélina, the text describes in lurid detail how she strips the victim and applies cosmetics to make her "virgin body" (2:64) more alluring. Having whetted readers' appetites, the author brings Durand to the scene and then virtuously annonces, "Let us let the half-lifted curtain fall back in place. . . . Let us not insult the noble victim to gain the applause of those depraved souls who get pleasure from the thought of virgin flesh throbbing under the bellowing excesses of hyena-eyed lust" (2:67). Clearly, he knew perfectly well how to exploit the misogynist potential of that image while pretending to do otherwise. The reduction of Angélina to a completely passive victim of male bourgeois lust and sadism also had unwelcome implications for the proletarian class her body represented. In spite of the vehement rhetoric in other parts of the text, *La révolte de Lyon* ultimately portrayed workers as incapable of exercising meaningful control over their own lives. In this respect, the message of the novel was less optimistic than even Jean-Baptiste Monfalcon's hostile representations.

La révolte de Lyon thus demonstrated how difficult it was to use the genre of the novel to challenge the conventions of bourgeois society. The novel's focus on individual characters, and its sentimental idealization of the domestic sphere as the arena in which true happiness is found, militated against a genuine rupture with that world. It is true that *La révolte de Lyon* was apparently too outspoken for France's middle-class reading public. Although its publisher, Moutardier, was a substantial firm whose 1835 catalogue listed works by the most popular authors of the day—Victor Hugo, Eugène Sue, Georges Sand, Félicité de Lamennais, Walter Scott, and Paul de Kock—this work seems to have been a commercial failure; it may have run afoul of the tightened censorship imposed in September 1835. The book appeared just before the transformation of commercial publishing in the late 1830s and early 1840s that allowed authors to reach a genuinely popular audience, as Eugène Sue's novels were to do. If Louis Couailhac was indeed its author, he drew the obvious lesson from this setback. In 1837, he published *Pitié pour elle*, a tale of middle-class provincial characters exposed to the seductions of Paris that was utterly devoid of concern for the lower classes.[44] The world's injustice to innocent women figured prominently in the plot—his heroine, like Emma Bovary, dies after her husband's behavior drives her to commit adultery—but Couailhac's clumsy prose and melodramatic plot demonstrated how badly a mediocre talent could mangle the same materials that Balzac and Flaubert used so effectively. Couailhac went on to become a successful journalist and the official secretary of Napoleon III's Imperial Senate.

The contrast between the book-length narratives of the Lyon insurrections as history and novel, and the representations of the same events in the press of the period, thus raises questions about the notion that newspapers were necessarily the vehicles of a dominant discourse and that the genres of high literature were more open to "counter-discourses," as Richard Terdiman has argued. The conventions governing the writing of both history and fiction worked to reinforce the message that there was no real alternative to the liberal bourgeois order. Although the point of view expressed in Jean-Baptiste Monfalcon's *Histoire des insurrections de Lyon* had been only one of many perspectives in the period's press, its apparent completeness and its supposedly scientific character gave it virtual hegemony after its publication. *La révolte de Lyon*, in which the very existence of Lyon's lively newspaper press is never mentioned, condemned the misery of the poor but denied them their specific identity and reinforced the notion that bourgeois patterns of life were the only ones possible.

Both Monfalcon's *Histoire des insurrections de Lyon* and *La révolte de Lyon* crystallized ways of representing bourgeois society that retained their power for a long time to come, a fact that reinforces the importance of the period during which they were written. Monfalcon's book provided bourgeois readers with detailed, concrete evidence of the threat from the "dangerous classes" in Europe's growing cities. At the same time, it reassured those readers of the moral justice of their resistance to working-class demands by demonstrating, apparently irrefutably, that the laws of economics made it impossible to change their condition. Finally, Monfalcon highlighted the danger posed by subversive groups such as the republican movement, whose propaganda risked setting off an explosion that would hurt all classes. Monfalcon thus laid out the bases for a conservative, class-conscious bourgeois identity that would be a major feature of French life well into the twentieth century. Similarly, *La révolte de Lyon* foreshadowed a long-lasting representation of French workers as downtrodden victims who would some day obtain the justice they deserved. Even more than the *Histoire des insurrections*, *La révolte de Lyon*, with its saintly workers and murderous bourgeois, purveyed exaggerated and emotionally laden social stereotypes. This novel itself appears to have had very little impact, but the way in which it portrayed French social conflicts became a staple of socialist and radical discourse as durable as Monfalcon's justification of the bourgeoisie.

The success of these two texts in using the events of the early 1830s to establish long-lasting patterns for the representation of French society underlines the significance of that revolutionary episode. But the narrowness and rigidity of these patterns also underlines the degree to which the space for contestation created by the press during the revolutionary period of 1830–35 was subsequently reduced. Unlike newspapers, histories and novels could stand alone; they were not part of a polyphonic "field," not in constant, overt dialogue with other texts representing the same events in differing ways. Furthermore, once published, they were static, congealed, incapable of responding to new events. Lyon's press had proposed to its readers a range of social identities, many of which were excluded completely from the totalizing narratives that replaced the newspapers. No book served, as the *Écho de la fabrique*, the *Papillon*, and the *Conseiller des femmes* had, to demonstrate that workers and women could speak for themselves, and none revealed, as the dialogue between the *Précurseur* and its rivals had, how varied and uncertain bourgeois identity could be. Furthermore, the press had accurately conveyed the fact that social identities were always in question, whereas both Monfalcon's history and *La révolte de Lyon* portrayed them as fixed and unchangeable. Examining the more stable narratives into which the revolutionary episode of the early 1830s was eventually compressed underlines the fact that newspapers truly were, as the author of the chapter on journalism in the *Nouveau Tableau de Paris* wrote in 1835, the only form of literature that had changed with the revolution, and the only one that reflected the "new and original *moeurs*" produced by that event.[45]

Conclusion

By 1835, when the September laws brought the press under control and ended the revolutionary cycle that had begun in July 1830, France—and, indeed, Europe as a whole—was no longer what it had been five years earlier. The July Revolution, the passage of the Reform Bill in England, the successful national movement in Belgium, the unsuccessful revolt in Poland, and nationalist demonstrations in Germany and Italy had demonstrated the fragility of the Restoration across the Continent. The liberal, democratic, and nationalist ideals inspired by the Revolution of 1789 were clearly still alive. At the same time, the debates over the "social question" in France and the passage of the New Poor Law in England had shown that a new set of issues was now also on the European agenda; not only political institutions but also the very structure of society was now in question. A language of social class had been added to the political vocabulary inherited from the 1790s. The combination of these political and social conflicts would characterize public debate in France and in Europe as a whole for the rest of the nineteenth century and well into the twentieth. Lyon between 1830 and 1835 was a microcosm in which all the political and social currents of this new era were clearly expressed. The vital role that its press played in this process enables us to follow its unfolding and its implications with unusual clarity.

As we have seen, the fact that the Lyon newspapers formed a "journalistic field" characterized by the conflictual interaction of its elements made them especially important in articulating the system of oppositions that structured the new political and social universe. Liberal, conservative, and radical journals, bourgeois and workers' papers, journals for women, and ostensibly nonpolitical cultural periodicals all offered definitions of identities that resonated with particular audiences and offered visions of what the public space of modernity should look like. Journalists emerged as critical figures in this new public world, which was localized in specific urban sites where public opinion took shape. Journalists wielded special influence through their ability to set the agenda for public discussion at these sites, but they paid for their privilege through their exposure to hazards ranging from duels to prison sentences; the tension surrounding their activities was a clear indication of the importance of the publications they shaped.

Newspapers were part of the liberal society and economy whose principles had been laid down by the revolutionary legislators of the 1790s. The elaborate system of press laws built up since 1789 was supposed to make press enterprises behave like rational, property-owning citizens who could be counted on to defend social order. At the same time, however, newspapers presented themselves as agents of progress and expressions of the dynamic, universalist ideals of 1789, and freedom of the press, safeguarded by journalists' right to public trials, was a hallmark of liberal regimes. Nowhere else were the contradictions inherent in liberalism's "bourgeois universalism" expressed so strongly as in the press, where rival interpretations of that heritage challenged each other on a daily basis. In the early 1830s, however, the press also became accessible to critics who stood outside the boundaries of liberalism—conservative legitimists, self-proclaimed "proletarians," women, and proponents of a cultural sphere that was immune to considerations of power and

Conclusion

money—who used it to express themselves and to underline the omissions and unexamined assumptions of liberal and bourgeois culture. The regular debate that went on among rival newspapers became one of the fundamental characteristics of nineteenth-century life; liberal society was a society whose values were constantly exposed to public questioning, and the press, even though it was central to the definition of liberalism, was at the same time a built-in element of instability.

One of the most powerful ways in which newspapers simultaneously structured and destabilized the liberal order was through their propagation of a repertoire of differentiated social identities. Until 1830, press debates had taken the form of a bipolar confrontation between adherents and opponents of the principles of 1789. The liberal press in particular had put itself forward as the voice of a supposedly universal public, from which only the adherents of the old regime were excluded. After 1830, the creation of newspapers claiming to speak for workers and women, on the one hand, and newspapers that denounced the triumphant liberals as a bourgeois caste that constituted a new aristocracy, on the other, threw this notion of the public as a single, unified entity into question. Some liberals, as we have seen, responded by embracing the notion of a bourgeois class with a vocation to govern society; others tried to translate liberal universalism into a new form, even while recognizing the existence of distinct social classes, such as the popular readership that the liberal *Précurseur* and the republican *Glaneuse* tried to appeal to with their pamphlet editions in 1834. In the years between 1830 and 1835, the press in Lyon demonstrated what a social universe structured by debates about the nature of the bourgeoisie, the relationship between workers and elites, and the roles of women would look like. The issues that were raised so forcefully in Lyon's press in those formative years were to remain central in French and European life for decades afterward.

Newspapers were central not only to the construction of new images of society and its constituent groups, but also to a new political culture. Public banquets promoted through the press and journalists' trials were ritualized occasions for affirming political ideas and opportunities to create new political realities. Although opposition groups in Lyon used these press-based events to challenge the Orleanist regime, in fact they were also helping to create a political culture that was capable of tolerating dissent, an important step in the creation of a stable liberal order whose legitimacy would be accepted even by those who did not immediately profit from it. As this new political culture took shape, the press in Lyon also had to confront directly the possibility of mass insurrection. Nowhere else in Europe was this threat so manifest as in Lyon, where journalists found themselves in the middle of uprisings twice in less than three years. The words they produced in the midst of these crises were critical in defining the significance of events and in containing the violence that had exploded in their midst. Blamed by the authorities for inciting these revolts, the newspapers in fact offered themselves as models of an alternative politics that was more regular and predictable than popular insurrection. Even though the press communicated the drama of Lyon's two insurrections through both its form and its content, it ultimately contributed to keeping the conflicts of the period within manageable boundaries.

As the revolutionary interval of the early 1830s gave way to greater stability, the writing of its story ceased to be journalism and instead became history and fiction. The representations of Lyon's revolutionary years in Jean-Baptiste Monfalcon's works and in France's first social novel owed a great deal to the city's press. Monfalcon was himself a journalist, and his vision of the city's experience had been shaped by his intimate involvement with its newspapers. His *Histoire des insurrections de Lyon* aimed to replace and thus silence the alternative narratives that had appeared in the press; at the same time, however, it provided a definitive portrait of the social and political reality that those newspapers had called into being. *La révolte de Lyon,* too, accepted the repertoire of social identities constructed through the press system of the early 1830s as an accurate representation of society, even as it occluded the role that newspapers had played in the events of those years. In symbolizing the suffering proletariat through the character of a victimized young woman, its anonymous author brought together themes that had suffused the workers' papers, women's journalism, and the radical democratic publications that had flourished after 1830. Like Monfalcon's history, the novel also deliberately omitted some themes that had been prominent in the press, particularly the emphasis on workers' capacities for acting on their own behalf that had been so strong in the *Écho de la fabrique* and its successors. Both history and novel, in differing ways, worked to stabilize the postrevolutionary situation, but the force and energy the press had exhibited during the crisis

Conclusion

years was only dimly reflected in those forms of publication. Alternative readings of the revolutionary crisis insinuated themselves even in the writings of Lyon's greatest defender of bourgeois supremacy, however. The contrast between Jean-Baptiste Monfalcon's widely circulated history and the recollections of the Lyon uprisings incorporated in his autobiography provide a fascinating example of how private experience could undermine public affirmations.

The close reading of the Lyon newspapers in the early 1830s thus sheds light on a number of larger historical questions about the nature of what has often been called bourgeois society, and about the way cultural patterns are constructed. The evidence presented here supports the claim that the years just after 1830 were the critical period in which references to both a self-conscious bourgeoisie and a self-conscious proletariat became major elements of public language in France, and thus underlines the importance of the Revolution of 1830. As we have seen, this was not because new social groups with distinct economic interests had suddenly emerged on the scene—the silk weavers' insurrection of 1786 had been quite similar to that of 1831—but because social conflicts were now being interpreted in new ways. Above all, the newspapers' ability to present those conflicts as clashes of abstract, anonymous groups—social classes—transformed the significance of these confrontations. In his study of how middle-class identity developed in Britain in this same period, Dror Wahrman has emphasized "the degree of freedom which in fact exists in the space between social reality and its representation."[1] Lyon's journalistic field offers a convincing demonstration of how one particular understanding of social reality—a particularly important one for the future of European society—was articulated. Similarly, we can see how the press played a major role in structuring the understanding of gender roles during this period.

Critical to this argument is the contention that the press does not simply record changes in the vocabulary of public debate, but that newspapers, more so than other forms of print, are active catalysts in changing social representations. Unlike other printed texts, newspapers are inherently part of what Pierre Bourdieu calls a "field." They necessarily define themselves by their rivalry with other newspapers, and compete to impose their visions of reality. Nineteenth-century newspapers—and in this respect the Lyon press of the 1830s is a model example—regularly

Conclusion

acted out the conflicts between political and social groups described in their own columns through the dialogue they conducted with one another; journalistic debates were symbolic representations of the larger clashes supposedly occurring in society. The close link between the appearance of Lyon's first workers' paper, the *Écho de la fabrique,* and its first explicitly class-conscious bourgeois paper, the *Courrier de Lyon,* in the weeks surrounding the November 1831 silk workers' insurrection is a striking example of how a change in the journalistic field could affect notions of social identity. The special characteristics of newspapers as a medium—not just the competition among them, but also their status as continuing series of publications, maintaining their identity over time, and their power to structure their audiences into cohesive "imagined communities" through shared and simultaneous reading—therefore distinguish them from the larger and more amorphous "print culture" of which they are a part. The tendency on the part of some recent historians to reduce the press to just one element in a larger spectrum of cultural representations, and often to privilege what strike them as more colorful and spontaneous forms of expression, obscures the special influence that periodical publications wield.

To be sure, the influence of the press varies from one context to another, and the model proposed here cannot simply be applied to other historical situations without taking their specificity into account. Following the overthrow of the July Monarchy in February 1848, Lyon and France both experienced another "media revolution" marked by the sudden appearance of numerous new publications expressing radical ideas. An examination of the newspapers from those years suggests, however, that in 1848 the vital arena for political and social debate in that revolutionary crisis was less the press than the clubs, sites where a participatory oral political culture manifested itself. In his chronicle of Lyonnais events in 1848, Jean-Baptiste Monfalcon, who had stressed so strongly the role of the newspapers in the 1830s, said little about the press. Instead, he concentrated his attention on these new forms of collective action, which he claimed systematically favored the rise of "demagogues of the lowest sort . . . who infect an ignorant population in the grip of passion with the most pernicious doctrines." He was especially appalled when the city's library, of which he had become the director, was taken over for club meetings, a literal displacement of print culture by this new

Conclusion

form of politics: "The peaceable refuge of study is taken away from its studious public; its marble tabletops are covered with bottles, its floor tiles stained with wine, its tables surrounded by men sent from all the city's clubs and organized in a sort of tribunal that terrifies the city with the excesses of its principles and the violence of its revolutionary motions."[2] Police reports in 1848, both in Lyon and Paris, also show a clear shift of attention from the newspapers, which were by then familiar phenomena, to clubs and public demonstrations.[3]

At the beginning of the twenty-first century, newspaper "fields" have obviously lost much of the hegemony over the representation of events and the structuring of public debate that they exercised in the early nineteenth century. In France, as elsewhere, the arena of public debate now takes in not only a wide range of other print media—books, which have by no means disappeared, and other forms of periodicals, such as newsweeklies and journals of opinion—but, even more important, television, radio, movies, and the rapidly evolving universe of the Internet.[4] The value of the case study provided here is to draw attention to the importance of the media as agents of political and social change; the challenge for scholars working on other periods is to identify those media, whatever their form, whose significance has been comparable to that of Lyon's "journalistic field" in the early 1830s.

Although the goal of this study has been to look in detail at one city's press during a specific period, in order to understand certain aspects of the general history of France and indeed of European society as a whole during the nineteenth century, and to offer a model for wider research on the historic role of communications media, I hope that it has also served to demonstrate the value of making a voyage to Lyon and meeting the remarkable journalists and periodicals who came to the fore during these five extraordinarily intense years of their city's history. Marius Chastaing, Eugénie Niboyet, Jean-Baptiste Monfalcon, Anselme Petitin, Théodore Pitrat, Adolphe Granier, and Léon Boitel have long been forgotten figures, even in the city where they lived and worked in the 1830s. They wrote for the moment, with little thought of achieving immortality through their words. In their own ways, however, they grappled with the same issues that inspired the great novelists and social thinkers of the period: how to put into words the living reality of the new social world around them, and how to induce readers to change it for the better. If we want to recover a sense of what it was like for a member of the middle classes to live through a nineteenth-century insurrection, for a manual worker to claim the right to justice and respect, for a talented woman to demand a public voice, we find powerful and often eloquent evidence in the obscure texts I have drawn on here. Thrust into the midst of events whose significance went far beyond the boundaries of their provincial city, these writers and their publications demonstrated that the different elements of their community could indeed find effective voices to speak for them. The originality and force of this local journalism reveals that literary creativity in this era was no monopoly of Paris, or of authors working in the period's high genres, and it reminds us that writers such as Balzac and Stendhal learned much from their anonymous journalistic contemporaries. In the 1830s, a voyage to Lyon was a way of grasping the implications of a society that was on the road to social and political democracy. Today, a return visit to the city's local periodicals is an opportunity to elaborate a more democratic view of literature and its relationship to human experience.

Appendix 1

Sophie Grangé, "Moi" and "A la femme"

Sophie Grangé's poems, "Moi" and "A la femme," both published in the first Lyon journal to solicit women's contributions, the *Papillon*, in 1832, are significant examples of the outspokenness of Lyon's women writers in the early 1830s. (See below for complete text of these poems.) Grangé emphasizes women's rights to pursue active projects and their moral and spiritual equality with men. Little is known about Grangé's life. Before 1830, she had published a small volume of verse, *Romances et poésies diverses, par Mlle. S* . . . (Lyon: Barret, 1826), celebrating the joys of *amitié*, primarily with other women, and her love of reading and study. In one of the poems included in that collection, "Vers sur son cabinet" ("Verses on her study"), she identified herself with the male heroes of the historical works she liked to read: "Je suis dans les combats les meilleurs capitaines, / Je partage leur gloire et prends part à leurs peines" ("In the battles I am the best leaders, / I share their glory and their sufferings") (11). Even before 1830, she thus claimed that, in imagination, women could do all the things that men could. She made a similar point in "Moi," which appeared in the *Papillon* on 10 July 1832 and in which she spoke of her dreams of "gloire, bonheur, amour" ("glory, happiness, love"). In her dreams, she claimed, these divinities inspired her so that "D'un drame souverain j'attaque tous les rôles,— / Et tous vont à ma taille!" ("In a drama of rulers I try all the roles—And all fit me perfectly!") And on her return to waking life, the memory of those dreams told her, "Je suis libre! et sans voir si la place est permise, / Sous le frac ou la robe, à l'aise dans ma mise, / Quand je suis lasse, je m'assieds!" ("I'm free! and without asking if I have permission, / Wearing a frock coat or a dress, comfortable in my clothes, / When I am tired, I sit down!") Grangé's "A la femme" is thus the work of a woman who imagined herself crossing gender boundaries, performing male roles and even wearing men's clothing. Despite its apparently apolitical content, the poem indicates the ways in which Lyon's press of the early 1830s permitted a questioning of conventional identities.

In "A la femme," published in the *Papillon* on 4 September 1832, Grangé made an even more explicit case for women's equality and independence. She insisted that women and men were equal in the sight of God, asked why baby girls were doted on when adult women who spoke up for themselves were scorned, and warned women against letting men limit their aspirations. She compared men's promises of love and respect to the deceptions used to lead sacrificial animals to their deaths, and concluded by exclaiming, "Levez-vous donc, mes soeurs . . ." ("Rise up, then, my sisters . . .").

Grangé's "A la femme" inspired at least two verse responses in the Papillon. The first, by one B. Jouvin, a resident of Grenoble, praised her outspokenness but doubted that most women would respond to her call for independence: "Tu l'appelles en vain cette esclave parée, / Elle est heureuse!—Car de sa chaîne dorée / La vieillesse et le temps n'ont pas montré l'acier!" ("In vain you call her a prettied-up slave, / She is happy—Because old age and time / Have not revealed the iron in her golden chain") (*Papillon*, 22 September 1832). A few weeks later, a male poet from Lyon, L. A. Berthaud, wrote that he had heard that Grangé intended to give up writing and urged her not to do so: ". . . Il est temps / Que les Sultanes soient en face des Sultans, / Que Diane ait son temple, et que la femme, libre, / Au vieux monde qui craque apporte l'équilibre!" (". . . It is time / That sultanas stand up to sultans, / That Diana should have her temple, and that woman, free / Should bring balance to an old world that crumbles!") (*Papillon*, 6 Oct. 1832).

Despite Berthaud's appeal, Grangé does not seem to have published any further work after "A la femme," at least under that name. She does not appear among the contributors to Eugénie Niboyet's *Conseiller des femmes* in 1833–34, and the Bibliothèque nationale lists no publications by her after 1832. In addition to the two poems published here, the *Papillon* published two short stories of unhappy love written by Grangé. In "Mariella," a young woman jilted by a faithless lover stabs him and poisons herself to prevent him from marrying someone else (*Papillon*, 11 August 1832). "Louise" featured a thoughtless young woman who breaks her promise to marry one suitor in favor of another. When the first man blows his brains out in front of her, she goes insane (*Papillon*, 25 August 1832). These prose works were much less effective than Grangé's poetry.

"Moi"

Souvent pendant la nuit, la paupière entr'ouverte
Et les regards errants sur la persienne verte,
J'attends que le jour pointe, et je compte en rêvant
Les heures qu'aux Terreaux sonne l'une après l'une
Le vieux cadran de fer qui fait tourner la lune
    Comme un astre vivant.

Alors je me recueille, et ma fraîche pensée
Rassemblant les débris d'une image passée,
Enfante comme un vers tous mes jours d'autrefois.
Alors je vous revois, rêves de jeune fille,
Gloire, Bonheur, Amour,—oh! ma sainte famille,
    Alors je vous revois.

Vous êtes là:—j'entends vos ailes frémissantes,
Vos paroles du ciel aux syllabes puissantes,
Et froler vos cheveux qui ne vieillisent pas,
Et je vois à vos fronts, comme deux soeurs jumelles
Deux étoiles briller, et brillantes comme elles,
    Des perles sous vos pas.

Et moi, le coeur empli de vos grandes paroles,
D'un drame souverain j'attaque tous les rôles,—
Et tous vont à ma taille!—Et je suis une enfant!—
Ce que c'est que rêver dans une nuit profonde
Qu'on a des ailes,—et qu'au dessus du vieux monde
    On s'élève en pensant.

Oh! que m'importe, à moi, ce monde qui gravite
Ou roule sur des chars? Hélas, tout meurt si vite!
Si vite tout s'enfuit des choses d'ici-bas!—

Je veux toucher les cieux pour mesurer la terre,
Ou me perdre si loin que jamais femme austère
    N'y poursuive mes pas.

Gloire, Bonheur, Amour, fantastiques génies,
Bercez-moi: j'aime tant vos douces harmonies,
Vos extases sans fin, votre vol vers les cieux.
Et j'aime que mes nuits de rêves toujours pleines,
S'embaument, ô mes dieux, à vos pures haleines,
    Et flambent à vos yeux.

Puis, que m'importe après que l'on dise à voix basse
Ou haute: "Voyez donc cette femme qui passe,
Un feutre noir au front et des bottes aux pieds!"—
Je suis libre! et sans voir si la place est permise,
Sous le frac ou la robe, à l'aise dans ma mise,
    Quand je suis lasse, je m'assieds!

A la femme

Puisque Dieu t'a voulue au milieu de la foule
Que son bras tout puissant fait mouvoir, et qu'il foule
    Au pressoir de la mort;
Là, pauvre et maladive, et grelottant la fièvre,
Ici reine du glaive, et du sang à la lèvre,
    Et jouant au remord.

Puisque lorsque tu vins, et frêle, et toute nue
Prendre au désert du monde une place inconnue,
    Tu gagnas en naissant
Ta part dans l'air qui vole, éternel héritage,
N'as-tu donc, ô ma soeur, pour vouloir davantage,
    Rien fait en grandissant!

Tu naquis:—et ton nom, au livre des baptêmes,
Fut inscrit. Ce jour là, les hurleurs d'anathèmes
    N'avaient pas encore de poumons:
C'étaient des chants, des fleurs pour toi, feuille éphémère

Encore pendue au sein de la tige, ta mère,
   Comme un chamois aux monts.

Alors il eût suffi d'une brise, d'une ondée,
Et tout mourait de toi!—hors une vague idée
   Un confus souvenir,
Une image incolore, un myrthe funéraire
Que l'on eût ébranché pour couronner le frère
   Qui pouvait te venir.

Et pour toi cependant, c'étaient de purs hommages,
Tels que pour le messie en apprenaient les mages;
   Chacun portait ses voeux
Au berceau de l'enfant venu dans la journée;
Et le veillard ému, passait, toute fanée,
   Sa main sur tes cheveux.

Et quoi donc! maintenant que ta forte pensée
(Soit qu'on cherche l'énigme ou future ou passée,
   Soit, si l'heure en a lui,—
Qu'on rêve poésie, et, sur la page blanche,
De vers impétueux on roule une avalanche),
   Creuse aussi loin que lui,

Quoi donc! l'homme viendrait te marquer la carrière,
Et formuler tout, loi, costume, prière!
   —Oh! le sang monte au coeur,
Et je le sens rougir ma lèvre débordée,
Dès que ma raison touche à cette morne idée,
   Lourde comme un vainqueur!

Oh! ne me dites pas que vos chaînes dorées
Sont légères:—Je sais que l'on vous a parées
   Et de soie et de fleurs;
Je sais qu'autour de vous la fadeur insulteuse,
Brode les mots d'amour, et la phrase menteuse
   Qui raille sous des pleurs.

Je sais que dans un bal, aux lueurs des bougies,
Lorsque l'homme s'enivre et de jeux et d'orgies,

Je sais qu'à nos genoux
Un instant il se plie, et que sa tête fière
S'éclipse une minute aux trombes de poussière
   Qui tournent avec nous.

Mais qu'importe!—Autrefois, au lieu de sacrifice,
C'est ainsi qu'on menait la tremblante génisse.
   Triste, sous un manteau
Dont les fleurs effaçaient la tenture pourprée,
Pas à pas, vers l'autel elle allait, adorée! . . .
   Puis—venait le couteau!

Levez-vous donc, mes soeurs,—Il est temps qu'on efface
L'empreinte de fer chaud qui brûla notre face.
   Il est temps que nos fronts
Enfin régénérés d'une infamante aumône,
Se montrent grands et fiers, et sur Lacédémone,
   Dévêtent leurs affronts.

Quoi donc toujours à nous les rubans, les poupées?
À l'homme le pouvoir, la foudre et les épées? . . .
   Hélas! Il l'a fallu,
Il a fallu que l'homme ainsi nous rapetisse:
Mais toujours nous garder sous sa haute justice!—
   Dieu ne l'a pas voulu!

Appendix 2

The *Écho de la fabrique*'s Anniversary Salute to the Victims of the 1831 Workers' Insurrection

The annual tribute of the *Écho de la fabrique* to the workers killed in the November 1831 insurrection, printed in a black-bordered box on the paper's front page, is one of the earliest examples of a rhetorical exercise that was to become central to the French labor movement: the celebration of the heroic martyrs of the labor movement, and the linking of their sacrifice to a predicted eventual triumph of their cause. Similar articles were published annually in the workers' press until the September Laws of 1835 barred such direct evocations of insurrection. This article originally appeared on 25 November 1832.

21, 22, 23 November 1831

"Drums, from our brothers' funeral procession!"

Will you go by unnoticed and bereft of all memory, deplorable days that November brings back? Will I be the only one to observe your sad anniversary? . . . I strain my ear and I don't hear the religious hymns you were promised! What has become of the priests? . . . Where is the censer? Only my profane voice, free of all fear, will ring out.

Lyon! My fatherland! Cover yourself with mourning . . . on those awful days, many of your children died . . . Say nothing against them . . . The parricide fury of Catalina, Caesar's ambition, did not make them take up arms . . . Nor was it the unthinking devotion to the ill-defined rights of a dead monarchy that made them leave their peaceful lives and summoned them to the battlefield. Frightful hunger, misery worthy of pity, were their heralds of arms . . . O, Lyon! in their distress, your children, unfortunate but still your citizens, did not raise the then-unstained flag of revolt or that tricolor standard, noble inheritance of republican France's great days—the glorious standard that, from the snows of Mount-Saint-Bernard, went to reflect the sun of the Orient in burning Egypt, and still glorious when, made sodden by the tears of betrayed Liberty, it was taken, protected by the imperial eagle, to carry its murderous caprices from capital to capital. They knew, these citizen-workers, that it could be justifiably unfurled only at the frontier, in the face of the foreigner. Nor did they raise the red flag of civil war, the banner of blood, signal of vengeance and proscription, but instead a black flag! . . . Lugubrious and sacred emblem, you were their only pennant. A short inscription served for your device: "Live working or die fighting!"

Rest in peace, victims of November! May the earth lie lightly on you! . . . Your blood has fertilized the ground from which the tree of proletarian emancipation shall grow . . . No glorious halo will surround your anonymous tombs . . . Ah! You would not have wanted a glory stained with the blood of your fellow citizens . . . Your memory will nevertheless not be forgotten in the history of the proletariat . . . The future is revealed! . . . I tell you . . . your descendants will no longer be the helots of civilization; in time they will consecrate to you a cenotaph as simple and beautiful as your life . . . The arts will embellish it. The David of those days will hang from the vault of the temple a memorial painting, and his genius will span the centuries between to retrace, on a canvas responsive to his brush, your three days in all their details. The first is the model of misery; the second is veiled; the palms of victory [and] the peaceful olive tree distinguish the third. Another Lebrun will consecrate to you his lyrical songs.

Salutations! Salutations to your remains!

Rest in peace, victims of November!

(Translation by Jeremy D. Popkin. Ellipses and emphases in the original.)

Notes

Introduction

1. *Journal des débats*, 8 December 1831. For typical historians' citations, see Pierre Rosanvallon, *Le Moment Guizot* (Paris: Gallimard, 1985), 297; and Louis Chevalier, *Classes laborieuses et classes dangereuses* (Paris: Librairie Générale Française, 1978), 595.

2. *Écho de la fabrique*, 6 November 1831. For a more detailed discussion of this paper, see below, Chapter 4.

3. Archives départementales du Rhône (hereafter ADR), Carton T 353, entry for 29 October 1831.

4. Charles Ledré, *La presse à l'assaut de la monarchie, 1815–1848* (Paris: Armand Colin, 1960); Claude Bellanger et al., *Histoire générale de la presse française*, vol. 2: 1815–71 (Paris: Presses Universitaires de France, 1969).

5. Two recent essays on Philipon and his colleagues are Elizabeth C. Childs, "The Body Impolitic: Press Censorship and the Caricature of Honoré Daumier," in Dean de la Motte and Jeannene Przyblyski, eds., *Making the News: Modernity and the Mass Press in Nineteenth-Century France* (Amherst: University of Massachusetts Press, 1999), 43–81; and James Cuno, "Violence, Satire, and Social Types in the Graphic Art of the July Monarchy," in Petra ten-Doesschate Chu and Gabriel P. Weisberg, eds., *The Popularization of Images: Visual Culture Under the July Monarchy* (Princeton, N.J.: Princeton University Press, 1994), 10–36. See also Robert J. Goldstein, *Censorship of Political Caricature in Nineteenth-Century France* (Kent, Ohio: Kent State University Press, 1989), 124–46. *La Caricature. Bildsatire in Frankreich, 1830–1835, aus der Sammlung von Kritter* (Göttingen: Kunstgeschichtliches Seminar der Universität Göttingen, 1980) gives detailed explanations of the pictures' subjects.

6. Evelyne Sullerot, *Histoire de la presse féminine en France, des origines à 1848* (Paris: Armand Colin, 1966); Laure Adler, *A l'aube du féminisme. Les premières journalistes, 1830–1850* (Paris: Payot, 1979); Cheryl A. Morgan, "Unfashionable Feminism? Designing Women Writers in the *Journal des femmes* (1832–1836)," in De la Motte and Przyblyski, eds., *Making the News*, 207–32.

7. David Pinkney, *The French Revolution of 1830* (Princeton, N.J.: Princeton University Press, 1972), 277.

8. See especially Peter Sahlins, *Forest Rites: The War of the Demoiselles in Nineteenth-Century France* (Cambridge, Mass.: Harvard University Press, 1994).

9. This point is stressed especially in Werner Giesselmann, *"Die Manie der Revolte." Protest unter der Französischen Julimonarchie, 1830–1848*, 2 vols. (Munich: R. Oldenbourg, 1993), 1:380, 2:583.

10. Ronald Aminzade, *Ballots and Barricades: Class Formation and Republican Politics in France, 1830–1871* (Princeton, N.J.: Princeton University Press, 1993), 209.

11. David Garrioch, *The Formation of the Parisian Bourgeoisie, 1690–1830* (Cambridge, Mass.: Harvard University Press, 1996), 266.

12. Philippe Perrot, *Fashioning the Bourgeoisie: A History of Clothing in the Nineteenth Century*, trans. Richard Bienvenu (Princeton, N.J.: Princeton University Press, 1994), 31–32; Bonnie G. Smith, *Ladies of the Leisure Class* (Princeton, N.J.: Princeton University Press, 1981).

13. On the spread of these terms, which stemmed originally from Roman times, see R. B. Rose, "Proletarians and Proletariat: Birth of a Concept, France 1789–1848," in Rose, *Tribunes and Amazons: Men and Women of Revolutionary France, 1789–1871* (Paddington, Australia: Macleay, 1998), 306–23.

14. On the transformation of workers' organizations, see William Sewell Jr., *Work and Revolution in France: The Language of Labor from the Old Regime to 1848* (Cambridge: Cambridge University Press, 1980); and Cynthia Truant, *The Rites of Labor: Brotherhoods of Compagnonnage in Old and New Regime France* (Ithaca, N.Y.: Cornell University Press, 1994).

15. Claire Goldberg Moses, "'Equality' and 'Difference' in Historical Perspective: A Comparative Examination of the Feminisms of French Revolutionaries and Utopian Socialists," in Sara E. Melzer and Leslie W. Rabine, eds., *Rebel Daughters: Women and the French Revolution* (Oxford: Oxford University Press, 1992), 231–54.

16. See especially Michèle Riot-Sarcey, *La démocratie à l'épreuve des femmes. Trois figures critiques du pouvoir, 1830–1848* (Paris: Albin Michel, 1994).

17. Fernand Rude, *Les révoltes des canuts, 1831–1834* (Lyon: Maspéro, 1982), 10. The label "shock city" is borrowed from the classic work of Asa Briggs, *Victorian Cities*. For a recent demonstration of the centrality of events in Lyon to changes in French social and political thinking during this period, see Michèle Riot-Sarcey, *Le réel de l'utopie. Essai sur le politique au XIX siècle* (Paris: Albin Michel, 1998).

18. Michel Chevalier, *A Lyon*, cited in Joseph Benoît, *Confessions d'un prolétaire, Lyon, 1871*, ed. Maurice Moissonnier (Paris: Éditions Sociales, 1968), 63n.

19. The Lyon uprisings have been the subject of several detailed monographs, particularly Fernand Rude, *Le mouvement ouvrier à Lyon de 1827 à 1832* (Paris: Domat-Montchrestien, 1944); Rude, *Révoltes des canuts;* Maurice Moissonnier, *Les canuts. "Vivre en travaillant ou mourir en combattant,"* 4th ed. (Paris: Messidor / Éditions Sociales, 1988); and Robert J. Bezucha, *The Lyon Uprising of 1834: Social and Political Conflict in the Early July Monarchy* (Cambridge, Mass.: Harvard University Press, 1974). For the casualty figures, see Rude, *Révoltes des canuts*, 43, 171.

20. John Bowring, *Second Report on the Commercial Relations Between France and Great Britain, Silks and Wine* (London: William Clowes and Sons, 1835); Louis-René Villermé, *Tableau de l'état physique et moral des ouvriers employés dans les manufactures de coton, de laine et de soie*, 2 vols. (Paris: Jules Renouard, 1840).

21. M. Roustan and C. Latreille, "Lyon contre Paris après 1830. Le mouvement de décentralisation littéraire et artistique," *Revue d'histoire de Lyon* 3 (1904), 24–42, 109–26, 306–18, 384–402.

22. On the circulation of Paris papers in Lyon, see Gilles Feyel, "La diffusion nationale des quotidiens parisiens en 1832," *Revue d'histoire moderne et contemporaine* 34 (1987), 50.

23. M. Lecocq and H.-J. Martin, "Le cas de Lyon," in Roger Chartier and Henri-Jean Martin, eds., *Histoire de l'édition française* (Paris: Fayard, 1990), 3:227–29.

24. Dean de la Motte and Jeannene Przyblyski, "Introduction," in De la Motte and Przyblyski, eds., *Making the News*, 1.

25. *A New History of French Literature*, ed. Denis Hollier (Cambridge, Mass.: Harvard University Press, 1989); the article on the Restoration press is by Tzvetan Todorov: "Freedom and Repression During the Restoration," 617–23.

26. Alexandre Saint-Cheron, "Du journalisme," *Revue encyclopédique*, August 1832, 533.

27. Richard Terdiman has reiterated the essentials of this argument in his recent contribution to De la Motte and Przyblyski: "Afterword: Reading the News," in *Making the News*, 351–76. NOTE: Throughout the present volume, all italics in quoted material are in the original source, not added by the author.

28. Richard Terdiman, *Discourse/Counter-Discourse: The Theory and Practice of Symbolic Resistance in Nineteenth-Century France* (Ithaca, N.Y.: Cornell University Press, 1985), 117–22, 127.

29. Robert Karl Manoff and Michael Schudson, "Reading the News," in Manoff and Schudson, eds., *Reading the News* (New York: Pantheon, 1987), 6.

30. Maurice Mouillaud and Jean-François Tétu, *Le journal quotidien* (Lyon: Presses Universitaires de Lyon, 1989), 22, 23.

31. Ibid., 58.

32. James W. Carey, "The Why and How of American Journalism," in Manoff and Schudson, eds., *Reading the News*, 150.

33. Honoré de Balzac, *Lost Illusions*, trans. Herbert J. Hunt (London: Penguin Books, 1971), 246–47.

34. Albert Camus, "Critique de la nouvelle presse," in Camus, *Actuelles. Chroniques, 1944–1948* (Paris: Gallimard, 1950), 31–32, originally published in *Combat*, 31 August 1944.

35. Terdiman, *Discourse/Counter-Discourse*, 123.

36. Manoff and Schudson, "Reading," in *Reading the News*, 5.

37. Jérôme Morin, *Du journalisme, à propos de la brochure intitulé. De l'enseignement du droit public en France, de M. Bellin, membre de la Société littéraire* (Lyon: L. Boitel, 1842), 42.

38. Marc Martin, *Trois siècles de publicité en France* (Paris: Éditions Odile Jacob, 1993), 82–83, 92. In any event, Martin relativizes the importance of Girardin, whose 1836 experiment was simply an extension of developments already under way since the late 1820s. Terdiman, for his part, relies heavily on the supposed shock effect of the 1836 "commercial revolution" in the French press.

39. [Pierre-Louis Roederer], "Essai analytique sur les diverses moyens établis pour la communication des pensées, entre les hommes en société," *Journal d'économie publique*, 30 brumaire An V (December 1796).

40. Mouillaud and Tétu, *Le journal quotidien*, 102–3.

41. Benedict Anderson, *Imagined Communities*, 2nd ed. (London: Verso, 1991), 35.

42. Otto Groth, *Die Zeitung*, 4 vols. (Mannheim: Bensheimer, 1928), 1:95.

43. Frédéric Barbier, "Une production multipliée," in Chartier and Martin, eds., *Histoire de l'édition française*, 3:118.

44. *Journal du commerce* (Lyon), 20 November 1831.

45. Printed in Percy Sadler, *Paris in July and August 1830* (Paris, 1830), 37–38.

46. "Bert," "La presse parisienne," in *Nouveau Tableau de Paris*, 6 vols. (Paris: Mme. Charles-Béchet, 1834–35), 5:131, 138.

47. Saint-Cheron, "Du journalisme," 535.

48. *Écho de la fabrique*, 23 September 1832. On the *Révolutions de Paris* as embodiment of the revolutionary process, see Pierre Rétat, "Forme et discours d'un journal révolutionnaire: Les *Révolutions de Paris*," in Claude Labrosse and Pierre Rétat, *L'instrument périodique* (Lyon: Presses Universitaires de Lyon, 1985), 139–66.

49. For this concept of the revolution as a process of creating a "new man," see Mona Ozouf, "La révolution française et l'idée de l'homme nouveau," in Colin Lucas, ed., *The Political Culture of the French Revolution* (Oxford: Pergamon, 1988), 213–32.

50. For an exploration of this issue in the French context, see Sewell, *Work and Revolution in France*.

51. For an elaboration of this notion of "media revolution," see Jeremy D. Popkin, "Media and Revolutionary Crises," in Popkin, ed., *Media and Revolution*, 12–30.

52. Mouillaud and Tétu, *Le journal quotidien*, 18.

53. William H. Sewell Jr., "Historical Events as Transformations of Structures: Inventing Revolution at the Bastille," *Theory and Society* 25 (1996), 844.

54. Ibid., 861.

55. Hans-Jürgen Lüsebrink and Rolf Reichardt, *The Bastille: A History of a Symbol of Despotism and Freedom* (Durham, N.C.: Duke University Press, 1997), 47.

56. Sandy Petrey, "Pears in History," *Representations*, no. 35 (1991), 69.

57. Pierre Bourdieu, "The Production of Belief: Contribution to an Economy of Symbolic Goods," in Pierre Bourdieu, *The Field of Cultural Production*, ed. Randal Johnson (New York: Columbia University Press, 1993), 78.

58. Pierre Bourdieu, "The Field of Cultural Production, or The Economic World Reversed," in Bourdieu, *Field of Cultural Production*, 30–31.

Chapter 1: Newspapers, Journalists, and Public Space

1. Pierre Bourdieu, "The Field of Cultural Production," in Bourdieu, *Field of Cultural Production*, 29–30.

2. "La Revue de Lyon," n.d. [Spring 1835], in Bibliothèque municipale de Lyon (hereafter BML), Fonds Coste, Estampe no. 775.

3. Archives nationales [Paris] (hereafter AN), F 18 495(A), prefect to minister of the interior, 4 January 1815.

4. Louis Trénard, *Lyon de l'Encyclopédie au préromantisme* (Paris: Presses Universitaires de France, 1958), 136–37.

5. On the ambivalent functions of the old-regime provincial press, see Claude Labrosse, "La region dans la presse regionale," in Jean Sgard, ed., *La presse provinciale au XVIIIe siècle* (Grenoble: Centre de Recherches sur les Sensibilités, 1983), 70–106; and Colin Jones, "The Great Chain of Buying: Medical Advertisement, the Bourgeois Public Sphere, and the Origins of the French Revolution," *American Historical Review* 101 (1996), 13–40.

6. Jeremy D. Popkin, "The Provincial Newspaper Press and Revolutionary Politics," *French Historical Studies* 18 (1993), 434–56.

7. *Journal de Lyon*, 31 December 1789.

8. Trénard, *Lyon*, 252–56, 348–53, 420–21.

9. On the role of the *Journal de Lyon* during the thermidorian period, see Renée Fuoc, *La réaction thermidorienne à Lyon, 1795* (Lyon: Les Éditions de Lyon, 1957), 41–42.

10. AN, F 18 495(A), De Cazes to prefect, 6 November 1815.

11. F. Baud, "La fondation et les débuts du *Précurseur*," *Revue d'histoire de Lyon* 13 (1914), 350–62. The paper was briefly revived in January 1823 and then suspended until 1826.

12. *L'Espiègle lyonnais*, 13 June 1824, copy in ADR, 4 M 449.

13. The attempts to close the paper down are documented in H. Valois, *Défense du Précurseur, journal de Lyon et du midi* (Lyon: Brunet, 1826), and in a letter from the minister of justice to minister of the interior, 28 November 1826, in AN, F 18 495(I). On the paper's subsequent legal troubles, see the prefect's letters to the Ministry of the Interior, 14 June 1827; 30 June 1827; 22 August 1827, in AN, F 18 495(I) and *Précurseur*, 13 May 1827; 1 July 1827; 22 August 1827.

14. AN, F 18 495(C), dossier *Champ de Mai, Journal du commerce de Lyon*.

15. Prefect to Ministry of Interior, 29 March 1828, in ADR, 4 M 449.

16. On the *Écho du Jour*, see Th. Pitrat to minister of the interior, n.d. [March 1828], in ADR, 4 M 449, and AN, F 18 495(D). For the *Gazette de Lyon*, see AN, F 18 495(F).

17. Ministry of the Interior, "Rapport," in AN, F 18 495(F), dossier *Gazette de Lyon*.

18. *Précurseur*, 1 January 1831.

19. *Journal du commerce*, 19 September 1830.

20. François-Nicolas Bavoux, *Développement de la proposition faite par M. Bavoux sur le cautionnement, le droit de timbre et le port des journaux et écrits périodiques* (Paris: Imprimerie Nationale, 1830), 8; *Résumé de la discussion générale de la proposition de M. Bavoux, relative aux journaux et ecrits périodiques* (Paris: Imprimerie Nationale, 1830).

21. Marcel Gauchet, "Right and Left," in Pierre Nora, ed., *Realms of Memory: The Construction of the French Past*, trans. Arthur Goldhammer (New York: Columbia University Press, 1996), 261.

22. *Glaneuse*, 9 July 1831.

23. *Précurseur*, 2 July 1831.

24. Sewell, *Work and Revolution*, 197–201.

25. *Précurseur du peuple*, [28 January 1834], in AN, CC 572.

26. Gasparin to minister of the interior, 9 December 1831 and 11 December 1831, in Archives municipales de Lyon (hereafter AML), Gasparin papers, 4 II 6 (on his initial efforts to control the *Précurseur* and his role in the founding of the *Courrier de Lyon*); Minister d'Argout to Gasparin, 13 August 1833 (congratulating him on influencing the choice of a new editor for the *Écho de la fabrique*) and 10 October 1833 (payment to *Journal du commerce*), in Gasparin papers, 4 II 9; *Mosaïque lyonnaise*, list of stockholders, 11 October 1834.

27. Bouvier-Dumolard, *Relation de M. Bouvier du Molart, ex-préfet du Rhône, sur les événemens de Lyon* (Lyon: Bureau du Journal du Commerce, 1832), 20, 29; Casimir-Périer to Gasparin, 21 December 1831, in AML, Gasparin papers, 4 II 2.

28. Gasparin to Casimir-Périer, 9 December 1831, in AML, Gasparin papers, 4 II 6.

29. Gasparin, report of 6 May 1834, in AML, Gasparin papers, 4 II 11.

30. On the denunciations of journalists as immoral, see Victoria Thompson, "*Splendeurs et misères des journalistes:* Imagery and the Commercialization of Journalism in July Monarchy France," *Proceedings of the Western Society for French History* 23 (1996), 361–68; and William Reddy, *The Invisible Code: Honor and Sentiment in Postrevolutionary France* (Berkeley and Los Angeles: University of California Press, 1997), chap. 5, "Condottieri of the Pen."

31. [Sébastien Kauffmann], *Biographie contemporaine des gens de lettres de Lyon* (Lyon and Paris: Chez Tous les Marchands de Nouveautés, 1826). Many of the same names also appear in *Biographie lyonnaise des auteurs dramatiques vivans, dits du terroir, rédigée dans la loge du portier des célestins; enrichie de quelques notes, par un bon enfant* (Lyon: Coque, n.d.). Both works omit the city's small number of woman authors.

32. Sébastien Kauffmann, *La Célestinade, ou La guerre des auteurs et des acteurs lyonnais, Poeme hério-comique en quatre chants* (Lyon: Laforgue, 1829), 4.

33. Anselme Petetin, in *Lyon vu de Fourvières. Esquisses physiques, morales et historiques* (Lyon: L. Boitel, 1833), xvii.
34. Kauffmann, *Célestinade*, 6, 78.
35. Petetin, in *Lyon vu de Fourvières*, x.
36. *National*, 5 June 1830.
37. Fuoc, *La réaction thermidorienne à Lyon*, 41.
38. ADR, T 353, entry of 22 May 1833.
39. Morin contract, 28 December 1828, in AN, CC 572.
40. AN, F 18 495 (F), dossier *Glaneuse*, contract of 29 September 1833.
41. The printer Léon Boitel, a major figure in Lyon's cultural life, served as legal figurehead for her two Lyon periodicals, the *Conseiller des femmes* and its successor, the *Mosaïque*. Niboyet's husband, Mouchon, was a member of the stockholders' committee for the latter paper.
42. *Précurseur*, 19 May 1830.
43. The paper had contracted with a Parisian journalist named Brait de la Mathe to function as its *gérant;* he evidently fulfilled his function by signing blank sheets of paper that were then used to print the official galley proof the paper had to submit to the local prefect. When other Lyon papers began to claim that Brait de la Mathe had died, the *Précurseur*, rather than refuting them, simply challenged them to prove their claims (5 November 1831). Brait de la Mathe's death was confirmed in the *Écho de la fabrique*, 20 November 1831, by which time the *Précurseur* had finally hired another *gérant*.
44. *Précurseur*, 26–27 December 1831.
45. *Précurseur*, 18 November 1831.
46. *Écho de la fabrique*, 9 September 1832.
47. Jean-Baptiste Monfalcon, *Souvenirs d'un bibliothécaire, ou Une vie d'homme de lettres en province* (Lyon: Nigon, 1853), 138.
48. Ibid., 134–35.
49. Ibid., 101.
50. *Précurseur*, 8 June 1831.
51. Favre to [Armand Carrel], 5 July 1833, in AN, CC 572.
52. Petetin, "Mémoire pour Anselme Petetin" [1835], in AN, CC 572.
53. On the significance of dueling in nineteenth-century French culture, see Robert A. Nye, *Masculinity and Male Codes of Honor in Modern France* (New York: Oxford University Press, 1993), esp. chap. 7. In 1832, supporters of the Duchesse de Berry, the mother of the Bourbon pretender who had been imprisoned after returning to France to launch a legitimist uprising, carried out a systematic nationwide campaign of challenges to journalists, including Monfalcon, who had reported her illegitimate pregnancy. Judges and government officials tried to convince the public that journalists should not have to defend their personal honor against "these new censors who don't mutilate books but kill authors," but most journalists were afraid to decline even these politically motivated challenges (*Gazette des Tribunaux*, 7 February 1833). In Lyon, Monfalcon worked out his own solution to this problem; granted the choice of conditions, he insisted on a single pair of pistols, only one of them loaded, and an exchange of shots at two paces. His aristocratic challenger's seconds talked their man out of an encounter that could hardly have failed to have a fatal result. Monfalcon, *Souvenirs*, 141–42.
54. Monfalcon, *Souvenirs*, 145, 191.
55. *Écho des travailleurs*, 8 January 1834.
56. Figures from AN, BB 20 56 (1831), BB 20 61 (1832), BB 20 66 (1833), BB 20 74 (1834). Determining the number of convictions is difficult, because a number of verdicts

delivered in absentia were later reversed when the defendant actually appeared in court; a number of trials involved multiple charges and penalties.

57. Théodore Pitrat, *Vote universel. Eléction à l'Assemblée nationale. Candidature* (n.d. [1848]), copy in BML, Pierre Charnier papers. This four-page declaration of Pitrat's candidacy for a seat in the National Assembly elected in April 1848 consists primarily of an account of the author's life.

58. Verne de Bachelard to minister of justice, 5 September 1833, in AN, BB 20 66.

59. Petetin to Monfalcon, 23 June [1834], tipped in to BML copy of Monfalcon, *Code moral des ouvriers, ou Traité des devoirs et des droits des classes laborieuses* (Paris and Lyon: Pélagaud Lesne et Crozet, 1836), sig. Rés. 434157. Monfalcon had a collection of letters to himself bound in at the back of this copy.

60. Cited in Riot-Sarcey, *La démocratie à l'épreuve des femmes*, 63. See also Niboyet's letters in Fonds Enfantin, carton 7815, Rapports d'Eugénie Niboyet et de Bottiau, July–December 1831.

61. On Chastaing's career, see below, Chapter 4.

62. Pitrat, *Vote universel*. In 1848, Pitrat ran on a democratic platform, but his election manifesto detailed his efforts on behalf of legitimism in earlier decades.

63. Anselme Petetin, "Des tendances socialistes imputées au gouvernement," *Revue de Lyon* (1849), 105.

64. Baron Amédée Girod de l'Ain, *Rapport fait à la Cour des Pairs. Affaire du mois d'avril 1834*, 10 vols. (Paris: Imprimerie Royale, 1834–35), 2:106.

65. Aimé Vingtrinier, "Notice nécrologique sur M. Léon Boitel," *Revue du Lyonnais* 2 (1855), 195; Eugène Vial, "La vie et l'oeuvre de Léon Boitel," *Revue du Lyonnais* 1 (1921), 109–21. The prefecture records in ADR, T 353, show Boitel listed at one time or another as the printer of the *Papillon*, the *Glaneuse*, the *Précurseur*, the *Conseiller des femmes*, *L'Industriel*, the *Écho de la fabrique*, the *Tribune prolétaire*, the *Mosaïque lyonnaise*, and the *Revue Lyonnaise*.

66. ADR, 4 M 450, 7 September 1833; 11 September 1833; 20 November 1834; Pitrat, *Vote universel*.

67. Jürgen Habermas, *Strukturwandel der Öffentlichkeit* (Frankfurt: Luchterhand, 1962).

68. See Joan Landes's feminist critique of Habermas in *Women and the Public Sphere in the Age of the French Revolution* (Ithaca, N.Y.: Cornell University Press, 1988), 39–45.

69. Jürgen Habermas, "The Public Sphere," cited in in Geoff Eley, "Nations, Publics, and Political Cultures: Placing Habermas in the Nineteenth Century," in Craig Calhoun, ed., *Habermas and the Public Sphere* (Cambridge, Mass.: MIT Press, 1992), 289.

70. On the significance of Habermas's distinction between the public and private dimensions of civil life, see Dena Goodman, "Public Sphere and Private Life: Toward a Synthesis of Current Historiographical Approaches to the Old Regime," *History and Theory* 31 (1992), 1–20.

71. Habermas, "The Public Sphere," cited in Eley, "Nations," 289.

72. On St.-Étienne in the early nineteenth century, see Aminzade, *Ballots and Barricades*, 141ff.

73. See John M. Merriman, *The Margins of City Life: Explorations on the French Urban Frontier, 1815–1851* (New York: Oxford University Press, 1991), 3.

74. Jean-Baptiste Monfalcon and A.-P.-I. Polinière, *Traité de la salubrité dans les grandes villes, suivi de l'hygiène de Lyon* (Paris: J. B. Ballière, 1846), 383.

75. Of the forty booksellers listed in the city almanac for 1831, two were located west

of the Saône, one was in the Croix-Rousse, one was in the nearby town of Villefranche, and the remaining thirty-six were in the *presqu'île*. See *Almanach historique et politique de la ville de Lyon et du département du Rhône, pour l'an de grace 1831* (Lyon: Rusand, 1831), 80–81.

76. *Précurseur*, 7 January 1831.

77. On the evolution of descriptive literature about Lyon, see Gilbert Gardes, *Le Voyage de Lyon* (Lyon: Horvath, 1993).

78. *Lyon vu de Fourvières*.

79. Pierre-Yves Saunier, "Haut-lieu et lieu haut: La construction du sens des lieux. Lyon et Fourvière au XIXe siècle," *Revue d'histoire moderne et contemporaine* 40 (1993), 202–27.

80. Léon Boitel, "La place des Célestins," in *Glaneuse*, 27 October 1831.

81. *Glaneuse*, 16 June 1831.

82. Lyon had four *cercles* at the time of the 1830 revolution. A fifth was created in 1834. AML, I (2) 40, no. 690. On the importance of the *cercle* in French bourgeois life during this period, see Maurice Agulhon, *Le cercle dans la France bourgeoise, 1810–1848* (Paris: Armand Colin, 1977). On the *cercles* of Lyon, see Catherine Pellissier, *Loisirs et sociabilités des notables lyonnais au XIXe siècle* (Lyon: Éditions Lyonnaises d'Art et d'Histoire, 1996), 158–69.

83. *Précurseur*, 7 January 1831.

84. *Conseiller des femmes*, 5 April 1834.

85. *Épingle*, 28 May 1835.

86. Edmond Goblot, *La barrière et le niveau* (1925; Paris: Presses Universitaires de France, 1967), 9.

87. AN, F 7 6998, 27 May 1827.

88. *Épingle*, 29 January 1835.

89. Police documents list nine lodges founded before 1830 that were still in existence in the early years of the July Monarchy. Two others were founded in the early 1830s. AML, I (2) 40, no. 467.

90. Léon Boitel, "Les tilleuls de Bellecour," in *Lyon vu de Fourvières*, 114.

91. AN, F 7 6998, 27 May 1827.

92. *Glaneuse*, 28 July 1831.

93. *Précurseur*, 31 December 1832.

94. ADR, 4 M 449, report of 1 June 1820.

95. "Un cabinet de lecture," *Glaneuse*, 18 August 1831.

96. *Glaneuse*, 15 September 1831.

97. *Précurseur*, 7 January 1831.

98. AML, I (2) 40, no. 690.

99. *Épingle*, 28 May 1835.

100. *Glaneuse*, 18 August 1831.

101. Agulhon, *Le cercle dans la France bourgeoise*.

102. Sermet and Auguste Baron, letter to Prefect Paulze d'Ivoy, 21 September 1830, in ADR, 4M 449.

103. *Écho de la fabrique*, prospectus, n.d. [October 1831]; *Écho des travailleurs*, prospectus, 5 October 1833. Auguste Baron, one of the backers of the unsuccessful 1830 project, was listed as a vendor for the *Écho de la fabrique*.

104. *Écho de la fabrique*, 6 November 1831.

105. *Cri du peuple*, 29 May 1831.

106. *Précurseur*, 26–27 October 1831.

107. *Glaneuse*, 9 January 1834.
108. *Précurseur du peuple*, "La liberté de la presse est un mensonge" and "La censure est rétablie," in AN, CC 572, d. "Imprimés divers concernant la participation de Petetin à l'affaire des crieurs publics."
109. Prunelle to Gasparin, 18 January 1834, in AML, Gasparin papers, 4 II 11.
110. *National*, 7 February 1834 and 28 January 1834. For the *Précurseur*'s sales claim, see *Précurseur du peuple*, "Lettre des Républicains de Lyon à M. le procureur du Roi," in AN, CC 572.
111. *Papillon*, 31 July 1832; 30 October 1833.
112. *Conseiller des femmes*, 4 January 1834.
113. *Conseiller des femmes*, 1 February 1834.

Chapter 2: The Press, Liberal Society, and Bourgeois Identity

1. Sarah Maza, "Luxury, Morality, and Social Change: Why There Was No Middle-Class Consciousness in Prerevolutionary France," *Journal of Modern History* 69 (1997), 201.
2. Goblot, *La barrière et le niveau*; Jean-Pierre Chaline, *Les bourgeois de Rouen. Une élite urbaine au XIXe siècle* (Paris: Presses de la Fondation Nationale des Sciences Politiques, 1982); Adeline Daumard, *Les bourgeois et la bourgeoisie en France depuis 1815* (Paris: Aubier, 1987); Jürgen Kocka, "The Middle Classes in Europe," *Journal of Modern History* 67 (1995), 783–806; Reddy, *Invisible Code*; Smith, *Ladies of the Leisure Class*.
3. Kocka, "Middle Classes," 785–88.
4. On the nineteenth-century secondary-education system and its curriculum, see Antoine Prost, *Histoire de l'enseignement en France, 1800–1967* (Paris: Armand Colin, 1967).
5. Anderson, *Imagined Communities*, 77.
6. See, for example, Roddey Reid, *Families in Jeopardy: Regulating the Social Body in France, 1750–1910* (Stanford, Calif.: Stanford University Press, 1993).
7. Patrice Higonnet, *Class, Ideology, and the Rights of Nobles During the French Revolution* (Oxford: Clarendon Press, 1981), esp. 28–36.
8. *Furet de Lyon*, 8 April 1832.
9. Pierre Rosanvallon, *Le Moment Guizot* (Paris: Gallimard, 1985).
10. Charles-Joseph Panckoucke, "Sur les journaux anglais," *Mercure de France*, 30 January 1790.
11. Jeremy Popkin, *Revolutionary News: The Press in France, 1789–1799* (Durham, N.C.: Duke University Press, 1990), 61–78, 143–68.
12. *National de 1834*, 27 January 1834.
13. *National de 1834*, 28 January 1834.
14. *Le Temps*, 2 December 1831 (Lyon, 29 November); reprinted in part in *National*, 3 December 1831.
15. *Écho des travailleurs*, 6 November 1833.
16. "Bert," "La presse parisienne," in *Nouveau Tableau de Paris*, 6 vols. (Paris: Mme. Charles-Béchet, 1834–35), 5:139.
17. *Précurseur*, 20 December 1833.
18. Charles E. Freedeman, *Joint-Stock Enterprise in France, 1807–1867* (Chapel Hill: University of North Carolina Press, 1979), 48, 52.
19. Contract between Jérôme Morin and stockholders, 28 December 1828, in AN, CC 572; *Précurseur*, 7 July 1831.

20. Gasparin to minister of the interior, 19 June 1832, in AML, Gasparin papers, 4 II 4.

21. Bouvier-Dumolard, *Relation*, 29. The paper in question was presumably the *Précurseur*.

22. Pitrat, *Vote universel;* ADR 4 M 450, report of 20 November 1834.

23. Monfalcon, *Souvenirs*, 5.

24. Ibid., 137. In July 1832, the president of the journal's board of directors wrote to a Parisian official to complain that pettyfogging officials were preventing the paper from selling some of the government bonds that formed its legal capital in order to cover its deficit. Letter of 12 July 1832, in AML, 4 II 1, Fonds Gasparin.

25. Jean-Baptiste Monfalcon, *Histoire des insurrections de Lyon, en 1831 et 1834, d'après des documents authentiques, précédée d'un essai sur les ouvriers en soie et sur l'organisation de la fabrique* (Lyon: Perrin, 1834), 145, 160–61.

26. AM Lyon, 4 II 10, Fonds Gasparin, Gasparin, 15 November 1832.

27. *Glaneuse*, 18 August 1831.

28. *Précurseur*, 30 August 1832. Not all of these were necessarily supporters of the paper's recently announced republican allegiance; as the editor Petetin frequently complained, there was a sizable minority of determined Orleanists who opposed him from within the stockholders' council.

29. Letters to Anselme Petetin in AN, CC 572, from Duc fils, 13 May 1833, from Pascal, 26 April 1833, etc. These documents do not furnish any evidence about the social composition of the readership in Lyon.

30. ADR, T 353, entries for 31 December 1831 and 6 November 1832; Monfalcon, *Souvenirs*, 132.

31. *Mosaïque*, 11 October 1834.

32. *Tribune prolétaire*, 10 May 1835.

33. AM Lyon, 4 II 10, Fonds Gasparin, "Rapport sur les différentes sociétés secrètes établies à Lyon," 6 April 1833.

34. AML, Gasparin papers, 4 II 11, note of 6 May 1834.

35. F. Baud, "La fondation et les débuts du Précurseur," *Revue d'histoire de Lyon* 13 (1914), 350–62. There was a short-lived effort to revive the paper in January 1823, but it was suppressed after a single issue. AN, F 18 495 (I), prefect's letter and order dated 25 January 1823.

36. AN F 18 495 I, 19 November 1826.

37. Contract between *Précurseur* stockholders and Jérôme Morin, 28 December 1828, in AN, CC 572.

38. *Précurseur*, 17 August 1826 (copy in AN, F 18 495[I])

39. *Précurseur*, 17 August 1826.

40. *Précurseur*, 17 August 1826.

41. Monfalcon, *Souvenirs*, 137; AN, F 18 495A, list of authorized periodicals in the *département* of the Rhône, 16 March 1829.

42. Th. Pitrat to minister of the interior, March 1828, in ADR, 4 M 449.

43. *Précurseur*, 17 August 1826.

44. *Précurseur*, 8 November 1828.

45. *Précurseur*, 28 July 1830.

46. *Précurseur*, 30 July 1830, ADR 4 M 449, dossier on enforcement of July Ordinances.

47. *Précurseur*, 31 July 1830.

48. ADR, 4 M 449, dossier on enforcement of July Ordinances; AN, F 18 495, prefect's

report of 30 July 1830; *Précurseur*, 30 and 31 July 1830. Pamphlet accounts included Trolliet, *Lettres historiques sur la révolution de Lyon, ou Une semaine de 1830* (Lyon: Targe, 1830), 4–12; and Claude Mornand, *Une semaine de révolution, ou Lyon en 1830* (Lyon: André Idt, 1831), 16, 25, 106; for Monfalcon's version, see *Histoire des insurrections de Lyon*, 143.

49. *Journal du commerce*, 3 August 1830.
50. *Précurseur*, 1 January 1831.
51. *Précurseur*, 1 January 1831.
52. *Précurseur*, 4 January 1831, "Letter from Paris."
53. *Précurseur*, 17 January, 6 February 1831.
54. *Précurseur*, 23–24 March 1831.
55. *Précurseur*, 29 April, 5 May 1831.
56. *Sentinelle*, 8 July 1831.
57. *Précurseur*, 6 June 1831.
58. *Précurseur*, 8 June 1831.
59. Monfalcon, *Souvenirs*, 101.
60. *Précurseur*, 18 November 1831.
61. *Précurseur*, 18 November 1831.
62. *Précurseur*, 23 and 24 October 1831. These articles were probably written by Jean-Baptiste Monfalcon, who at this point still had hopes of being named the paper's editor-in-chief.
63. *Précurseur*, 26–27 October 1831; 6 November 1831.
64. *Précurseur*, 25 October 1831. This article was by the paper's former editor, Morin.
65. *Journal du commerce*, 28 October 1831 (sympathetic to workers), 6 November 1831 (critical of them).
66. *Courrier de Lyon*, prospectus, n.d. [late December 1831].
67. *Précurseur*, 26–27 December 1831.
68. *Précurseur*, 3 December 1831; AML, Gasparin papers, 4 II 6, letter of 9 December 1831.
69. *Précurseur*, 9 September 1832.
70. *Précurseur*, 13–14 June 1832.
71. *Précurseur*, 29 August 1832.
72. *Précurseur*, 4 November 1832. On the social and political significance of subsidies to municipal theaters, which often amounted to more than spending on education, see William B. Cohen, *Urban Government and the Rise of the French City: Five Municipalities in the Nineteenth Century* (New York: St. Martin's Press, 1998), 127–36.
73. *Précurseur*, 10 March 1833.
74. *Précurseur*, 2 and 4 February 1833.
75. Carrel to Petetin, letter of 5 March 1833, in AN, CC 572.
76. *Précurseur du peuple*, "La liberté de la presse est un mensonge," in AN, CC 572.
77. Petetin, "Mémoire pour Anselme Petetin," in AN, CC 572, p. 58.
78. *National*, 8 February 1834.
79. *Glaneuse*, 23 April 1833; 14 April 1833; 25 April 1833. The *Catéchisme républicain* was drawn up by the Paris Société des droits de l'homme.
80. Monfalcon, *Souvenirs*, 101. The prefect Gasparin reported his efforts to support this initiative to the interior ministry soon after his arrival in Lyon; Casimir-Périer responded by praising the idea but reminding him that money was tight. AML, 4 II 6, Fonds Gasparin, reports of 9 and 11 December 1831 and 4 II 2, letter of 21 December 1831.

81. *Journal du commerce*, 24 November 1833.
82. *Courrier de Lyon*, 15 January 1832.
83. *Courrier de Lyon*, 17 February 1833.
84. *Courrier de Lyon*, 30 October 1833.
85. *Courrier de Lyon*, 24 February 1833.
86. *Précurseur*, 10 March 1833; 29 April 1833.
87. Monfalcon, *Souvenirs*, 144.
88. Monfalcon, *Histoire des insurrections de Lyon*, 145n, 159n.
89. Monfalcon, *Code moral*, 713.
90. *Gazette du Lyonnais*, 1 May 1833.
91. *Cri du peuple*, 29 May 1831; *Réparateur*, 28 August 1833.
92. *Cri du peuple*, 20 July 1831.
93. *Cri du peuple*, 13 October 1831.
94. *Réparateur*, 30 August 1833, *Gazette du Lyonnais*, 23 February 1834.
95. Georges Ribe, *L'opinion publique et la vie politique à Lyon lors des premières années de la seconde Restauration* (Paris: Sirey, 1957), 62–63.
96. *Précurseur*, 6 November 1831.
97. *Courrier de Lyon*, 21 August 1834.
98. *Précurseur*, 16 April 1834; Monfalcon, *Histoire des insurrections de Lyon*, 316–19. The Paris republican paper *National de 1834* (16 April 1834) reported that "the civil population . . . was treated as suspect by the soldiers who had supposedly been sent to its defense."
99. Bouvier-Dumolard, *Relation*, 20.
100. Reports of 9 and 11 December 1831, in AML, Gasparin papers, 4 II 6.
101. Minister d'Argout to Gasparin, letter of 19 September 1833, in AML, Gasparin papers, 4 II 9.
102. Report of 6 May 1834 in AML, Gasparin papers, 4 II 11.
103. Monfalcon, *Code moral*, 713.
104. *Précurseur*, 21 February 1834; Monfalcon, *Histoire des insurrections de Lyon*, 309–10; Petetin, "Mémoire," in AN, CC 572, p. 66.
105. *Censeur*, prospectus, 20 November 1834.
106. *Censeur*, 27 February 1848; *Courrier de Lyon*, 29 February 1848; Jean-Baptiste Monfalcon, *Annales de la ville de Lyon, ou Histoire de notre temps*, 2 vols. (Lyon: n.p., 1849–50), 1:18.

Chapter 3: Reshaping Journalistic Discourse: The Alternative Press in Lyon

1. *Glaneuse*, 1 September 1831.
2. *Conseiller des femmes*, 23 November 1833.
3. *Glaneuse*, 16 June 1831; *Papillon*, 22 February 1835.
4. *Papillon*, 25 January 1834; 13 March 1834.
5. *Quatrième procès de la Glaneuse*, n.d. [March or April 1832].
6. *Papillon*, 28 July 1832.
7. On Boitel, see Eugène Vial, "La vie et l'oeuvre de Léon Boitel," *Revue du Lyonnais* 1 (1921), 109–21.
8. *Conseiller des femmes*, prospectus.
9. *Papillon*, prospectus, 30 June 1832.
10. *Glaneuse*, 18 September 1831.
11. *Glaneuse*, 18 August 1831.

Notes to Pages 111–123

12. *Glaneuse*, 18 September 1831.
13. *Glaneuse*, 4 August 1831. On its significance, see Maurice Agulhon, *Marianne into Battle: Republican Imagery and Symbolism in France, 1787–1880* (Cambridge: Cambridge University Press, 1981), 18, 21–22.
14. *Glaneuse*, 7 December 1831.
15. *Glaneuse*, 29 April 1832; *Papillon*, 4 September 1832.
16. Louis-Auguste Berthaud, *Asmodée, Satire* (Lyon: André Idt, n.d. [1833]). Berthaud's work was inspired by a Parisian model, Joseph Barthélémy's republican *Némesis*. Arrested in Lyon in March 1833, Berthaud moved to Paris and started a similar publication, *L'homme rouge, satire hebdomadaire*, co-edited with Veyrat. AN, BB 20 66 (entry for 25 March 1833).
17. *Glaneuse*, 25 December 1831.
18. Gabriel Perreux, *Au temps des sociétés secrètes* (Paris: Rieder, 1931), 104.
19. Musée Gadagne (Lyon), Fonds Justin Godart, Journal de Joseph Bergier, 9 October 1833.
20. AML, Gasparin papers, II 4 11, 6 May 1834.
21. *Glaneuse*, 16 June 1831.
22. *Glaneuse*, 16 June 1831.
23. *Glaneuse*, 9 July 1831; 28 July 1831.
24. On the *poissard* genre, see Pierre Frantz, "Travestis poissards," *Revue des sciences humaines* 190 (1983), 7–20; and Alexander Hull, "The First Person Plural Form: *Je Parlons*," *French Review* 62 (1988), 242–48. On the *Père Duchesne*, see Jeremy D. Popkin, *Revolutionary News*, 151–68.
25. *Glaneuse*, 31 July 1831.
26. *Glaneuse*, 28 March 1833.
27. *Glaneuse*, 13 July 1831.
28. *Glaneuse*, 28 October 1832.
29. *Procès de la Glaneuse, contenant les douze articles incriminés* (Lyon: Aux Bureaux de la Glaneuse, 1833), 43; *Association républicaine pour la liberté individuelle et la liberté de la presse. Procès de la Glaneuse, journal républicain de Lyon* (Paris: Auffray, [1833]), 5, 8.
30. *Conseiller des femmes*, 28 December 1833.
31. *Quatrième Procès de la Glaneuse*, n.d. [March or April 1832]. A copy of this pamphlet is tipped in to the Bibliothèque nationale's collection of the paper for 1831.
32. *Glaneuse*, 22 March 1832.
33. Terdiman, *Discourse/Counter-Discourse*, 165.
34. *Glaneuse*, 25 November 1831.
35. *Glaneuse*, 22 December 1831.
36. *Glaneuse*, 29 April 1832.
37. *Glaneuse*, 11, 14, 16, 18 April 1833. On the background of the "Catechism," see Bezucha, *Lyon Uprising*, 85–86.
38. Sophie Grangé, "Louise," in *Papillon*, 25 August 1832.
39. Claire G. Moses, *French Feminism in the Nineteenth Century* (Albany, N.Y.: SUNY Press, 1984), 63–70.
40. *Espiègle*, 13 June 1824.
41. *Papillon*, 3 July 1832.
42. *Papillon*, prospectus, 30 June 1832.
43. *Papillon*, 14 July 1832.
44. *Papillon*, 1 September 1832.

45. *Papillon*, 4 September 1832. For the full text of Grangé's poem, see Appendix 1. Grangé had published two small collections of poetry in 1826. An article in the *Papillon* just a month later lamented that she had decided to give up writing. Whether she published subsequently under a married name has been impossible to determine.

46. *Papillon*, Clara Francia, "A elle," 11 August 1832; 19 March 1833; 3 July 1832.

47. *Papillon*, 31 July 1834; 28 August 1834.

48. *Papillon*, 10 August 1834, by "Mlle. J.D." [probably Jane Dubuisson, who also contributed to the *Conseiller des femmes*].

49. Bibliothèque municipale de Lyon, Fonds Coste, Ms. 1130, letter of 7 October 1833. The *Papillon* urged support for the *Conseiller* on 30 October 1833.

50. Older studies of Eugénie Niboyet have been superseded by Riot-Sarcey, *La démocratie à l'épreuve des femmes*, which includes a listing of archival sources on Niboyet's life. See also Laure Adler, *A l'aube du féminisme. Les premières journalistes, 1830–1850* (Paris: Payot, 1979); Lucette Czyba, "L'oeuvre lyonnaise d'une ancienne saint-simonienne: Le *Conseiller des femmes* (1833–1834) d'Eugénie Niboyet," in J. R. Derré, ed., *Regards sur le Saint-Simonisme et les Saint-Simoniens* (Lyon: Presses Universitaires de Lyon, 1986), 103–41; Laura Strumingher, "Mythes et réalités de la condition féminine à travers la presse féministe lyonnaise des années 1830," *Cahiers d'histoire* 21 (1976), 409–24; Maximilien Buffenoir, "Le féminisme à Lyon avant 1848," *Revue de l'histoire de Lyon* 7 (1906), 348–58; and Sullerot, *Histoire de la presse féminine en France*, 186–87.

51. Most of Ulliac Trémadeure's articles in the *Conseiller des femmes* appeared under the pseudonym of Ulliac Dudrezène. The Bibliothèque nationale catalog lists numerous publications by her, dealing with children's education and popularizing scientific and historical topics, and ranging in date from 1821 to the 1860s. Her articles in the *Conseiller des femmes* show that she deserves to be considered among the most radical of the French feminists in the early 1830s. In her autobiography, *Souvenirs d'une vieille femme*, 2 vols. (Paris: E. Maillet, 1861), she tried to distinguish herself from "women who no longer wanted to be part of their sex" (2:70), but her articles of the early 1830s had a stronger tone. As she wrote in her autobiography, "I was not going to accept with resignation the mockery that people thought they could indulge in without punishment."

52. *Conseiller des femmes*, prospectus, n.d. [1833].

53. Niboyet, the daughter of a doctor, had been born in 1796 to a prominent Protestant family. She had participated in the Saint-Simonian movement in 1831, before leaving Paris for Lyon. After the demise of the *Conseiller des femmes*, she was active in various movements for the disadvantaged. In 1844, she founded France's first pacifist periodical, and she resumed her advocacy of women's rights during the Revolution of 1848, editing a Parisian newspaper, *La Voix des femmes*, and serving as president of the Club des femmes. Riot-Sarcey, *Démocratie*, 51, 162–66.

54. *Conseiller des femmes*, 2 November 1833.

55. *Conseiller des femmes*, 2 November 1833.

56. *Conseiller des femmes*, 16 November 1833.

57. *Conseiller des femmes*, 7 December 1833.

58. *Conseiller des femmes*, 5 April 1834.

59. *Papillon*, 1, 4, 8 September 1832.

60. *Conseiller des femmes*, 28 December 1833.

61. *Conseiller des femmes*, 7, 14 December 1833.

62. *Conseiller des femmes*, 15 February 1834.

63. *Conseiller des femmes*, 8 March 1834.

64. *Conseiller des femmes*, 11 January 1834.

65. *Conseiller des femmes*, 2 November 1833
66. Riot-Sarcey, *Démocratie*, 138.
67. Claire Goldberg Moses, "'Equality' and 'Difference' in Historical Perspective: A Comparative Examination of the Feminisms of French Revolutionaries and Utopian Socialists," in Melzer and Rabine, eds., *Rebel Daughters*, 231.
68. *Écho des travailleurs*, 27 November 1833.
69. Eugénie Niboyet to Eugène de Lamerlière, 7 October 1833, in BML, Fonds Coste, Ms. 1130.
70. *Écho de la fabrique*, 27 October 1833.
71. *Écho des travailleurs*, 27 November 1833; *Courrier de Lyon*, 16 December 1833.
72. *Conseiller des femmes*, 11 January 1834.
73. *Papillon*, 30 October 1833.
74. *Conseiller des femmes*, 7 June 1834.
75. *Conseiller des femmes*, 14 December 1833.
76. Smith, *Ladies of the Leisure Class*, esp. chap. 4.
77. Amédée Roussillac, "Revue bibliographique," in *Revue du Lyonnais* 1 (1835), 406–7.
78. Monfalcon to Boitel, 7 January 1850, in BML, Fonds Charavay, Ms. 613.
79. *Papillon*, 15 January 1835.
80. *Mosaïque*, 11 October 1834.
81. *Papillon*, 15 January 1835. On the cultural significance of the *Revue du Lyonnais*, see Pierre-Yves Saunier, "Représentations sociales de l'espace et histoire urbaine. Les quartiers d'une grande ville française, Lyon au XIXe siècle," *Social History / Histoire sociale* 29 (1996), 42.

Chapter 4: Echoes of the Working Classes

1. The first English publication along similar lines, the *Trades' Newspaper*, had appeared in 1825. Iorwerth Prothero, *Artisans and Politics in Early Nineteenth-Century London: John Gast and His Times* (Folkestone: William Dawson and Son, 1979), 183.

2. The *Écho de la fabrique* (1831–34) was the "stem paper" from which all the others descended. When the April 1834 insurrection led to its closing down, it was continued under the title *Indicateur* in 1834–35. The *Écho des travailleurs* was founded by dissident contributors to the *Écho de la fabrique* in October 1833 and ceased publication in March 1834. After the April 1834 uprisings, it reappeared as the *Tribune prolétaire* (1834–35) and subsequently, for a few issues, as the *Union des travailleurs* (1835) and the *Nouvel Écho de la fabrique* (1835). After a five-year hiatus, *Écho des ouvriers* appeared in 1840–41, followed by *Le Travail* (1841) and *Écho de la fabrique de 1841*, which gave way in 1845 to a monthly, the *Tribune lyonnaise, Revue politique, sociale, industrielle, scientifique et littéraire des travailleurs*, which Chastaing continued to publish until 1851. The most thorough account of these later papers is Mary Lynn McDougall, "After the Insurrections: The Workers' Movement in Lyon, 1834–1852" (Ph.D. diss., Columbia University, 1973), 128–30. Among historians of the Lyon insurrections, Fernand Rude put relatively little emphasis on the role of the *Écho*, stressing instead the importance of the silk weavers' mutualist organization. Maurice Moissonnier gave the paper greater attention, insisting on its importance in creating a working-class consciousness. Moissonnier, *Canuts*, 72–73. The most detailed account of the paper to date is that of Robert Bezucha, who emphasizes above all what he sees as the paper's eventual identification with the republican movement. Bezucha, *Lyon Uprising*, 98–116. See also Jean Gaumont, *Le commerce*

véridique et social, 1835–1838, et son fondateur Michel Derrion, 1803–1850 (Amiens: Imprimerie Nouvelle, 1935), 42–51 (on the *Indicateur*). The *Écho de la fabrique* is available in a modern facsimile edition (Paris: Edhis, 1973, 2 vols.). This reprint does not include its rivals and continuators.

3. *Écho des travailleurs*, 6 November 1833.
4. *Écho de la fabrique*, 28 October 1832.
5. *Écho des travailleurs*, 9 November 1833.
6. Prospectus in BML, Charnier papers.
7. *Écho de la fabrique*, 9 September 1832.
8. *Écho de la fabrique*, 13 May 1832; 27 May 1832.
9. *Écho de la fabrique*, 9 September 1832.
10. *Écho de la fabrique*, 6 October 1833.
11. *Écho de la fabrique*, 27 April 1834.
12. The term "proletarian" was used fairly frequently during the 1790s but dropped out of favor until after the Revolution of 1830, when it began to take on its modern connotation. Rose, "Proletarians and Proletariat," in Rose, *Tribunes and Amazons*, 306–23. On "travailleur," see Maurice Tournier, "*Travailleur* aux prises avec l'histoire," in Tournier, *Des mots sur la grève* (Paris: Klincksieck, 1992), 127–33.
13. *Écho des travailleurs*, 2 November 1833.
14. *Écho des travailleurs*, 9 November 1833.
15. *Écho de la fabrique*, prospectus (October 1831).
16. On the association between geographic and social marginality in French urban life in this period, see Merriman, *The Margins of City Life*.
17. *Écho de la fabrique*, prospectus, n.d. [October 1831]. The paper's first issue appeared on 30 October 1831.
18. *Écho de la fabrique*, 9 September 1832.
19. *Tribune prolétaire*, 15 February 1835.
20. The most detailed account of the paper's origins is in *Tribune prolétaire*, 10 May 1835.
21. The history of the *mutuellistes* is the main thread of Fernand Rude's *Le mouvement ouvrier à Lyon de 1827 à 1832*.
22. Bowring, *Second Report*, 36, 37. See also Villermé, *Tableau*, 1:354.
23. *Écho de la fabrique*, 18 August 1833.
24. Villermé, *Tableau*, 1:367.
25. *Écho de la fabrique*, 28 October 1832.
26. Bezucha, *Lyon Uprising*, 117.
27. H. Rivière, "Fabricant non mutuelliste," in *Écho des travailleurs*, 23 November 1833.
28. Monfalcon, *Histoire des insurrections de Lyon*, 152.
29. *Indicateur*, 5 July 1835.
30. *Indicateur*, 12 April 1835. The paper listed donations as small as 25 centimes.
31. *Furet de Lyon* (15 January 1832–15 April 1832). This paper was edited by a republican agitator, Joseph Beuf, who seems to have been an isolated figure in the city's political affairs. A report on his trial in June 1832 commented, "Beuf is one of those hotheads who push all ideas to their most extreme consequences. A fanatical royalist in 1815, he became a *leveling* republican in 1830." AN, BB 20 61, 18 June 1832.
32. Jacques Rancière, *La nuit des prolétaires* (Paris: Fayard, 1981).
33. *Écho de la fabrique*, 5 August 1832. Pierre-Jean de Béranger, then at the height of

his fame, was a popular poet whose works mixed glorification of Napoleon with democratic sentiments.

34. *Tribune lyonnaise*, 20 April 1848; *Écho de la fabrique*, 18 December 1831.

35. ADR, État civil microfilm 730606, Lyon Nord et Ouest, entry for 7 thermidor An VII; *Écho de la fabrique*, 3 March 1832.

36. Marius Chastaing, *Vingt-deux jours de captivité* (Lyon: Bureau de la Tribune Lyonnaise, 1849), 5.

37. *Écho de la fabrique*, 5 August 1832; *Tribune lyonnaise*, 20 April 1848.

38. ADR, 4 M 263, dossier re Loge de la Bienveillance (1835–38). Chastaing, identified as "homme de lettres," is on the membership list for late 1835 and claimed to have drafted the lodge's statutes himself. He had evidently quit the group by April 1836, although he remained strongly committed to the Masonic movement and published several pamphlets on the subject, including *Astrée. Discours Maçonnique sur la justice* (Lyon: Isidore Deleuze, 1838).

39. Marius Chastaing, *Astréologie, ou Remède aux causes du malaise social* (Lyon: Rodanet, 1848), 6.

40. Truant, *Rites of Labor*.

41. On this point, see the secret police report of 6 April 1833 in AML, Gasparin papers, 4 II 10.

42. Sigaud, "Défense des fabricans d'étoffes de soie, qui ne sont pas mutuellistes," *Écho des travailleurs*, 20 November 1833.

43. Rude, *Mouvement social*, 324n.

44. *Tribune prolétaire*, 10 May 1835.

45. *Tribune prolétaire*, 10 May 1835.

46. *Indicateur*, 28 September 1834.

47. Charles Tilly, *The Contentious French: Four Centuries of Popular Struggle* (Cambridge, Mass.: Harvard University Press, 1986), 310.

48. *Écho de la fabrique*, 3 February 1833.

49. Ministry of Commerce to Gasparin, 1 August 1833, in AML, Gasparin papers, 4 II 1.

50. *Écho de la fabrique*, 25 December 1831.

51. *Courrier de Lyon*, 27 January 1833.

52. Monfalcon, *Histoire des insurrections de Lyon*, 150–54.

53. *Précurseur*, 28 November 1831; 7–8 May 1834; 30 December 1834 (re prosecution of *Tribune prolétaire*).

54. *Globe*, 31 October 1831.

55. *Le Temps*, 18 April 1834 (Lyon, 14 April). The anonymous article was in fact by J.-B. Monfalcon, former editor of the *Courrier de Lyon*.

56. *Écho de la fabrique*, 17 March 1833. I have not been able to locate this translation, if in fact it existed.

57. *Écho de la fabrique*, 23 February 1834.

58. *Écho de la fabrique*, 28 October 1832; 7 April 1833. No winning term was ever selected.

59. *Écho de la fabrique*, 19 May 1833.

60. Letter to *Écho de la fabrique*, 30 September 1832, in AML, Gasparin papers, 4 II 7. This letter, arguing for the legitimacy of Gasparin's request, is in his secretary's handwriting; the prefect presumably did not want to give the newspaper a piece of paper with his signature on it.

61. *Écho de la fabrique*, 27 May 1832.
62. *Indicateur*, 23 November 1834.
63. *Écho de la fabrique*, 30 December 1832; 20 January 1833.
64. *Écho de la fabrique*, 25 November 1832.
65. *Indicateur*, 7 June 1835.
66. *Courrier de Lyon*, 27 January 1833.
67. *Tribune prolétaire*, 10 May 1835.
68. AN, CC 558, d. *Écho de la fabrique;* police report, 6 April 1833, in AML, Fonds Gasparin, 4 II 10.
69. The changeover was announced in *Écho de la fabrique*, 18 August 1833.
70. *Écho de la fabrique*, 5 and 12 January 1834; *Écho des travailleurs*, 8 January 1834.
71. Letter from Favier to Pierre Charnier, 26 February [1835], in BML, Charnier papers.
72. *Écho des travailleurs*, 13 November 1833.
73. A. Vidal, in *Écho de la fabrique*, 11 March 1832.
74. *Écho de la fabrique*, 9 September 1832.
75. *Écho des travailleurs*, 6 November 1833.
76. *Indicateur*, 12 December 1834.
77. *Nouvel Écho de la fabrique*, August 1835.
78. *Écho de la fabrique*, 6 November 1831.
79. *Écho des travailleurs*, "Prospectus," 5 October 1833.
80. *Écho de la fabrique*, 13 May 1832; *Écho des travailleurs*, 14 December 1833.
81. *Écho de la fabrique*, 19 May 1833.
82. *Écho des travailleurs*, 25 January 1834.
83. *Écho des travailleurs*, 14 December 1833.
84. *Écho de la fabrique*, 15 January 1832; *Écho des travailleurs*, 22 February 1834.
85. *Écho des travailleurs*, 30 November 1833.
86. *Écho des travailleurs*, 1 February 1834.
87. *Écho de la fabrique*, 27 January 1833.
88. *Écho des travailleurs*, 8 March 1834.
89. *Écho de la fabrique*, 8 December 1833.
90. *Écho des travailleurs*, 16 November 1833.
91. *Écho de la fabrique*, 17 November 1833.
92. *Écho des travailleurs*, 27 November 1833.
93. The project's historical importance is underlined in Ellen Furlough, *Consumer Cooperation in France: The Politics of Consumption, 1834–1930* (Ithaca, N.Y.: Cornell University Press, 1991), 20–21.
94. Derrion, in *Indicateur*, 11 January 1835. On Derrion and the *Commerce véridique*, see Jean Gaumont, *Commerce véridique*.
95. *Tribune prolétaire*, 19 April 1835.
96. *Écho de la fabrique*, 14 October 1832.
97. *Écho de la fabrique*, 9 December 1832.
98. *Écho des travailleurs*, 5 October 1833.
99. *Globe*, 31 October 1831.
100. *Écho de la fabrique*, 26 August 1832; 5 May 1833.
101. *Écho des travailleurs*, 7 December 1833.
102. *Écho des travailleurs*, 15 March 1834.
103. *Tribune prolétaire*, 26 April 1835.
104. *Écho de la fabrique*, 9, 16 September 1832.

105. *Tribune prolétaire*, 9 November 1834.
106. *Tribune prolétaire*, 23 November 1834; *Indicateur*, 9 November 1834.
107. *Écho de la fabrique*, 6 January 1833.
108. *Écho de la fabrique*, 26 August 1832.
109. *Écho de la fabrique*, 30 December 1832.
110. *Écho de la fabrique*, 6 January 1833.
111. *Écho de la fabrique*, 23 February 1834.
112. *Indicateur*, 23 November 1834.

113. Several members of the *Écho de la fabrique*'s oversight commission were arrested after the April 1834 uprising, but no evidence of their involvement in the insurrection was found, and charges against them were dismissed. Girod de l'Ain, *Rapport*, 10:141–47; AN, CC 558.

114. *Écho de la fabrique*, 28 October 1832.
115. *Écho de la fabrique*, 28 October 1832.

116. *Indicateur*, 8 February 1835; *Tribune prolétaire*, 15 February 1835. The *Indicateur* hastened to distance itself from the disputed article in a subsequent issue (22 February 1835).

117. *Tribune prolétaire*, 23 November 1834.

118. *Écho de la fabrique*, 25 November 1832; *Écho des travailleurs*, 23 November 1833.

119. *Courrier de Lyon*, 27 January 1833.

120. Monfalcon, *Histoire des insurrections de Lyon*, 153.

121. 120. AML, Gasparin papers, 4 II 4, 2 November 1832; Comte d'Argout to Gasparin, Gasparin papers, 4 II 9, 13 August 1833. Gasparin's letter detailing his attempted involvement in the replacement of Chastaing has not survived.

122. AML, Prunelle to Gasparin, 14 November 1833, in Gasparin papers, 4 II 13; *Écho des travailleurs*, 8 February 1834.

123. See *Tribune prolétaire*, 7 December 1834; 4 January 1835; and *Indicateur*, 18 January 1835; 5 July 1835.

124. *Écho de la fabrique*, 21 October 1832.

125. Gasparin to minister of the interior, 30 September 1832, in AML, Gasparin papers, 4 II 4.

126. *Écho des travailleurs*, 6 November 1833.

127. Giesselmann, *Revolte*, 2:1008–10, 1025.

Chapter 5: Creating Events: Press Banquets and Press Trials in the July Monarchy

1. Daniel Boorstin, *The Image* (New York: Penguin Books, 1962), 22–23.

2. Tilly, *The Contentious French*, 10.

3. On early nineteenth-century English political banquets, see James A. Epstein, *Radical Expression: Political Language, Ritual, and Symbol in England, 1790–1850* (New York: Oxford University Press, 1994), esp. chap. 5, "Rituals of Solidarity: Radical Dining, Toasting, and Symbolic Expression."

4. On nineteenth-century French political banquets, see Rebecca Spang, "'La fronde des nappes': Fat and Lean Rhetoric in the Political Banquets of 1847," in Carrol F. Coates, ed., *Repression and Expression: Literary and Social Coding in Nineteenth-Century France* (New York: Peter Lang, 1996), 167–78; and Spang, *The Invention of the Restaurant: Paris and Modern Gastronomic Culture* (Cambridge, Mass.: Harvard University Press, 2000), especially chaps. 4 and 8.

5. *Écho de la fabrique*, 4 November 1832.

6. *Procès-verbal de la fête Maç∴offerte par les* ▫▫ *réunies des O∴de Lyon et de la Croix-Rousse, au F∴Odilon-Barrot, le 4e j∴du 7e m∴de l'an de la V∴L∴. 5832, 4 September 1832 (ère vulg) dans le local des enfans d'Hiram et de la sincère amitié* (Lyon, 1832).

7. *Précurseur*, 9 September 1829.

8. *Écho des travailleurs*, 6, 16 November 1833.

9. *Glaneuse*, 4 August 1831.

10. *Précurseur*, 31 August 1832; AML, Gasparin papers, 4 II 4, 31 August 1832; *Précurseur*, 1 October 1832; *Glaneuse*, 2 October 1832; AML, Gasparin papers, 4 II 4, 26 September 1832.

11. *Écho de la fabrique*, special number following 28 October 1832 issue; AML, Gasparin papers, 4 II 4, 29 October 1832. The diners gathered in two groups, one in the suburb of Vaise and the other in the Perrache section of Lyon.

12. AML, Gasparin papers, 4 II 10, 5 May 1833. Tickets to this banquet were sold partly to raise money for a pending trial involving the *Précurseur;* probably not all the purchasers really intended to attend.

13. AML, I 36, no. 208.

14. *Écho de la fabrique*, 21 October 1832.

15. *Procès-verbal de la Fête Maç*, 10.

16. *Glaneuse*, 2 October 1832.

17. *Courrier de Lyon*, 1 September 1832.

18. *Précurseur*, 31 August 1832.

19. *Glaneuse*, 2 October 1832.

20. *Écho de la fabrique*, special number following 28 October 1832 issue.

21. *Glaneuse*, 2 October 1832.

22. *Glaneuse*, 2 October 1832.

23. *Courrier de Lyon*, 4 October 1833.

24. *Courrier de Lyon*, 1, 2, 9 September 1832.

25. *Précurseur*, 16 September 1832.

26. *Précurseur*, 5 September 1832.

27. *Écho des travailleurs*, 6 November 1833.

28. See, in addition to Gasparin's letters, the reports from the commander of the *gendarmerie* unit stationed in the city in AN, F 7 6782, for the dates of 29 and 30 September 1832; 28 April 1833; and 10 May 1833.

29. AML, Gasparin papers, 4 II 4, 26 July 1832.

30. AML, Gasparin papers, 4 II 4, 30 September 1832.

31. Prunelle to Gasparin, in AML, Gasparin papers, 4 II 11, 10 March 1834.

32. AML, Gasparin papers, 4 II 4, 2 November 1832.

33. AML, Gasparin papers, 4 II 10, letters of 21 April 1833 and 24 April 1833.

34. *Précurseur*, 29 April 1833.

35. AML, Gasparin papers, 4 II 10, 30 April 1833.

36. *Précurseur*, 5 May 1833.

37. AML, Gasparin papers, 4 II 10, 9 May 1833.

38. *Courrier de Lyon*, 30 April 1833; 10 May 1833.

39. *Glaneuse*, 10 May 1833.

40. AML, Gasparin papers, 4 II 10, 10, 11 May 1833.

41. AML, Gasparin papers, 4 II 10, 10 May 1833.

42. *Gazette du Lyonnais*, 11 May 1833.

43. Otto Kirchheimer, *Political Justice* (Princeton, N.J.: Princeton University Press, 1961), 6, 52, 426.

44. This critical weakness of the regime's repressive apparatus is underlined in Giesselmann, *Die Manie der Revolte*, 449.

45. AN, BB 20 61, cover letter dated 12 September 1832, referring to acquittals in cases against the *Précurseur;* there is an almost identical remark in BB 20 66, session of 4 December 1833, after a trial involving the *Glaneuse*. Judges were equally scandalized by Lyon juries' decisions in many nonpolitical cases, such as the light sentences given to several defendants who had infected prepubescent girls with venereal disease. AN, BB 20 74, December 1834.

46. Kirchheimer, *Political Justice*, 18.

47. ADR, 4 M 449, Varenare to prefect, 27 November 1830.

48. See, for example, the discussion of press coverage of the "Chicago Seven" trial in Juliet Dee, "Constraints on Persuasion in the Chicago Seven Trial," in Robert Hariman, ed., *Popular Trials: Rhetoric, Mass Media, and the Law* (Tuscaloosa: University of Alabama Press, 1990), 86–113.

49. On trials as means of communication, see Robert Hariman, "Introduction," in Hariman, ed., *Popular Trials*, 3.

50. François Guizot, "Des conspirations et de la justice politique," in Guizot, *Mélanges politiques et historiques* (1821; Paris: Michel Lévy Frères, 1869), 193.

51. Harry S. Bryant Jr., "*Ce n'est pas moi qui ai cherché cette publicité:* Pamphlet Coverage of Political Trials During the Early July Monarchy, 1830–1835," *Proceedings of the Western Society for French History* 24 (1997), 423–29. A number of press-trial pamphlets (but none of those from Lyon) are reprinted in *Les révolutions du XIXe siècle*, vol. 10: *La presse républicaine devant les tribunaux, 1831–1834* (Paris: Edhis, 1974).

52. On the English press-trial tradition, see Epstein, *Radical Expression*, 30–33.

53. Pidansat de Mairobert, attrib., *Journal historique de la révolution opérée dans la constitution de la monarchie françoise, par M. de Maupeou, chancelier de France*, 7 vols. (London: "John Adamson," 1774–76), 5:28, 39, 40, 42, 46–47 (January 24, 27, 28, 29, and February 1, 1774).

54. For a recent account of one of the most spectacular of these trials, see Morris Slavin, *The Hébertistes to the Guillotine* (Baton Rouge: Louisiana State University Press, 1994).

55. Isser Woloch, *Jacobin Legacy: The Democratic Movement Under the Directory* (Princeton, N.J.: Princeton University Press, 1970), 56–63.

56. For a discussion of the regulations on courtroom reporting, see *Le National* (Paris), 17 October 1833.

57. For details of the period's press laws, see Irene Collins, *The Government and the Newspaper Press in France, 1814–1881* (London: Oxford University Press, 1959), 22–23, 37, 55, 62–63.

58. *Glaneuse*, 18 April 1832.

59. *Cri du peuple*, 14 September 1831.

60. *Précurseur*, 11 June 1832.

61. *Précurseur*, 14 June 1832.

62. Perreux, *Sociétés secrètes*, 58–63.

63. AML, I (2) 59, 14 June 1832.

64. *Précurseur*, 29 June 1832.

65. The legitimist editor Théodore Pitrat had already been prosecuted in March 1832. AN, BB 20 61.

66. The politically influential banker Laffitte also intervened with him. AN, 271 AP 2 (Papiers Odilon-Barrot), item B.6.4, letter of Lafayette to Odilon-Barrot, 15 August 1832.

67. *Précurseur*, 29 August 1832.

68. *Précurseur*, 30 August 1832.
69. *Précurseur*, 1, 2 September 1832.
70. Anon., *Procès du "Précurseur," avec la plaidoirie de M. Odilon-Barrot* (Lyon: Bureau du Précurseur, 1832), 25, 56.
71. *Précurseur*, 2 September 1832.
72. *Précurseur*, 3–4 September 1832.
73. *Réparateur*, 13, 15 June 1834.
74. *Glaneuse*, 13 January 1833; 28 March 1833.
75. *Précurseur*, 27–28 March 1833.
76. *Précurseur*, 27–28 March 1833.
77. *Glaneuse*, 19 May 1833.
78. *Précurseur*, 31 March 1833.
79. *Procès de la Glaneuse, contenant les douze articles incriminés; le compte rendu des audiences du 11 et du 17 mai, avec tous les incidens; la défense de M. Adolphe Granier, et les plaidoiries recueillies en entier de MM. Dupont, Michel-Ange Perier et Charassin* (Lyon: Aux Bureaux de la Glaneuse, 1833), 26.
80. AML, Gasparin papers, 4 II 10, 30 May 1833; AN, F 18 495 (I), documents dated 4 June, 9 June, 25 June, 11 November, 28 November 1833, 7 March 1834; CC 572, minutes of *Précurseur* shareholders' assembly, 6 September 1833.
81. *Courrier de Lyon*, 16 May 1833.
82. *Courrier de Lyon*, 27 February 1833.
83. *Courrier de Lyon*, 28 February 1833.
84. Monfalcon, *Souvenirs*, 140.
85. AML, Gasparin papers, 4 II 10, 22 May 1833.
86. AN, BB 20 61, session of 15 March 1832.
87. Collins, *Government and the Newspaper Press*, 82.

Chapter 6: *Textualizing Insurrection: The Press and the Lyon Revolts of 1831 and 1834*

1. For the casualty figures, see Fernand Rude, *Les révoltes des canuts*, 43, 171.
2. *Gazette du Lyonnais*, 4 December 1831.
3. Paul Ricoeur, "Événement et sens," in Jean-Luc Petit, ed., *L'événement en perspective* (Paris: Éditions de l'École des Hautes Études en sciences sociales, 1991), 41.
4. *Globe*, 9 December 1831.
5. Boorstin, *The Image*, 20.
6. For an analysis of press coverage of the Paris insurrection of 5–6 June 1832 that reaches similar conclusions, see Thomas Bouchet, "'Un événement grave vient de jeter la consternation dans Paris. . . .' La presse française face aux 5 et 6 juin 1832," in Hans-Jürgen Lüsebrink and Jean-Yves Mollier, eds., *Presse et événement. Journaux, gazettes, almanachs (XVIII–XIXe siècles)* (Bern: Peter Lang, 2000), 27–44.
7. For an example from the early seventeenth century, see William Beik, *Urban Protest in Seventeenth-Century France* (Cambridge: Cambridge University Press, 1997), 100–113.
8. On the 1744 revolt, see Maurice Garden, *Lyon et les lyonnais au XVIIIe siècle* (Paris: Société d'Édition "Les Belle-Lettres," 1970), 584–88; and E. Pariset, *Histoire de la fabrique lyonnaise* (Lyon: A. Rey, 1901), 195–202; the most detailed account of the 1786 uprising is Louis Trénard, "La crise sociale lyonnaise à la veille de la Révolution," *Revue d'histoire moderne et contemporaine* 2 (1955), 5–45.
9. *Gazette de Leyde*, 1 September 1786.

Notes to Pages 196–211

10. For a more intensive analysis of the difference between press coverage of the 1786 and 1831 insurrections, see Jeremy D. Popkin, "Texte et insurrection. La presse et les insurrections de Lyon en 1786 et 1831," in Hans-Jürgen Lüsebrink and Jean-Yves Mollier, eds., *La perception de l'événement dans la presse*, 47–65.

11. On Lyon's press during the revolutionary decade, see Trénard, *Lyon de l'Encyclopédie au préromantisme*, 252–57, 348–53, 420–21; and Trénard, "Les périodiques lyonnais dans les luttes révolutionnaires, 1792–1794," in Pierre Rétat, ed., *La Révolution du journal* (Paris: Éditions du CNRS, 1989), 297–305.

12. Granier's self-serving version was published as part of the public dispute about responsibility for the Lacombe poster that followed the reoccupation of the city and the arrest of those charged with leading the insurrection. It appears in the form of a small brochure, entitled *Lacombe*, tipped into the Bibliothèque nationale's collection of the *Glaneuse* after the issue of 18 December 1831.

13. *Journal du commerce*, 7 October 1831.
14. *Précurseur*, 23 October 1831.
15. *Précurseur*, 16 October 1831.
16. *Précurseur*, 24 October 1831.
17. *Précurseur*, 23 October 1831; 18 November 1831.
18. *Journal du commerce*, 23 October 1831; 6 November 1831.
19. *Glaneuse*, 6 November 1831.
20. *Glaneuse*, 6 November 1831.
21. *Précurseur*, 25 October 1831.
22. *Journal du commerce*, 9 November 1831. The Bristol riot was covered extensively in the Paris press, mostly by way of translations from the London newspapers.

23. *Écho de la fabrique*, 6 November 1831. The rumor of a pending insurrection circulated widely in the city; the Saint-Simonians' official correspondent, Peiffer, reported it to the movement's leaders in Paris. Peiffer, letter to Pereire jeune, 8 November 1831, in Bibliothèque de l'Arsenal, Fonds Enfantin, c. 7606.

24. *Écho de la fabrique*, 20 November 1831.
25. *Gazette du Lyonnais*, 5 November 1831.
26. *Courrier de Lyon*, 17 February 1834; 21 February 1834; *Gazette du Lyonnais*, 19 February 1834.
27. *Écho de la fabrique*, 23 February 1834; 16 March 1834.
28. *Glaneuse*, 23 March 1834.
29. *Précurseur*, 17 February 1834; 2–3 April 1834; 8 April 1834.
30. Petetin, "Mémoire," AN CC 572, 62, 66. Gasparin confirmed this visit in his testimony to the Cour des Pairs after the uprising. Girod de l'Ain, *Rapport fait à la Cour des Pairs*, 10:80–81.
31. *Le Temps*, 16 April 1834 (Lyon, 10 April)
32. *Gazette du Lyonnais*, 6 April 1834; *Réparateur*, 9 April 1834.
33. *Conseiller des femmes*, 22 February 1834; *Papillon*, 20 February 1834.
34. Popkin, *Revolutionary News*, 136–40.
35. *Précurseur*, 15 April 1834.
36. *Précurseur*, 22 November 1831.
37. Broadsheet, tipped in with issue of 22 November 1831 in *Précurseur*, collection in Bibliothèque municipale de Lyon.
38. *Journal du commerce* and *Gazette du Lyonnais*, reprinted in *National* (Paris), 27 November 1831.
39. *Précurseur*, 25 November 1831.

40. *Précurseur,* 25 November 1831; *Écho de la fabrique,* 27 November 1831; *Journal du commerce,* 30 November 1831.
41. *Écho de la fabrique,* 27 November 1831.
42. *Écho de la fabrique,* 12 February 1832.
43. *Gazette du Lyonnais,* 25, 27 November 1831.
44. *Le Temps,* 2 December 1831. The *Temps* of the early 1830s was not the same enterprise as the paper of the same name that dominated the serious press of the Third Republic era.
45. *Précurseur,* 15 April 1834.
46. *Le Temps,* 13 April 1834 (Lyon, 9 April, 1 P.M. and 4 P.M.); 16 April 1834 (Lyon, 10 April).
47. "Journée du 12 avril," *Réparateur,* 16 April 1834.
48. *Réparateur,* 17 April 1834; *Précurseur,* 16 April 1834.
49. *Journal des débats,* 25, 26 November 1831.
50. *National,* 25, 27 November 1831.
51. *Le Temps,* 25 November 1831 (Lyon, 21 November); 2 December 1831 (Lyon, 29 December 1831).
52. On the significance of Monfalcon's repeated reuse of the story, see Chapter 7.
53. *Journal des Débats,* 8 December 1831.
54. Saint-Marc Girardin, *Souvenirs et reflexions politiques d'un journaliste* (Paris: Michel Levy, 1859), 142–43.
55. *Journal des Débats,* 27 June 1831, reprinted in Girardin, *Souvenirset réflexions,* 122.
56. *Le Temps,* 13 April 1834 (Lyon, 9 April) and 16 April 1834 (Lyon, 10 April).
57. For a brief account of the trial, see Paul Bastid, *Les grands procès politiques de l'histoire* (Paris: Fayard, 1962), 307–18. The Lyon defendants objected to the disruptive courtroom strategy adopted by the leaders of the Paris republican movement.
58. *Précurseur,* 15 April 1834.
59. *Réparateur,* 9 April 1834.
60. *Conseiller des femmes,* 19 April 1834.
61. *Journal du commerce,* 2 May 1834.
62. "Combat du Pont Morand, le 22 9.bre 1831," BML, Fonds Coste no. 716. This crudely drawn illustration shows armed workers on the east bank of the Rhône firing at silk merchants' establishments on the presqu'île, with the Croix-Rousse hill in the background. The difference in scale between the figures on the lower right and the much smaller ones on the lower left reflects either the artist's haste or his clumsiness.
63. In a composition in circular format, "Événements de Lyon 21, et 22 Novembre 1831. Place des Bernardines" (BML, Fonds Coste 716 bis) the viewer looks over the shoulders of armed insurgents firing at troops. "Événements de Lyon, Barrière de la Croix-Rousse 21 et 22 November 1831" (BML, Fonds Coste 715), whose inscription says it was for sale at locations in both Lyon and La Croix-Rousse, shows uniformed soldiers in flight. A woodcut of the fighting in 1834, included in a popular almanac, shows combatants on both sides in an even-handed fashion. (BML, Fonds Coste 718 bis)
64. On the conventions that governed popular prints of the Revolution of 1830, see Michael Marrinan, "Historical Vision and the Writing of History at Louis-Philippe's Versailles," in Chu and Weisberg, eds., *Popularization of Images,* 115. Like the print of the "Combat du Pont Morand" reproduced here, the image Marrinan analyzes shows insurgents in the foreground storming a bridge.
65. Étienne Stanislas, a journeyman dyer, was later tried and acquitted on charges of having killed eight National Guards during the fighting. Moissonnier, *Canuts,* 134. He is

probably the open-shirted, dark-skinned man standing alone at bottom center of "Combat du Pont Morand." Jean-Baptiste Monfalcon drew on conventional stereotypes of blacks as savages in his description of "this hideous Negro . . . choosing his victims from the Pont Morand, and, with his eyes aflame, his mouth foaming, his arms covered with blood, giving a barbarous cry and jumping for joy every time his lead, skillfully aimed, toppled a dragoon or an artilleryman," but his description does place Stanislas exactly where he is shown in this print. (Monfalcon, *Histoire des insurrections de Lyon*, 82.) Louis Blanc, although he did not share Monfalcon's conservative views, reproduced parts of this passage in the section of his *Histoire des dix ans* dealing with Lyon (Louis Blanc, *The History of Ten Years, 1830–1840*, trans. Walter R. Kelly, 2 vols. [Philadelphia: Lea and Blanchard, 1848], 2:527). In a later work, Monfalcon mentions popular support for Stanislas at his trial (Monfalcon, *Code moral*, 403). Jean-Claude Romand, a participant in the 1831 uprising who claimed to have composed the slogan "Vivre en travaillant ou mourir en combattant," recalled that Stanislas "had stood alone on the Pont Morand against the opposing forces." Out of respect for such courage, Romand chose to be chained next to him when the two of them had been arrested and were being transported to prison. Jean-Claude Romand, *Confession d'un malheureux. Vie de Jean-Claude Romand, forçat libéré, écrite par lui-même, et publiée par M. Edouard Servan de Sugny, procureur du roi près le Tribunal de première instance de Nantua* (Paris: Au Comptoir des Imprimeurs Unis, 1846), 122.

66. ADR, T 380 contains police records concerning engravings and lithographs sold in Lyon during this period, but there are no documents relating to depictions of the insurrections of the 1830s. For the exchange between Casimir-Périer and Gasparin, see AML, Gasparin papers, 4 II 6, 10 December 1831, Gasparin to Casimir-Périer, and 4 II 2, Casimir-Périer to Gasparin, 15 December 1831.

67. Pamphlets concerning the 1831 insurrection include *Histoire de Lyon pendant les journées des 21, 22 et 23 novembre 1831, contenant les causes, les conséquences et les suites de ces déplorables événements*, ed. Auguste Baron (Lyon: A. Baron, 1832) (pro-worker); Bouvier Dumolard, *Rélation de M. Bouvier du Molart, ex-préfet du Rhône, sur les événemens de Lyon* (Lyon: Bureau du Journal du Commerce, 1832); [Michel Chevalier, ed.], *Religion Saint-Simonienne. Événemens de Lyon* (Paris: Everat, 1832) (Saint-Simonian); Jean-François Mazon, *Événemens de Lyon, ou Les trois journées de novembre 1831* (Lyon: Guyot, 1831) (relatively neutral); and Collomb, *Détails historiques sur les journées de Lyon, et les causes qui les ont précédées* (Lyon: Charvin, 1832) (pro-worker). Titles published after the 1834 uprising include [Anon.], *Historique des événemens de Lyon, du neuf au quatorze avril 1834* (Lyon: Rossary, 1834) (excerpts from *Courrier de Lyon*); *La vérité sur les événemens de Lyon au mois d'avril 1834* (legitimist); [Eugénie Niboyet], *Précis historique sur les événemens de Lyon* (Lyon: L. Boitel, 1834); Adolphe Sala, *Les ouvriers lyonnais en 1834, esquisses historiques* (Paris: Hivert, 1834) (legitimist, but outspokenly sympathetic to the workers); Terson, *Un St.-Simonien au peuple de Lyon, à l'occasion des événemens d'avril 1834* (Lyon: Durval, 1834).

68. *La vérité sur les événemens de Lyon*, 224.

69. [Eugénie Niboyet], *Précis historique*, 6–7, 14. The Harvard University library's copy of this pamphlet includes a note in nineteenth-century handwriting attributing it to Niboyet.

70. *Courrier de Lyon*, 19 April 1834.

71. *Courrier de Lyon*, 5 July 1833.

72. *Courrier de Lyon*, 16 February 1834.

Notes to Pages 231–242

Chapter 7: From Newspapers to Books: The Recasting of Revolutionary Narrative

1. *Le Temps*, 13 April 1834 (Lyon, 9 April).
2. General Aymard to Monfalcon, letter of 14 June 1834, in BML, Rés. 434157; Monfalcon to Gasparin, letter of 20 June 1834, in AML, Gasparin papers, 4 II 8.
3. Monfalcon to Gasparin, letter, 8 [June] 1834, in AML, Gasparin papers, 4 II 8.
4. On the place of history in the period's culture, see Linda Orr, *Headless History: Nineteenth-Century French Historiography of the Revolution* (Ithaca, N.Y.: Cornell University Press, 1990), 16–21; and Michael Marrinan, *Painting Politics for Louis-Philippe: Art and Ideology in Orléanist France, 1830–1848* (New Haven, Conn.: Yale University Press, 1988), 22–25.
5. Monfalcon to Gasparin, letter of 16 May 1834, in AML, Gasparin papers, 4 II 8.
6. On the relationship of early nineteenth-century French history to the period's paradigm of positivist science, see Hayden White, *Tropics of Discourse: Essays in Cultural Criticism* (Baltimore, Md.: Johns Hopkins University Press, 1978), 123–25.
7. Monfalcon, *Histoire des insurrections de Lyon*, 159n; readers might have inferred Monfalcon's identity from the fact that the editors of all the other Lyon newspapers he referred to were identified by their full names.
8. *Courrier de Lyon*, 17 February 1834.
9. *Précurseur*, 3 September 1834; *Réparateur*, 24 July 1834.
10. Monfalcon's research efforts are documented in letters inserted in his personal copy of his *Code moral des ouvriers* in BML, sig. Rés. 434157, and his letters to the prefect, Gasparin, in AML, Gasparin papers, 4 II 8.
11. Jean-Baptiste Monfalcon, *Histoire monumentale de la ville de Lyon*, 9 vols. (Paris: Firmin Didot, 1866), 1:xiv.
12. [Jean-Baptiste Monfalcon], "Maladies des Tisserands," in P. Patissier, *Traité des maladies des artisans* (Paris: J.-B. Ballière, 1822), 391, 394, 400; Monfalcon claimed that this was a reprint of an article first written in 1820. On Monfalcon's deliberate effort to use scientific publications to advance his career, see his *Souvenirs*, 26, 28–29.
13. Monfalcon, "Maladies," in Patissier, *Traité*, 393.
14. Monfalcon, *Souvenirs*, 188.
15. Bezucha, *Lyon Uprising*, 186.
16. Bruno Benoît, *L'identité politique de Lyon. Entre violences collectives et mémoire des élites, 1786–1905* (Paris: L'Harmattan, 1999), 60. Benoît argues, however, that the federalist uprising of 1793 and its brutal repression were much more important than the conflicts of the nineteenth century in this process. In fact, 1793 was hardly ever mentioned in the press during the early 1830s, whose importance Benoît may well have underrated.
17. Villermé, *Tableau*, 352–68.
18. Blanc copied Monfalcon's passage about the role of the "negro, named Stanislas" during the November 1831 insurrection. Blanc, *History of Ten Years*, 1:527.
19. Jean-Baptiste Monfalcon, *Histoire de la ville de Lyon*, 2 vols. (Lyon: Guilbert and Dorier, 1847), 2:1159–85; *Histoire monumentale de la ville de Lyon*, 9 vols. (Paris: Firmin Didot, 1866), 3:276–301.
20. Paris and Lyon: Pélagaud Lesne and Crozet, 1836.
21. *Revue du Lyonnais* 3 (1836), 253.
22. Monfalcon, *Code moral*, 424, 502, 503.
23. Report to the Royal Academy of Lyon, 17 May 1836, in *Code moral*, xli.
24. Lyon: J. Nigon, 1853. Monfalcon had a limited number of copies reproduced by a procedure known as *autographie*, which was used in the nineteenth century to duplicate

manuscript documents. The text is handwritten rather than typeset, and the work was never sold commercially. The title reflects the fact that, in 1841, Monfalcon, an ardent bibliophile, had been named librarian of one of Lyon's two public collections, launching him on a new career that he pursued to the end of his life in 1874. After 1834, he had abandoned journalism and devoted only a small portion of his time to his obligations as a doctor.

25. Monfalcon, *Souvenirs*, 5, 13.
26. Monfalcon, *Souvenirs*, 29.
27. Monfalcon, *Souvenirs*, 93–98.
28. Monfalcon, *Souvenirs*, 112–13.
29. *Le Temps*, 2 December 1831, letter from Lyon, 29 November 1831, mentions a "doctor from the Hôtel-Dieu" without naming him. In the issue of 4 December 1831 (letter from Lyon, 1 December 1831), Monfalcon is named. Although these letters from Lyon were published anonymously, Monfalcon was the author of both. Monfalcon mentioned the incident again in one of his news bulletins about the April 1834 insurrection (*Le Temps*, 16 April 1834, letter of 12 April 1834). He included the story in his *Code moral* (p. 377) and later in his *Histoire monumentale de Lyon*, 3:282.
30. Monfalcon, *Souvenirs*, 115.
31. Joseph Benoît, *Confessions d'un prolétaire, Lyon, 1871*, ed. Maurice Moissonnier (Paris: Éditions Sociales, 1968), 49.
32. Ibid., 42, 69–70; Romand, *Confession d'un malheureux*, 25, 41–42, 80, 101. The authenticity of Romand's text, published in 1847 and seen through the press by Edouard Servan de Sugny, royal prosecutor at the court in the town of Nantua, is obviously questionable, and the latter half of the book is a stereotypical account of his moral redemption in prison. Nevertheless, the tone of his recollections of the 1820s and 1830s is convincing, and some details, such as his reference to "Stanislas the Negro," cited in Chapter 6, correspond to other sources about the insurrection.
33. For a recent survey of French writings about poverty and social inequality in this period, see Giovanna Procacci, *Gouverner la misère. La question sociale en France, 1789–1848* (Paris: Seuil, 1993).
34. Martyn Lyons cites Étienne-Léon Lamothe-Langon's *Le gamin de Paris*, published in 1833, as the first work he identified with a bourgeois villain; like *La révolte de Lyon*, its plot involved the seduction of an innocent girl. Lamothe-Langon was a former emigré whose hostility to the bourgeoisie was an expression of his aristocratic background. Martyn Lyons, *Le triomphe du livre. Une histoire sociologique de la lecture dans la France du XIXe siècle* (Paris: Promodis, 1987), 120.
35. *Revue républicaine*, 5:128 (1835).
36. *La révolte de Lyon en 1834, ou La fille du prolétaire*, 2 vols. (Paris: Moutardier, 1835), 1:ix, 170–71.
37. Reid, *Families in Jeopardy*, 140.
38. "Le Canut," in Louis Couailhac, *Les sept contes noirs* (Paris and Lyon: Bohaire, 1832), 123–42. The story was serialized in the Lyon silk workers' newspaper, *Écho de la fabrique*, on 14 April, 21 April, 28 April, and 5 May 1833. Couailhac was born in Lille and educated at the Lycée Henri IV in Paris. He apparently spent some time in Lyon in the early 1830s, but left before the April 1834 uprising. By 1852, he had pledged loyalty to Napoleon III's regime, which later rewarded him with the post of *sécretaire-rédacteur* to the Imperial Senate. See the articles on him in the *Dictionnaire de biographie française* and the *Grande encyclopédie*.
39. Michel Ragon, *Histoire de la littérature prolétarienne en France* (Paris: Albin Michel, 1974).

40. The *Revue républicaine*'s reviewer, who wrote as if he knew the author's identity, assumed that the author of the "Aperçu" had also written the novel. There is no indication in the work itself to justify this assumption.

41. "Hte. D.," "Des machines, et de leur influence sur la production et les salaires," *Revue républicaine* (1835) 5:11, 14.

42. Reid, *Families*, 182–85.

43. Monfalcon, *Histoire des insurrections de Lyon*, 33.

44. Louis Couailhac, *Pitié pour elle*, 2 vols. (Brussels: Société Typographique Belge, 1837).

45. "Bert," in *Nouveau Tableau de Paris*, 5:131.

Conclusion

1. Dror Wahrman, *Imagining the Middle Class: The Political Representation of Class in Britain, c. 1780–1840* (Cambridge: Cambridge University Press, 1995), 6.

2. Jean-Baptiste Monfalcon, *Annales de la ville de Lyon, ou Histoire de notre temps*, 2 vols. (Lyon: n.p., 1849–50), 1:46, 47.

3. See the daily police reports on the press and clubs in Lyon in 1848 in AN, BB 30 327.

4. For an interesting examination of the effect computers and the Internet have had on one form of French writing—the keeping of diaries—see Philippe Lejeune, *"Cher écran." Journal personnel, ordinateur, internet* (Paris: Seuil, 2000).

Bibliography

Archival Sources

Sources in Lyon

Archives départementales du Rhône (ADR)

4 M 212	April 1834 uprising
4 M 262	"Associations politiques"
4 M 263	Freemasons
4 M 287	"Crieurs et afficheurs publics"
4 M 449–50	"Affaires de presse"
T 338	"Cabinets de lecture"
T 339	"Libraires"
T 353	Newspaper registrations, 1828–1841
T 357	"Journaux 1828–1850"
Microfilme 730606	État civil (Chastaing)

Archives municipales de Lyon (AML)

I (2) 36	Police politique
I (2) 40	Police
I (2) 46	Police
I (2) 59–60	Press
4 II 1–13	Gasparin papers
39 II 1–2	Monfalcon paper

Bibliothèque municipale de Lyon
    Pierre Charnier papers
    Ms. 613, 2429, 2464, 5751, Rés. 434157: correspondence of J.-B. Monfalcon
    Fonds Coste (illustrations)

Musée Gadagne (Lyon)
    Fonds Justin Godart   Carnets de Joseph Bergier

Sources in Paris

Archives nationales (AN) (Paris)
    271 AP   Odilon-Barrot papers
    BB 20 56, 61, 66, 74   Cour d'assises (Rhône)
    F 7 6698   Associations (Rhône)
    F 7 6782   Gendarmerie reports (Rhône)
    F 18 495 A–K   Newspaper registration (Rhône)
    CC 558, 572   Cour des pairs (1834 uprising)

Bibliothèque de l'Arsenal (Paris)
    Fonds Enfantin   Letters of Eugénie Niboyet and Peiffer

Bibliothèque Historique de la Ville de Paris
    Ms. carton 42 48   Letters of Eugénie Niboyet

Bibliothèque Marguerite-Durand (Paris)
    Dossier Niboyet   Letters of Eugénie Niboyet

Newspapers and Periodicals

Publications in Lyon

Asmodée (1833)
Censeur (1834–)
Conseiller des femmes (1833–34)
Courrier de Lyon (1832–)
Cri du peuple (1831)
Écho de la fabrique (1831–34)
Écho de la fabrique de 1841 (1841)
Écho des ouvriers (1840)
Écho des travailleurs (1833–34)
Épingle (1835)
Furet de Lyon (1832)
Gazette du Lyonnais (1831–)
Glaneuse (1831–34)
Indicateur (1834–35)
Journal du commerce (1823–)
Nouvel Écho de la fabrique (1835)
Papillon (1832–35)
Précurseur (1821–22, 1823, 1826–34)
Précurseur du peuple (1834)
Réparateur (1833–)
Revue de Lyon (1835–36)
Revue du Lyonnais (1835–)

Sentinelle nationale (1831)
Tribune lyonnaise (1845–51)
Tribune prolétaire (1834–35)
Union des travailleurs (1835)

Publications in Paris

Gazette des tribunaux
Globe
Homme rouge
Journal des débats
National
National de 1834
Revue encyclopédique
Revue républicaine
Le Temps
Tribune politique et littéraire

Printed Primary Sources

Almanach historique et politique de la ville de Lyon et du départment du Rhône, pour l'an de grace 1831. Lyon: Rusand, 1831.
Association républicaine pour la liberté individuelle et la liberté de la presse. Procès de la Glaneuse, journal républicain de Lyon. Paris: Auffray, 1833.
Associations nationales en faveur de la presse patriote. Extrait du National. 11 July 1833.
Aux amis de la verité. Lyon: Brunet, 10 December 1831.
Baune, Eugène. *Essai sur les moyens de faire cesser la détresse de la fabrique.* Lyon: Baron, 1832.
Bavoux, François-Nicolas. *Développement de la proposition faite par M. Bavoux sur le cautionnement, le droit de timbre et le port des journaux et écrits périodiques.* Paris: Imprimerie Nationale, 1830.
Benoît, Joseph. *Confessions d'un prolétaire, Lyon, 1871.* Edited by Maurice Moissonnier. Paris: Éditions Sociales, 1968.
Bert. "La presse parisienne." In *Nouveau Tableau de Paris.* 6 vols., 5:131–45. Paris: Mme. Charles-Béchet, 1834–35.
[Beuf, Joseph]. *Procès et défense de Joseph Beuf, prolétaire.* Lyon, 1832.
Biographie des accusés d'avril, de leurs défenseurs, des pairs, juges du procès, etc. Paris: Collibert, 1835.
Biographie lyonnaise des auteurs dramatiques vivans, dits du terroir, rédigée dans la loge du portier des célestins; enrichie de quelques notes, par un bon enfant. Lyon: Chez tous les libraires, Coque, n.d.

Bouvier-Dumolard. *Rélation de M. Bouvier du Molart, ex-préfet du Rhône, sur les événements de Lyon.* Lyon: Bureau du Journal du Commerce, 1832.

Bowring, John. *Autobiographical Recollections of Sir John Bowring, with a Brief Memoir by Lewin B. Bowring.* London: Henry S. King and Company, 1877.

———. *Second Report on the Commercial Relations Between France and Great Britain, Silks and Wine.* London: William Clowes and Sons, 1835.

Chastaing, Marius. *Astrée. Discours Maçonnique sur la justice.* Lyon: Isidore Deleuze, 1838.

———. *Astréologie, ou Remède aux causes du malaise social.* Lyon: Rodanet, 1848.

———. *Vingt-deux jours de captivité.* Lyon: Bureau de la Tribune Lyonnaise, 1849.

[Chevalier, Michel, ed.] *Religion Saint-Simonienne. Événemens de Lyon.* Paris: Everat, 1832.

Collomb. *Details historiques sur les journées de Lyon, et les causes qui les ont précédées.* Lyon: Charvin, 1832.

Couailhac, Louis. *Les sept contes noirs.* Paris and Lyon: Bohaire, 1832.

Dupin, Charles. *Aux chefs d'atelier composant l'association des mutuellistes lyonnais.* Paris: Lacombe, 1834.

Favre, Jules. *6e procès du Précurseur. Défense de M. Anselme Petetin.* Lyon: Babeuf, 1833.

Ferton, J. *Moyens de défense qui devaient être présentée à la Cour d'Assises du Rhône, le 12 mai 1834.* N.p., n.d.

———. *Ordre public et amnistie.* Lyon: Perret, 1834.

Girardin, Saint-Marc. *Souvenirs et reflexions politiques d'un journaliste.* Paris: Michel Levy, 1859.

Girod de l'Ain, Baron Amédée. *Rapport fait à la Cour des pairs. Affaire du mois d'avril 1834.* 15 vols. Paris: Imprimerie Royale, 1834–35.

Grangé, Sophie. *Romances et poésies diverses, par Mlle. S. . . .* Lyon: Banet, 1826.

Granier, J. Adolphe. *Charles X, ou La leçon au roi tyran.* Lyon: André Idt, 1830.

———. *Pamphlet.* Lyon: Perret, 1831.

———. *Procès de la "Glaneuse."* Lyon: Glaneuse, 1833.

Histoire de Lyon pendant les journées des 21, 22 et 23 novembre 1831, contenant les causes, les conséquences et les suites de ces déplorables événements, éditeur Auguste Baron. Lyon: A. Baron, 1832.

Historique des événemens de Lyon, du neuf au quatorze avril 1834. Lyon: Rossary, 1834.

Kauffmann, Sébastien. *Biographie contemporaine des gens de lettres de Lyon.* Lyon and Paris: Chez Tous les Marchands de Nouveautés, 1826.

———. *La Célestinade, ou La guerre des auteurs et des acteurs lyonnais.* Lyon: Laforgue, 1829.

L[amothe]-L[angon], Baron Étienne-Léon de. *Une semaine de l'histoire de Paris.* Paris: Mame and Delaunay-Vallée, 1830.

Lyon vu de Fourvières. Esquisses physiques, morales et historiques. Lyon: L. Boitel, 1833.

Bibliography

Mazon, Jean-François René. *Événemens de Lyon, ou Les trois journées de novembre 1831*. Lyon: Guyot; Paris: Dentu, 1831.
M. Lacombe. Lyon: André Idt, 1831.
Monfalcon, Jean-Baptiste. *Annales de la ville de Lyon, ou Histoire de notre temps*. 2 vols. Lyon, 1845–50.
———. *Code moral des ouvriers, ou Traité des devoirs et des droits des classes laborieuses*. Paris and Lyon: Pélagaud Lesne and Crozet, 1836.
———. *Histoire de la ville de Lyon*. 2 vols. Lyon: Guilbert and Dorier, 1847.
———. *Histoire des insurrections de Lyon, en 1831 et 1834, d'après des documents authentiques, précédée d'un essai sur les ouvriers en soie et sur l'organisation de la fabrique*. Lyon: Perrin, 1834.
———. *Histoire des marais, et des maladies causées par les émanations des eaux stagnantes*. Paris: Bechet Jeune, 1824.
———. *Histoire monumentale de la ville de Lyon*. 9 vols. Paris: Didot, 1866–69.
———. "Maladies des tisserands." In Philippe Patissier, *Traité des maladies des artisans, et de celles qui résultent des diverses professions*, 384–400. Paris: J.-B. Ballière, 1822.
———. *Souvenirs d'un bibliothécaire, ou Une vie d'homme de lettres en province*. Lyon: Nigon, 1853.
Monfalcon, Jean-Baptiste, and A.-P.-I. Polinière. *Traité de la salubrité dans les grandes villes, suivi de l'hygiène de Lyon*. Paris: J.-B. Ballière, 1846.
Morin, Jérôme. *Du journalisme, à propos de la brochure intitulée. De l'enseignement du droit public en France, de M. Bellin*. Lyon: L. Boitel, 1842.
———. *Histoire de Lyon depuis la révolution de 1789*. 3 vols. Paris: Furne, 1845.
Mornand, Claude. *Une semaine de révolution, ou Lyon en 1830*. Lyon: André Idt, 1831.
Moyens de défense qui devaient être présentés à la cour d'Assises du Rhône, le 12 mars 1834, par le citoyen J. Ferton, l'un des gérans de la Glaneuse. Lyon: Jérome Perret, 1834.
[Niboyet, Eugénie]. *Précis historique sur les événemens de Lyon*. Lyon: L. Boitel, 1834.
Le Parfaite Silence, Société maçonnique de Saint-Jean d'Écosse, au rite Écossais et Français, constituée à Lyon. Lyon: L. Boitel, 1834.
Pointe, J.-P. "Fragment pour servir à l'histoire de Lyon, pendant les événements du mois d'avril 1834." *Revue du Lyonnais* 3 (1836), 216–32.
Principes d'un vrai républicain. Réception de plusieurs membres dans la société des droits de l'homme. Lyon: Charvin, n.d.
Procès de la Glaneuse, contenant les douze articles incriminé. Lyon: Aux Bureaux de la Glaneuse, 1833.
Procès du Précurseur, avec la plaidoirie de M. Odilon-Barrot. Lyon: Bureau du Précurseur, 1832.
Procès et défense de Joseph Beuf, prolétaire, condamné à trois ans et demi de prison et 2,500 fr. d'amende par la cour d'assises de Rhône. Lyon: Chez les Principaux Libraires, 1832.

Procès-verbal de la fête Maç∴offerte par les □□ *réunies des O∴de Lyon et de la Croix-Rousse, au F∴Odilon-Barrot, le 4e j∴du 7e m∴de l'an de la V∴L∴ 5832, 4 September 1832 (ère vulg) dans le local des enfans d'Hiram et de la sincère amitié* (Lyon, 1832).
Publication du "Populaire." *Procès de la Glaneuse, journal républicain de Lyon.* Paris: n.p., 1834.
Quelques événemens du jour, Fragment du manuscrit trouvé à Lyon dans une vielle armoire de sacristie, et destiné à une prochaine publication a Paris sous le titre de Lettres Contemporaines. . . . Paris: Delaunay, 1829.
Réponse d'un homme de lettres à une agression brutale du Journal du Commerce de Lyon. Paris: Delaunay, 1829.
Résumé de la discussion générale de la proposition de M. Bavoux, relative aux journaux et ecrits périodiques. Paris: Imprimerie Nationale, 1830.
La révolte de Lyon en 1834, ou La fille du prolétaire. 2 vols. Paris: Moutardier, 1835.
Rittiez, F. *Lettre du citoyen Rittiez aux patriotes de 1830.* Paris: Prévôt, 1831.
[Roederer, Pierre-Louis]. "Essai analytique sur les diverses moyens établis pour la communication des pensées, entre les hommes en société." *Journal d'économie publique,* 30 brumaire An V [December 1796].
Romand, Jean-Claude. *Confession d'un Malheureux. Vie de Jean-Claude Romand, Forçat libéré, écrite par lui-même, et publiée par M. Eduard Servan de Sugny, procureur du roi près le Tribunal de première instance de Nantua.* Paris: Au Comptoir des Imprimeurs-Unis, 1846.
Sadler, Percy. *Paris in July and August 1830.* Paris: Baudry, 1830.
Saint-Cheron, Alexandre. "Du journalisme." *Revue encyclopédique,* August 1832, 533.
Sala, Adolphe. *Les ouvriers lyonnais en 1834, esquisses historiques.* Paris: Hivert, 1834.
Seynes, Théodore. *Lyon 1831.* Lyon: Charvin, 1831.
Souscription pour les ouvriers blessés. Lyon: Pitrat, 1831.
Sugny, Edouard Servande. *Ma vie judiciaire.* Lyon: Veuve Ayné, 1847.
Terson, Jean. *Un St.-Simonien au peuple de Lyon, à l'occasion des événemens d'avril 1834.* Lyon: Durval, 1834.
Trémadeure, Sophie Ulliac. *Souvenirs d'une vieille femme.* 2 vols. Paris: E. Maillet, 1861.
Trolliet, L. F. *Lettres historiques sur la révolution de Lyon, ou Une semaine de 1830.* Lyon: Targe, 1830.
Valois, H. *Défense du Précurseur, journal de Lyon et du midi.* Lyon: Brunet, 1826.
Vente pour impot. Par récidive et au profit de la meilleure des républiques. Lyon: Pitrat, 1834.
La vérité sur les événemens de Lyon au mois d'avril 1834. Paris: Dentu; Lyon: Chambet, 1834.
Villermé, Louis-René. *Tableau de l'état physique et moral des ouvriers employés dans les manufactures de coton, de laine et de soie.* 2 vols. Paris: Jules Renouard, 1840.

Bibliography

Select Bibliography of Secondary Sources

N.B.: *This Select Bibliography has been limited to works cited concerning nineteenth-century history, the history of Lyon, the history of the press, and theoretical issues relevant to this book.*

Adler, Laure. *A l'aube du féminisme. Les premières journalistes, 1830–1850.* Paris: Payot, 1979.
Agulhon, Maurice. *Le cercle dans la France bourgeoise, 1810–1848.* Paris: Armand Colin, 1977.
Allen, James Smith. *In the Public Eye: A History of Reading in Modern France, 1800–1930.* Princeton, N.J.: Princeton University Press, 1991.
Aminzade, Ronald. *Ballots and Barricades: Class Formation and Republican Politics in France, 1830–1871.* Princeton, N.J.: Princeton University Press, 1993.
Anderson, Benedict. *Imagined Communities.* 2nd edition. London: Verso, 1991.
Bastid, Paul. *Les grands procès politiques de l'histoire.* Paris: Fayard, 1962.
Baud, F. "La fondation et les débuts du Précurseur." *Revue d'Histoire de Lyon* 13 (1914), 350–62.
Bellanger, Claude, et al. *Histoire générale de la presse française.* 5 vols. Paris: Presses Universitaires de France, 1969.
Benoît, Bruno. *L'identité politique de Lyon. Entre evidences collectives et mémoire des elites, 1786–1905.* Paris: L'Harmattan, 1999.
Bezucha, Robert J. *The Lyon Uprising of 1834: Social and Political Conflict in the Early July Monarchy.* Cambridge, Mass.: Harvard University Press, 1974.
Biré, E. *La presse royaliste de 1830 à 1852. Alfred Nettement, sa vie et ses oeuvres.* Paris: Victor Le Coffre, 1901.
Blanc, Louis. *The History of Ten Years, 1830–1840, or France Under Louis Philippe.* Translated by Walter R. Kelly. 2 vols. Philadelphia: Lea and Blanchard, 1848.
Boorstin, Daniel. *The Image.* New York: Penguin Books, 1962.
Bouchet, Thomas. "'Un événement grave vient de jeter la consternation dans Paris. . . .' La presse française face aux 5 et 6 juin 1832." In Hans-Jürgen Lüsebrink and Jean-Yves Mollier, eds., *Presse et événement. Journaux, gazettes, almanachs (XVIIIe–XIXe siècles)*, 27–44. Bern: Peter Lang, 2000.
Bourdieu, Pierre. *The Field of Cultural Production.* Edited by Randal Johnson. New York: Columbia University Press, 1993.
Brisac, Marc. *Lyon et l'insurrection polonaise de 1830–1831.* Lyon: Revue d'Histoire de Lyon, 1909.
Brooks, Peter. *Reading for the Plot: Design and Intention in Narrative.* New York: Alfred A. Knopf, 1984.
Bryant, Harry S., Jr. "*Ce n'est pas moi qui ai cherché cette publicité:* Pamphlet Coverage of Political Trials During the Early July Monarchy, 1830–1835." *Proceedings of the Western Society for French History* 24 (1997), 423–29.

Buffenoir, Maximilien. "Le féminisme à Lyon avant 1848." *Revue d'Histoire de Lyon* 7 (1906), 348–58.

———. "Le 'Précurseur' et la révolution de Juillet." *Revue d'Histoire de Lyon* 6 (1907), 351–62.

Calhoun, Craig, ed. *Habermas and the Public Sphere*. Cambridge, Mass.: MIT Press, 1992.

La Caricature. Bildsatire in Frankreich, 1830–1835, aus der Sammlung von Kritter. Göttingen: Kunstgeschichtliches Seminar der Universität Göttingen, 1980.

Chaline, Jean-Pierre. *Les bourgeois de Rouen. Une élite urbaine au XIXe siècle*. Paris: Presses de la Fondation Nationale des Sciences Politiques, 1982.

Chartier, Roger, and Henri-Jean Martin, eds. *Histoire de l'édition française t. III. Le temps des éditeurs*. Paris: Fayard, 1990.

Chevalier, Louis. *Classes laborieuses et classes dangereuses*. Paris: Librairie Générale Française, 1978.

Chu, Petra ten-Doesschate, and Gabriel P. Weisberg, eds. *The Popularization of Images: Visual Culture Under the July Monarchy*. Princeton, N.J.: Princeton University Press, 1994.

Cohen, William B. *Urban Government and the Rise of the French City: Five Municipalities in the Nineteenth Century*. New York: St. Martin's Press, 1998.

Collins, Irene. *The Government and the Newspaper Press in France, 1814–1881*. London: Oxford University Press, 1959.

Czyba, Lucette. "L'oeuvre lyonnaise d'une ancienne Saint-Simonienne: *Le conseiller des femmes* (1833–34) d'Eugénie Niboyet." In J. R. Derré, ed., *Regards sur le Saint-Simonisme et les Saint-Simoniens*, 103–141. Lyon: Presses Universitaires de Lyon, 1986.

Daumard, Adeline. *Les bourgeois et la bourgeoisie en France depuis 1815*. Paris: Aubier, 1987.

De la Motte, Dean, and Jeannene Przyblyski, eds. *Making the News: Modernity and the Mass Press in Nineteenth-Century France*. Amherst: University of Massachusetts Press, 1999.

Droux, Georges. *La chanson lyonnaise. Histoire de la chanson à Lyon. Les Sociétés Chansonnières*. Lyon: Revue d'Histoire de Lyon, 1907.

Dutacq, F. "Les journées lyonnaises de novembre 1831." *Revolution de 1848* 29 (1932), 68–103.

Epstein, James A. *Radical Expression: Political Language, Ritual, and Symbol in England, 1790–1850*. New York: Oxford University Press, 1994.

Favre, Jules. *Plaidoyers politiques et judiciaires*. 2 vols. Paris: Plon, 1882.

Feyel, Gilles. "La diffusion nationale des quotidiens parisiens en 1832." *Revue d'histoire moderne et contemporaine* 34 (1987), 31–65.

Freedeman, Charles E. *Joint-Stock Enterprise in France, 1807–1867*. Chapel Hill: University of North Carolina Press, 1979.

Furlough, Ellen. *Consumer Cooperation in France: The Politics of Consumption, 1834–1930*. Ithaca, N.Y.: Cornell University Press, 1991.

Garden, Maurice. *Lyon et les lyonnais au XVIIIe siècle.* Paris: Société d'Édition "Les Belle-Lettres," 1970.
Gardes, Gilbert. *Le voyage du Lyon.* Lyon: Horvath, 1993.
Garrioch, David. *The Formation of the Parisian Bourgeoisie, 1690–1830.* Cambridge, Mass.: Harvard University Press, 1996.
Gaumont, Jean. *Le commerce véridique et social, 1835–1838, et son fondateur Michel Derrion, 1803–1850.* Amiens: Imprimerie Nouvelle, 1935.
Giesselmann, Werner. "Die Manie der Revolt." *Protest unter der Französischen Julimonarchie, 1830–1848.* 2 vols. Munich: R. Oldenbourg, 1993.
Goblot, Edmond. *La barrière et le niveau.* Paris: Presses Universitaires de France, 1967.
Godart, Justin. "Le compagnonnage à Lyon." *Revue d'histoire de Lyon* 2 (1903), 425–69.
———. *La révolution de 1830 à Lyon.* Paris: Presses Universitaires de France, 1930.
Goldstein, Robert J. *Censorship of Political Caricature in Nineteenth-Century France.* Kent, Ohio: Kent State University Press, 1989.
Goodman, Dena. "Public Sphere and Private Life: Toward a Synthesis of Current Historiographical Approaches to the Old Regime." *History and Theory* 31 (1992), 1–20
Groth, Otto. *Die Zeitung.* Volume 1. Mannheim: Bensheimer, 1928.
Habermas, Jürgen. *Strukturwandel der Öffentlichkeit.* Neuwied and Berlin: Luchterhand, 1962.
Hariman, Robert, ed. *Popular Trials: Rhetoric, Mass Media, and the Law.* Tuscaloosa: University of Alabama Press, 1990.
Historique des événements de Lyon, du neuf au quatorze avril 1834. Lyon: Rossary, 1834.
Johnson, Christopher H. *Utopian Communism in France: Cabet and the Icarians, 1839–1851.* Ithaca, N.Y.: Cornell University Press, 1974.
Jones, Gareth Stedman. *Languages of Class: Studies in English Working Class History, 1832–1982.* Cambridge: Cambridge University Press, 1983.
Kirchheimer, Otto. *Political Justice.* Princeton, N.J.: Princeton University Press, 1961.
Kleinclausz, A., et al. *Histoire du Lyon.* Lyon: Pierre Masson, 1952.
Kocka, Jürgen. "The Middle Classes in Europe." *Journal of Modern History* 67 (1995), 783–806.
Latreille, A. "Un salon littéraire à Lyon, 1830–1860, Mme. Yéméniz." *Revue d'histoire de Lyon* 2 (1903), 21–47.
Ledré, Charles. *La presse à l'assaut de la monarchie, 1815–1848.* Paris: Armand Colin, 1960.
Lyons, Martyn. *Le triomphe du livre. Une histoire sociologique de la lecture dans la France du XIXe siècle.* Paris: Promodis, 1987.
Manoff, Robert Karl, and Michael Schudson, eds. *Reading the News.* New York: Pantheon, 1987.

Marrinan, Michael. *Painting Politics for Louis-Philippe: Art and Ideology in Orléanist France, 1830–1848*. New Haven, Conn.: Yale University Press, 1988.

Martin, Marc. *Trois siècles de publicité en France*. Paris: Éditions Odile Jacob, 1993.

Marty, Corinne. "Les bibliothèques publiques de la ville de Lyon du XVIIIème au XIXème siècles, d'après les papiers d'un érudit lyonnais, Jean-Baptiste Monfalcon (1792–1874)." Mémoire de maîtrise d'histoire, Université Jean-Moulin Lyon III, 1988.

Matlock, Jann. *Scenes of Seduction: Prostitution, Hysteria, and Reading Difference in Nineteenth-Century France*. New York: Columbia University Press, 1994.

Maza, Sarah. "Luxury, Morality, and Social Change: Why There Was No Middle-Class Consciousness in Prerevolutionary France." *Journal of Modern History* 69 (1997), 199–229.

McDougall, Mary Lynn. "After the Insurrections: The Workers' Movement in Lyon, 1834–1852." Ph.D. diss., Columbia University, 1973.

Merriman, John M. *The Margins of City Life: Explorations on the French Urban Frontier, 1815–1851*. New York: Oxford University Press, 1991.

Merriman, John M., ed. *1830 in France*. New York: Franklin Watts, 1975.

Moissonnier, Maurice. *Les canuts. "Vivre en travaillant ou mourir en combattant."* 4th edition. Paris: Messidor / Éditions Sociales, 1988.

Momblet, E. "Leon Boitel, fondateur de la Revue du lyonnais." *Revue du Lyonnais*, 3rd ser. (1866), 1:15–30.

Moses, Claire Goldberg. "'Equality' and 'Difference' in Historical Perspective: A Comparative Examination of the Feminisms of French Revolutionaries and Utopian Socialists." In Sara E. Melzer and Leslie W. Rabine, eds., *Rebel Daughters: Women and the French Revolution*, 231–54. Oxford: Oxford University Press, 1992.

———. *French Feminism in the Nineteenth Century*. Albany, N.Y.: SUNY Press, 1984.

Moss, Bernard H. *The Origins of the French Labor Movement, 1830–1914*. Berkeley and Los Angeles: University of California Press, 1978.

Mouillaud, Maurice, and Jean-François Tétu. *Le journal quotidien*. Lyon: Presses Universitaires de Lyon, 1989.

Nora, Pierre, ed., *Realms of Memory: The Construction of the French Past*. Translated by Arthur Goldhammer. New York: Columbia University Press, 1996.

Nye, Robert A. *Masculinity and Male Codes of Honor in Modern France*. New York: Oxford University Press, 1993.

Orr, Linda. *Headless History: Nineteenth-Century French Historiography of the Revolution*. Ithaca, N.Y.: Cornell University Press, 1990.

Pariset, E. *Histoire de la fabrique lyonnaise*. Lyon: A. Rey, 1901.

Pelissier, Catherine. *Loisirs et sociabilités des notables lyonnais au XIXe siècle*. Lyon: Editions Lyonnaises d'Art et d'Histoire, 1996.

Bibliography

Perreux, Gabriel. *Au temps des sociétés secrètes.* Paris: Rieder, 1931.

———. "L'espirit public dans les départements au lendemain de la révolution de 1830: La region lyonnaise." *Revolution de 1848* 33 (1936–38), 98–106.

Perrod, Pierre Antoine. *Jules Favre, avocat de la liberté.* Lyon: La Manufacture, 1988.

Perrot, Philippe. *Fashioning the Bourgeoisie: A History of Clothing in the Nineteenth Century.* Translated by Richard Bienvenu. Princeton, N.J.: Princeton University Press, 1994.

Petrey, Sandy. "Pears in History." *Representations,* no. 35 (1991), 52–71.

Pilbeam, Pamela M. *Republicanism in Nineteenth-Century France, 1814–1871.* New York: St. Martin's Press, 1995.

Pinkney, David M. *The French Revolution of 1830.* Princeton, N.J.: Princeton University Press, 1972.

Popkin, Jeremy D. "Un grand journal de province à l'époque de la révolution de 1830. *Le précurseur de Lyon,* 1826–1834." In Michel Biard, Annie Crepin, and Bernard Gainot, eds., *Hommages à Jean-Paul Bertaud.* Paris: forthcoming.

———. "Media and Revolutionary Crises." In Jeremy D. Popkin, ed., *Media and Revolution,* 12–30. Lexington: University of Press of Kentucky, 1995.

———. *News and Politics in the Age of Insurrection: Jean Luzac's "Gazette de Leyde."* Ithaca, N.Y.: Cornell University Press, 1989.

———. *Revolutionary News: The Press in France, 1789–1799.* Durham, N.C.: Duke University Press, 1990.

———. "Texte et insurrection: La presse et les insurrections de Lyon en 1786 et 1831." In Hans-Jürgen Lüsebrink and Jean-Yves Mollier, eds., *Presse et événement. Journaux, gazettes, almanachs (XVIIIe–XIXe siècles),* 45–63. Bern: Peter Lang, 2000.

Procacci, Giovanna. *Gouverner la misère. La question sociale en France, 1789–1848.* Paris: Seuil, 1993.

Prost, Antoine. *Histoire de l'enseignement en France, 1800–1967.* Paris: Armand Colin, 1967.

Prothero, Iorwerth J. *Artisans and Politics in Early Nineteenth-Century London: John Gast and His Times.* Folkestone: William Dawson and Son, 1979.

Ragon, Michel. *Histoire de la littérature prolétarienne en France.* Paris: Albin Michel, 1974.

Reddy, William M. *The Invisible Code: Honor and Sentiment in Postrevolutionary France, 1814–1848.* Berkeley and Los Angeles: University of California Press, 1997.

Rancière, Jacques. *La nuit des prolétaires.* Paris: Fayard, 1981.

Reid, Roddey. *Families in Jeopardy: Regulating the Social Body in France, 1750–1910.* Stanford, Calif.: Stanford University Press, 1993.

Rémusat, Charles François Marie de. *Mémoires de ma vie.* Edited by Charles Pouthas. 5 vols. Paris: Plon, 1959.

Les révolutions du XIXe siècle, vol. 10: *La presse républicaine devant les tribunaux, 1831–1834.* Paris: Edhis, 1974.

Ribe, Georges. *L'opinion publique et la vie politique à Lyon lors des premières années de la seconde Restauration.* Paris: Sirey, 1957.

Ricoeur, Paul. "Événement et sens." In Jean-Luc Petit, ed., *L'événement en perspective.* Paris: Éditions de l'École des Hautes Études en Sciences Sociales, 1991.

Riot-Sarcey, Michèle. *La démocratie à l'épreuve des femmes. Trois figures critiques du pouvoir, 1830–1848.* Paris: Albin Michel, 1994.

———. "Histoire et autobiographie. *Le vrai livre des femmes* d'Eugénie Niboyet." *Romantisme* 17 (1987), 59–68.

———. *Le réel de l'utopie. Essai sur la politique au XIXe siècle.* Paris: Albin Michel, 1998.

Rosanvallon, Pierre. *Le Moment Guizot.* Paris: Gallimard, 1985.

Rose, R. B. *Tribunes and Amazons: Men and Women of Revolutionary France, 1789–1871.* Paddington, Australia: Macleay, 1998.

Roustan, M., and C. Latreille. "Lyon contre Paris après 1830. Le mouvement de décentralisation littéraire et artistique." *Revue d'histoire de Lyon* 3 (1904), 24–42, 109–26, 306–18, 384–402.

Rude, Fernand. *Les canuts en 1789. Doléances des maîtres-ouvriers fabricants en étoffe d'or, argent et de soie de la ville de Lyon.* Lyon: Fédérop, 1976.

———. "L'insurrection ouvrière de Lyon en 1831 et le rôle de Pierre Charnier." *Revolution de 1848* 34 (1938), 18–49, 65–117; 35 (1938), 140–79.

———. *Le mouvement ouvrier à Lyon.* Lyon: Fédérop, 1977.

———. *Le mouvement ouvrier à Lyon de 1827 à 1832.* Paris: Domat-Montchrestien, 1944.

———. *Les révoltes des canuts, 1831–1834.* Lyon: Maspero, 1982.

Sahlins, Peter. *Forest Rites: The War of the Demoiselles in Nineteenth-Century France.* Cambridge, Mass.: Harvard University Press, 1994.

Saunier, Pierre-Yves. "Haut-lieu et lieu haut. La construction du sens des lieux. Lyon et Fourvière aux XIXe siècle." *Revue d'histoire moderne et contemporaire* 40 (1993), 202–27.

———. "Représentations sociales de l'espace et histoire urbaine. Les quartiers d'une grande ville française, Lyon au XIXe siècle." *Social History / Histoire sociale* 29 (1996), 23–52.

Sewell, William H., Jr. "Historical Events and Transformations of Structures: Inventing Revolution at the Bastille." *Theory and Society* 25 (1996), 841–81.

———. *Work and Revolution in France: The Language of Labor from the Old Regime to 1848.* Cambridge: Cambridge University Press, 1980.

Sheridan, George J. "The Political Economy of Artisan Industry: Government and the People in the Silk Trade of Lyon, 1830–1870. " *French Historical Studies* 11 (1979), 215–38.

Smith, Bonnie G. *Ladies of the Leisure Class.* Princeton, N.J.: Princeton University Press, 1981.

Smith, Timothy B. "Public Assistance and Labor Supply in Nineteenth-Century Lyon." *Journal of Modern History* 68 (1996), 1–30.

Spang, Rebecca. "'La fronde des nappes': Fat and Lean Rhetoric in the Political Banquets of 1847." In Carrol F. Coates, ed., *Repression and Expression: Literary and Social Coding in Nineteenth-Century France*, 167–78. New York: Peter Lang, 1996.

———. *The Invention of the Restaurant: Paris and Modern Gastronomic Culture*. Cambridge, Mass.: Harvard University Press, 2000.

Stein, Lorenz von. *Geschichte der sozialen Bewegung in Frankreich von 1789 bis auf unsere Tage*. 3 vols. Hildesheim: Georg Olms Verlag, 1959.

Strumingher, Laura. "Mythes et réalitites de la condition féminine à travers la presse féministe lyonnaise des années 1830." *Cahiers d'histoire* 21 (1976), 409–24.

Sullerot, Evelyne. *Histoire de la presse féminine en France, des origines à 1848*. Paris: Armand Colin, 1966.

Terdiman, Richard. *Discourse/Counter-Discourse: The Theory and Practice of Symbolic Resistance in Nineteenth-Century France*. Ithaca, N.Y.: Cornell University Press, 1985.

Thompson, Victoria. "*Splendeurs et misères des journalistes:* Imagery and the Commercialization of Journalism in July Monarchy France." *Proceedings of the Western Society for French History* 23 (1996), 361–68.

Tilly, Charles. *The Contentious French: Four Centuries of Popular Struggle*. Cambridge, Mass.: Harvard University Press, 1986.

Tournier, Maurice. *Des mots sur la grève*. Paris: Klincksieck, 1992.

Trénard, Louis. "La crise sociale lyonnaise à la veille de la Révolution." *Revue d'histoire moderne et contemporaine* 2 (1955), 5–45.

———. *Lyon de l'Encyclopédie au préromantisme*. Paris: Presses Universitaires de France, 1958.

———. "Les périodiques lyonnais dans les luttes révolutionnaires, 1792–1794." In Pierre Rétat, ed., *La Révolution du journal*, 297–305. Paris: Éditions du CNRS, 1989.

Truant, Cynthia. *The Rites of Labor: Brotherhoods of Compagnonnage in Old and New Regime France*. Ithaca, N.Y.: Cornell University Press, 1994.

Varille, Mathieu. *Les journées d'Avril 1834 à Lyon*. Lyon: Noirclerc and Fénétrier, 1923.

Vermorel, Jean. *Un préfet du Rhône sous la Monarchie de Juillet, M. de Gasparin*. Lyon: Audin, 1933.

Vial, Eugène. "La vie et l'oeuvre de Léon Boitel." *Revue du Lyonnais* 1 (1921), 109–21.

Vingtrinier, Aimé. "Léon Boitel." *Revue du Lyonnais* 2 (1855), 193–98.

Wahrman, Dror. *Imagining the Middle Class: The Political Representation of Class in Britain, c. 1780–1840*. Cambridge: Cambridge University Press, 1995.

Weill, Georges. *Histoire du parti republicain en France, 1814–1870*. New edition. Paris: Felix Alcan, 1929.

White, Hayden. *Tropics of Discourse: Essays in Cultural Criticism*. Baltimore, Md.: Johns Hopkins University Press, 1978.

Index

Page numbers in *italics* indicate illustrations.

"A la femme" (by Sophie Grangé), 112, 274–76
Agulhon, Maurice (historian), 61
"Aide-toi, le ciel t'aidera," 30
alternative press. *See* press, alternative
Ami du peuple (by Marat), 20
Aminzade, Ronald (historian), 5
Anatomy of Revolution (by Brinton), 5
Anderson, Benedict (scholar), 14, 69
Annales school of history, 236
"Aperçu sur la question du prolétariat," 251–53. See also *La Révolte de Lyon*
Argout, Comte d' (minister), 164
Asmodée (periodical), 106, 113, 291 n. 16
Association Lyonnais pour la Liberté de la Presse, 185
Association pour la Liberté de la Presse, 185
Athénée des femmes, 65, 127–28

Babeuf, Gracchus, 182
Balzac, Honoré de, 12, 41, 259, 270
banquets (in support of press), 159–60, 167–80, 186–87, 189, 191, 265
Barrière et le niveau, La (by Goblot), 56
Bastille, 18
Benoît, Joseph (silkworker), 246
Bellecour, place, 53–54, 56–57, 63, 127
Berger (worker-journalist), 159
Berryer, Pierre-Antoine (lawyer), 187
Berthaud, L. A. (journalist), 272, 291 n. 16. See also *Asmodée*
Bertholon, César (republican militant), 112, 119, 134
Beuf, Joseph (journalist), 294 n. 31
Bezucha, Robert (historian), 230, 237

Biographie des gens de lettres de Lyon, 41
Blanc, Louis (socialist theorist), 153, 176, 239
Boitel, Léon (publisher), 48, 55, 57–58, 109–10, 120, 133–34, 269
booksellers, 285–86 n. 75
Boorstin, Daniel (scholar), 169, 195
Bourbon dynasty, 4, 34, 67, 83, 183
Bourdieu, Pierre (sociologist), 12, 19, 24, 267
"bourgeois monarchy," 1, 76–77, 88, 102
bourgeois press. *See* press, bourgeois
bourgeoisie, 5, 17, 50, *58*, 68–70, 93–95, 97–98
bourgeois family, 256–58
bourgeois identity, 5, 17, 46, 68, 73–74, 82, 98, 102–3, 107–8, 112
bourgeois ideology, 2, 10, 157
bourgeois society, 22, 67, 69–70
  and government, 101
  in Lyon, 52
  as press stockholders, 79–80, 134
  women and, 127, 132
  See also Monfalcon, Jean-Baptiste; press, bourgeois; public sphere; silk industry
Bouvery, Henri-Joseph (worker-journalist), 157
Bouvier-Dumolard (prefect), 77, 101, 209, 211, 225
Bowring, John (English Utilitarian), 7, 140, 147
Bret, Alexander (journalist), 176
Brinton, Crane (historian), 4
broadsheets, 221–24, 302 nn. 62–64
Brotteaux, les, 222–23

Cabet, Étienne (socialist), 157
cabinets de lecture, 56–57, 59–60, 86, 108
cafés, 59–60, 139, 146, 151
Camus, Albert (author), 12
canuts, 147. See also silk workers
Carey, James W. (scholar), 11
Caricature, La (periodical), 8, 26, 95–96, 118
Carlists, 33, 80, 211
Carrel, Armand (journalist), 45–46, 94
Casimir-Périer (politician), 38, 88, 90, 115, 224
Catéchisme républicain (April 1833), 95
Catholicism, 34, 36, 51, 62, 100, 130, 154
caution money, 72, 81, 99, 107, 113, 157, 183–84
Célestinade, La (by Kauffmann), 41
Célestins, place des, 54–55
Censeur, 103, 148
Cercle de commerce (Lyon), 57
Chamber of Deputies, 34, 59, 71, 86, 101
Chamber of Peers, 115
Charles X (King of France), 15, 45, 116
Chastaing, Marius (journalist), 43–44, 47–48, 77, 141, 143, 152–53, 156, 158, 160–61, 269, 295 n. 38
  and Écho de la fabrique, 44, 141–42, 149, 176
  and Écho des travailleurs, 141–42, 151
  See also Écho de la fabrique; Écho des travailleurs; press, working-class
chefs d'atelier, 136, 140–42. See also Écho de la fabrique, silk workers
Chevalier, Michel (Saint-Simonian), 6
"Combat du Pont Morand" (broadsheet), 222–23
Commerce véridique, 155
Communist Manifesto (by Marx and Engels), 137, 252
compagnonnages, 5, 137, 144
Condorcet, Marquis de, 128
Confessions d'un malheureux (by Romand), 246
Confessions d'un prolétaire (by Benoît), 246
conseil des prud'hommes, 107, 140, 144, 148, 160

Conseiller des femmes, 34, 36–37, 65, 73, 79, 106–11, 113–14, 124–132, 258
  and insurrection of 1834, 208, 221
  reactions to, 130
  See also Niboyet, Eugénie; press, alternative; press, women's
Constituent Assembly, 143
Constitutionnel (newspaper), 59, 81
cooperative movement, 155
Cordeliers, place des, 53
Cormenin, Louis-Marie de (deputy), 93
Couailhac, Louis (author), 249, 251, 257, 259, 305 n. 38
Couderc, Jean (deputy), 83, 186
Cour des Pairs, 47
Courrier de Lyon (newspaper) 25, 35, 77, 97–103, 163, 228, 236, 238
  and bourgeoisie, 97–99, 101–3
  creation of, 33, 82, 95, 97
  finances of, 78
  format of, 36
  and Lyon insurrections, 224–25
  and Orleanist government, 37, 38, 77, 97
  political orientation of, 33, 35, 77, 97
  and press banquets, 176–77, 179
  and press trials, 188–89
  relations with other papers, 92, 130, 146, 148–49, 163, 206
  stockholders of, 79
  See also Monfalcon, Jean-Baptiste; press, bourgeois
Cri du peuple (newspaper), 32–33, 61–62, 79, 99–100, 114, 184. See also Pitrat, Théodore; press, legitimist;
Croix-Rousse, 2, 52–54, 222–23
Croix-Rousse, place de la, 202
"cultural field," definition of, 19, 24–25, 264, 267–68
cultural history, 8–10

Daumard, Adeline (historian), 68
"De la coalition des chefs d'atelier à Lyon" (by Favre), 251
Declaration of the Rights of Man and Citizen, 92, 182
Derrion, Michel (journalist), 155
Dictionnaire des sciences médicales, 243

Index

Directory, 72, 171
Dubroca, Jules (Fourierist), 156
Dubuisson, Jane (journalist), 125, 128
dueling, 46, 149, 284 n. 53

Écho de la fabrique (newspaper), 34, 62, 92, 98, 135–66, 172–73, 174–78, 293 n. 2
  banquet in honor of, 172–73, 174–77
  and Marius Chastaing, 44, 142–43, 149–50
  and chefs d'atelier, 140–42
  and *conseil de prud'hommes*, 148
  content of, 37, 73, 150–51, 152–63
  creation of, 138–40, 144–45
  and *Écho des travailleurs*, 35–36, 46, 150
  and Fourier, 156–57, 160
  impact of, 20, 147, 149
  and insurrections in Lyon, 2, 158–59, 161–63, 205–6, 211, 213, 222, 225, 277–78 (doc.)
  readership of, 79, 146, 165
  relations with government, 38, 146, 148, 164
  relations with other papers, 130, 146, 148, 156, 163–64
  and Saint-Simonians, 155–56, 157, 160
  and Society of Mutual Duty, 140–41, 144–45, 149–50, 157
  trial of, 189
  and workers, 137, 139, 141, 144, 147, 150–51, 165, 237
  See also Chastaing, Marius; *Écho des travailleurs*; *Indicateur*; press, workers'
Écho des travailleurs, 35, 38, 98, 135–36, 138, 141, 150, 152–56, 161, *162*, 164–65
  content of, 152–56, 161, 164–65
  creation of, 35, 150
  and *Écho de la fabrique*, 46, 177
  relations with other papers, 129–30
  See also *Tribune prolétaire*
Écho du jour (newspaper), 31
éducation mutuelle, 143
Enfantin, Prosper (Saint-Simonian), 121, 129

Épingle (periodical), 25, 27, 106
Estates-General, 28

Falconnet, Joachim (journalist), 139–40, 142, 159
Favre, Jules (lawyer), 45–46, 164, 251, 253
feminist press. See press, women's
Femme libre, La (periodical), 121
First Restoration, 28
Foucault, Michel (scholar), 74
Fourier, Charles (socialist), 152, 154–57, 160
Fourvière hill, 51–52
Freedeman, Charles (historian), 74
Fulchiron, Jean-Claude (politician), 83, 213
Furet de Lyon (periodical), 106, 294 n. 31
Furet, François (historian), 39

Galois, Antoine-Louis Christophe (journalist), 82
Garnier-Pagès, Étienne, banquet in honor of, 173–78
Garrioch, David (historian), 5
Gauchet, Marcel (scholar), 32
Gasparin, Adrien-Étienne-Pierre de (prefect), 37–38, 78, 93, 101–2, 114, 134, 148, 164, 174, 177–79, 189, 231–32, 238
  and Lyon press, 37–38, 78, 93, 101–2, 114, 134, 148, 164
  and Jean-Baptiste Monfalcon, 231–32, 238
  and press banquets, 174, 177–79
  and press trials, 189
Gazette de France, 19
Gazette de Leyde, 195–97
Gazette de Lyon, 31
Gazette des tribunaux, 153
Gazette du Lyonnais, 25, 33, 47, 80, 87, 99–100, 206, 211, 213, 216. See also Pitrat, Théodore; press, legitimist
Gazette universelle (Lyon), 29–31, 83, 85
gérant, 43–44, 74, 183–84
Germinal (by Zola), 256
Giesselmann, Werner (historian), 165
Gilardin, Alphonse, 186

Girardin, Émile de (journalist), 39, 46
Girardin, Saint-Marc (journalist), 1–2, 194, 217–20, 227–28
Glaneuse (newspaper), *96*, 106–20, 131, 197–202, 236
  content of, 23, 36, 95, 112–20, 204
  creation of, 32
  and gender imagery, 110–11
  and Adolphe Granier (editor), 43, 109, 188, 199
  and insurrections in Lyon, 119, 197–202, 207, 212
  popular edition of, 35, 63
  and press banquets, 165, 173–74, 178
  and press trials, 184, 187–88
  readership of, 78, 79, 107
  relations with other papers, 89, 95, 158
  and republicanism, 112–13, 119–20, 207
  See also press, alternative; press, republican
Globe (newspaper), 81, 146, 156, 216
Goblot, Edmond (scholar), 56, 68
"Grand Croisade Contre la Liberté" (caricature), *96*
Grandville (caricaturist), 26, 95
Grangé, Sophie (poet), 112, 123–24, 271–76
Granier, Adolphe (journalist), 42–43, 109, 119, 188, 199, 201, 269. See also *Glaneuse*; insurrection (Lyon, November 1831)
Groth, Otto (scholar), 14
Guillotière, La, 22
Guizot, François (politician), 88, 181

Habermas, Jürgen (scholar), 49–51
Higonnet, Patrice (historian), 70
Histoire générale de la presse française (by Bellanger et al.), 3
Holbach, Baron d', 251, 253
Hollier, Denis (scholar), 9
Hôtel-de-Ville (Lyon), 52, 62–63, 119, 199–200, 205
Hôtel-Dieu (Lyon), 44, 217

Indicateur, 25, 135, 148, 155, 159–60, 190. See also *Écho de la fabrique*

insurrection: Lyon, 1786, 195–97
insurrection: Lyon, November 1831, 1, 6, *200*
  broadsheets and, 221–24, *223*, 302 nn. 62–64
  casualties in, 6, 193–94
  commemoration of, in press, 159, 161–63, *162*, 277–78
  and *Courrier de Lyon*, 225
  and *Écho de la fabrique*, 205–6
  and Saint-Marc Girardin, 1, 194, 218–19, 227–28
  and *Gazette du Lyonnais*, 206
  and *Glaneuse*, 119, 197–202, 212
  and *Journal des débats*, 1, 194, 218
  and *Journal du commerce*, 212–13
  and Lyon press, 170, 194, 197–206, 208–20, 226
  Jean-Baptiste Monfalcon's writings on, 217–18, 231–39, 244–48
  pamphlets and, 224, 303 n. 67
  and Paris press, 215–19, 220
  and *Précurseur*, 91, 93, 202–5, 210–13
  workers' autobiographies and, 246
insurrection: Lyon, April 1834, 4, 6, 51, 193–94, 206–9, 214–15, 219–21, 224–26, 235–38, 246
  casualties in, 6, 193–94
  causes of, 171, 191
  pamphlets and, 224, 303 n. 67
  Jean-Baptiste Monfalcon's writings on, 231, 235–38, 242
  Anselme Petetin and, 102
  press, role in, 206–9, 214–15, 219–21, 224–26
  repression of press after, 133, 164
  in *La Révolte de Lyon*, 250–51, 255–56
  trial of participants, 47, 48, 102
  workers' press and, 158
  See also *Courrier de Lyon*; Gasparin, Adrien-Étienne-Pierre de; Monfalcon, Jean-Baptiste
insurrection: Paris, June 1832, 4, 119, 185

Jacobin regime, 29, 42
Jacobin club, 171
Jesuits, 116
Journal de Lyon, 28, 42, 196

Journal de Paris, 19
Journal des débats, 1, 2, 81, 194, 215–18, 220
Journal des femmes, 125
Journal du commerce, 25, 29–30, 35, 81–82, 91, 97, 202–3
  creation of, 29
  format of, 36, 87
  and Antoine Louis-Christophe Galois (publisher), 82
  government, 38
  and Lyon insurrections, 211–12, 221
  political orientation of, 31, 91, 114
  readership of, 61, 78
  See also press, bourgeois; press, liberal
journalism. *See* press
journalists, 40–49
July Days of 1830. *See* Revolution of 1830
July Monarchy, 32, 69
  and press, 32, 182–84, 191
  See also liberalism; Revolution of 1830
July Ordinances (1830), 45

Kauffmann, Sebástien (journalist), 41
Kirchheimer, Otto (scholar), 180
Kocka, Jürgen (scholar), 68

Lacombe affiche, 199, 212
Lafayette, General (politician), 185
  banquet in honor of, 172
Lamarlière, Eugène de (journalist), 124–25
Lamarque, General, 93
Lamennais, Félicité de (author), 34
"Le Canut, histoire contemporaine" (by Couailhac), 249, 251, 257
Le Chapelier law, 144
Ledré, Charles (historian), 3
legitimism. *See* Carlists, *Cri du peuple*; *Gazette du Lyonnais*; press, legitimist; Pitrat, Théodore; *Réparateur*
lettres de cachet, 182
liberalism, 29–31, 69–71, 75, 83, 95, 100–101. *See also* bourgeoisie; press, bourgeois; press, liberal
Liberation (in 1944), 12
literary theory, 10, 260
Lost Illusions (by Balzac), 12

Louis XV (king of France), 116
Louis-Philippe (king of France), 63, 115, 117, 170, 191, 241
Louis le Grand, place de. *See* Bellecour, place
Lüsebrink, Hans-Jürgen (scholar), 18
Lyon, 6–8, 51–55
  public sphere of, 55–65
  See also insurrection: Lyon, November 1831; insurrection: April 1834; press in Lyon; silk industry; silk workers
Lyon vu de Fourvières (book), 41, 55–56, 58, 120

Maignaud, Louise (journalist), 125, 127, 137
Manoff, Robert (scholar), 11
Marat, Jean-Paul (journalist), 182, 199, 254
Martin, Marc (scholar), 13
Marx, Karl (socialist), 17, 251–52
Marxist tradition, 17, 50, 68, 95, 138, 236
"Marseillaise, La," 115
Masonic lodges, 57, 143, 172, 174, 295 n. 38
Mathon de la Cour (journalist), 42
Maupeou "coup," 182
Maza, Sarah (historian), 68
"media revolution," 17–18, 23, 38, 68, 105, 226, 268
Merriman, John (historian), 52
Moissonier, Maurice (historian), 230
Moment Guizot, Le (by Rosanvallon), 71
Monfalcon, Jean-Baptiste (journalist and author), 42, 99, 101, 133, 163–64, 230–48, 258, 269, 304–5 n. 24, 305 n. 29
  correspondent for *Le Temps*, 217–18, 220
  early life, 243
  edits *Courrier de Lyon*, 44–45, 77, 233–34, 236
  and insurrections in Lyon, 214–15, 217–18, 220, 230–42, 244–47, 267
  involvement in duels, 46
  portrait of, *240*
  and *Précurseur*, 90, 203, 243–44
  relations with other journalists, 47–48, 102, 141, 146

Monfalcon, Jean-Baptiste (*continued*)
　and Revolution of 1848, 268–69
　WORKS:
　Code moral des ouvriers, ou Traité des devoirs et des droits des classes laborieuses, 241, 247–48
　Histoire des insurrections de Lyon, 21, 87, 146, 230–32, 235–39, 241–42, 246–48, 260, 266
　Souvenirs d'un bibliothécaire, 44–45, 46, 235, 242–47, 304–5 n. 24
　See also *Courrier de Lyon*; *Précurseur*; press, bourgeois
Moniteur, Le (newspaper), 71
Morin, Jérôme (journalist), 40–41, 43, 77, 86–87, 89–90, 176, 244
Mosaïque de Lyon (periodical), 38, 79, 134
Moses, Claire (historian), 129
Mouillaud, Maurice (scholar), 11, 18
Moutardier (publisher), 259
Mystères de Paris, Les (by Sue), 248–49, 256–57

Napoleonic Code, 128
Napoleonic period, 29, 115, 144, 182
Napoleonic regime, 26, 72, 117, 171
National, Le (newspaper), 42, 63–64, 81, 94, 215–16, 220
National Assembly, 28, 71, 171
National de 1834 (newspaper), 219
négociants, 79, 83
New History of French Literature (by Hollier), 9
Niboyet, Eugénie, 43, 46, 77, 79, 109, 121, 125–26, 129–32, 224, 269
　early life, 47, 125, 292 n.53
　editor of *Mosaïque*, 134
　ideas of, 126, 221, 224
　later career, 125, 292 n.53
　See also *Conseiller des femmes*; press, women's
Nodier, Charles (author), 124
Nouveau Tableau de Paris, 15–16, 261
November 1831, insurrection of. *See* insurrection: Lyon, November 1831

Odilon-Barrot, H. C. (politician), banquet in honor of, 168–69, 172, 174–77, 186

pamphlets (in Lyon), 224, 289 n. 48, 303 n. 67
Panckoucke, Charles (publisher), 71
Panckoucke publishing house, 243
Papillon, 25, 27, 34–36, 64, 106–10, 112–14, 120–25, 127–31, 133, 134
　and insurrection of 1834, 208
　See also press, alternative; press, women's
parliamentary debates, 31
Paroles d'un croyant (by Lamennais), 14
Pelzin, Alexandre (journalist), 42
Père Duchesne (periodical), 116, 199
Périer, Michel-Ange (journalist), 200
Perrache Prison, 44, 47, 234
Petetin, Anselme (journalist), 41–42, 77, 90–94, 185–88, 209–12, 269
　becomes editor of *Précurseur*, 44, 88, 90–92
　and insurrections in Lyon, 209–12, 214, 221
　opposition to in Lyon, 45–46
　political views of, 48, 93–94, 157, 203
　prosecutions of, 102, 185–88
Petrey, Sandy (scholar), 19
Philipon, Charles (jounalist), 3, 26, 95, 118
Phrygian bonnet, 111
Pinkney, David (historian), 4
Pitié pour elle (novel), 259
Pitrat, Théodore, 43, 47–49, 77, 99–100, 269. See also *Cri du peuple*; *Gazette de Lyon*; *Gazette du Lyonnais*; press, legitimist
poissarde style, 115–16
"Pourquoi je suis républicain" (by Bertholon), 112, 119
Précurseur, 20, 43–46, 76, 81–103, 114, 168–69, 173–78, 202–4, 209–12, 221
　ceases publication, 103, 148
　and *Courrier de Lyon*, 92, 95, 97
　creation of, 29, 30, 83
　format of, 30, 36, 86
　impact of, 31
　and insurrection of November 1831, 202–4, 210–12
　and insurrection of April 1834, 209, 221

and Jean-Baptiste Monfalcon, 90, 203, 217, 234, 236, 243
and Jérome Morin (editor), 43, 86–87, 89–90
and Anselme Petetin (editor), 44–46, 88, 90, 203
political views of, 77, 81, 83–85, 87, 93–94, *96*, 98, 120, 157
popular edition of, 35, 63, 99
and press banquets, 168–69, 172–78
prosecutions of, 93, 102, 184–89
and public opinion in Lyon, 60, 83–84, 88–89, 98
readership of, 60, 72, 78, 85, 99
relations with government, 30, 37, 93, 100–102
relations with other newspapers, 83, 92, 95, 97, 146, 157–59
and Revolution of 1830, 31, 40, 86–87
and Saint-Simonians, 88
stockholders of, 77, 81, 83–85, 87, 93–94, *96*
and workers, 62, 90–92
See also Association Lyonnaise pour la liberté de la presse; *Censeur*; Monfalcon, Jean-Baptiste; Morin, Jérôme; Petetin, Anselme; press, bourgeois; press, liberal; press, republican
Précurseur du peuple, 35, 99
prefect, 28, 30–32, 37, 63, 76, 97, 103, 172, 181, 189. *See also* Bouvier-Dumolard, Gasparin
Prefecture, 25, 42, 51
presqu'île (Lyon), 52–54
press, 2–3, 8–22, 65, 226–28, 264–70
and cultural history, 8–15
and revolution, 3, 15–22, 226–28
and social identity, 17–21
See also banquets; *Conseiller des femmes*; *Courrier de Lyon*; *Écho de la fabrique*; *Écho des travailleurs*; *Gazette du Lyonnais*; *Glaneuse*; *Journal du commerce*; *Papillon*; *Précurseur*; *Réparateur*; trials
press, alternative, 33–34, 105–14, 120, 129, 131, 133–34, 208
and bourgeois society, 107–8
definition of, 105–7

editors of, 109–10
and gender, 110–11
and politics, 112–14
See also *Asmodée*; Boitel, Léon; *Conseiller des femmes*; *Épingle*; *Glaneuse*; Lamarlière, Eugène de; Niboyet, Eugénie; *Papillon*
press, bourgeois, 64, 67, 69–82, 98–99, 102–3
definition of, 69–77, 81
editors of, 77–78
and government, 100–102
Lyonnais, 81–82
Parisian, 81
and press laws, 71–72, 74–76
readers of, 78–81
See also bourgeoisie; *Courrier de Lyon*; *Journal du commerce*; liberalism; Monfalcon, Jean-Baptiste; *Précurseur*
press, legitimist, 25, 29–33, 95, 97–103, 179, 187, 206–8. *See also Cri du peuple*; *Gazette de Lyonnais*; Pitrat, Théodore; *Réparateur*
press, liberal, 29–31, 70, 81–92, 97–99, 101–3. *See also Courrier de Lyon*; *Journal du commerce*; *Précurseur*; press, bourgeois
press, in Lyon, 7–8, 23–37, *27*, 265–70
during French Revolution, 28–29
and government, 37–38
popular, 63–64
reading of, 57–62, *58*
during Restoration, 29–31
after Revolution of 1830, 31–37
press, Parisian, 1–2, 7, 81, 214–20, 227. *See also Caricature*; Carrel, Armand; *Constitutionnel*; Girardin, Emile de; *Globe*; *Journal des débats*; *National*; *Presse, La*; *Revue encyclopédique*; *Revue républicaine*
press, republican, 32, 35, 93–95, *96*, 119–20, 173, 179, 207. *See also Glaneuse*; *Précurseur*; *Sentinelle*
press, women's, 3, 34, 36–37, 64–65, 110, 113, 120–34, 261. *See also Conseiller des femmes*; Grangé, Sophie; Niboyet, Eugénie; *Papillon*

press, workers', 34, 98, 135–66, *190*, 205.
   See also Chastaing, Marius; *Écho de la fabrique*; *Écho des travailleurs*
Presse à L'assaut de la monarchie, La (by Ledré), 3
Presse, La (newspaper), 39
Prix Montyon, 241
Prunelle, Gabriel (politician), 63, 75, 83, 164, 238
"public sphere," 49–51, *53*, 54–65, 84, 106, 131–33, 183, 227
   Habermas's definition of, 49–51
   newspaper reading in, 57–61
   workers and women and, 61–65, 127–29, 147, 165–66

Ragon, Michel (scholar), 250
Rancière, Jacques (scholar), 143
Reddy, William (historian), 68
Reichhardt, Rolf (historian), 18
Reid, Roddey (scholar), 248, 256
Réparateur (newspaper), 25, 33, 80, 120, 130, 187, 215, 221, 233. *See also* press, legitimist
"Republican Catechism" (1833), 119
republicanism, 113, 158, 172, 177. *See also* press, republican
Restoration, 3, 20
   and press, 3, 15, 29–30, 72, 183
   See also Revolution of 1830
Révolte de Lyon en 1834, ou La fille du prolétaire, La, 21, 230, 248–61, 266
   authorship of, 249
   and insurrection of April 1834, 255
   plot of, 249–250, 256–58
revolution, 3, 15–22. *See also* insurrection: Lyon, November 1831; insurrection: Lyon, April 1834; Revolution of 1789; Revolution of 1830; Revolution of 1848
Revolution of 1789, 4–5, 16–17, 19, 28, 39, 84–85, 144, 152, 227
   legacy, 24, 226, 264
Revolution of 1830, 3–5, 17, 23, 39, 70
   in Lyon, 24, 31, 40, 86–87
   and press, 20, 182–84
   See also July Monarchy; liberalism
Revolution of 1848, 4, 17, 171, 191, 226

Révolutions de Paris, Les (newspaper), 16, 159
Revue de Lyon, La (periodical), 25–26, 110
Revue du Lyonnais (periodical), 134, 226
Revue encyclopédique (periodical), 16
Revue républicaine (periodical), 252
Richomme, Fanny (journalist), 125
Ricoeur, Paul (scholar), 194–95
Riot-Sarcey, Michèle (historian), 129
Robespierre, Maximilien (politician), 158
Roederer, Pierre-Louis (politician), 13–14
Romand, Jean-Claude (author), 246, 305 n. 32
romantic movement, 109
Rosanvallon, Pierre (historian), 71
Rude, Fernand (historian), 230

Saint-Étienne, 52
Saint-Georges neighborhood (Lyon), 61
Saint-Jean, place, 208
Saint-Simonian movement, 47, 129, 152, 160
   Eugénie Niboyet and, 47, 107, 121, 125
   and Lyon, 6, 88, 121, 216
   and press, 16, 155–56
Say, J. B. (economist), 157
Schudson, Michael (scholar), 11
Second Empire, 103, 249
Senones (journalist), 43
Sentinelle, Le (newspaper), 32–33, 79, 89, 202
"September Laws" (1835), 3–4, 118, 142–43, 164, 191, 225, 259, 263
"Serment de l'Hôtel-de-Ville (de Lyon)" 200
Sewell, William (historian), 18
silk industry, 6, 52, 54, 140–44
   Jean Baptiste Monfalcon and, 73, 141, 218, 233–34, 237–39
   press and, 148
silk workers, 1, 6, 136–37, 140–42, 144, 147–48, 153, 202, 204, 234, 237–38, 249. See also *canuts*; *Écho de la fabrique*; *Écho des travailleurs*; press, workers'
Sismondi, Léonard Simonde de (economist), 157
Smith, Bonnie (historian), 68, 132

Index

social identities, 16–22, 265
socialism, 5, 154–57, 160, 254–55, 260
société de lecture, 56
Société des Amis du peuple, 158
Société des droits de l'homme, 113, 154
Société des Ferrandiniers, 141
Société des Mutuellistes, 62, 140–41, 144–45, 148–49
société en commandite, 74, 75
Society of Mutual Duty. See *Société des Mutuellistes*
Society of the Rights of Man. See *Société des droits de l'homme*
Spang, Rebecca (historian), 172
Stanislas, Étienne (Stanislas the Negro), 222, *223*, 302–3 n. 65
Stendhal (author), 270
Strukturwandel der Öffentlichkeit (by Habermas), 49
Sue, Eugéne (author), 248–49, 256–57, 259

tariff (in Lyon silk trade), 91–92, 101, 140, 203, 205
Temps, Le (newspaper), 73, 81, 146, 214–15, 217–18, 220, 231, 236
Terdiman, Richard (scholar), 10–12, 76, 118, 260
Terreaux, place de, 53–54, 60, 62, 208
Tétu, Jean-François (scholar), 11, 18
"The New Tom Thumb" (conte), 117
thermidorian reaction, 72

Thiers, Adolphe (politician), 117
Third Republic, 103
Tilly, Charles (scholar), 145, 170
"To Woman" (poem, by Sophie Grangé), 123, 274–76
trials, press-related, 47, 167–70, 173, 180–89, 191, 265, 299 n. 45
Tribunal de commerce, 76
Tribune, La (newspaper), 207
Tribune lyonnaise (periodical), 135, 142
Tribune prolétaire (periodical), 25, 135, 138, 142, 155–57, 160–61
Tuileries palace, 116

Ulliac Trémadeure, Sophie (author), 125, 128, 292 n. 52
Ultras, 29. *See also* legitimists
Union des travailleurs (periodical), 142
Utilitarians, 154

Vidal, Antoine (journalist), 142–43
Villèle ministry, 30, 83, 183
Villermé, Louis-René (author), 7, 141, 239
Voix des Femmes, La (by Niboyet), 125

Wahrman, Dror (historian), 267
"Why I am a Republican." *See* "Pourquoi je suis Républicain"

Zenger trial, 181
Zola, Émile (novelist), 256